T0342651

WATER
SUPPLY

Water Supply
4th Edition
TAFE NSW

Head of content management: Dorothy Chiu
Content manager: Chee Ng / Sandy Jayadev
Content developer: Samantha Brancatisano / Eleanor Yeoell
Project editor: Natalie Orr
Cover designer: Chris Starr (MakeWork)
Text designer: Linda Davidson (Original design by Norma Van Rees)
Project designer: Linda Davidson
Editor: Pete Cruttenden
Permissions/Photo researcher: Liz McShane
Proofreader: Duncan Campbell-Avenell
Indexer: Mary Russell
Typeset by KnowledgeWorks Global Ltd

Any URLs contained in this publication were checked for currency during the production process. Note, however, that the publisher cannot vouch for the ongoing currency of URLs.

Third edition published in 2016 by Cengage Learning Australia.

Previous editions published by Pearson Australia (a division of Pearson Australia Group Pty Ltd).

© 2021 Cengage Learning Australia Pty Limited

Copyright Notice
This Work is copyright. No part of this Work may be reproduced, stored in a retrieval system, or transmitted in any form or by any means without prior written permission of the Publisher. Except as permitted under the *Copyright Act 1968,* for example any fair dealing for the purposes of private study, research, criticism or review, subject to certain limitations. These limitations include: Restricting the copying to a maximum of one chapter or 10% of this book, whichever is greater; providing an appropriate notice and warning with the copies of the Work disseminated; taking all reasonable steps to limit access to these copies to people authorised to receive these copies; ensuring you hold the appropriate Licences issued by the Copyright Agency Limited ("CAL"), supply a remuneration notice to CAL and pay any required fees. For details of CAL licences and remuneration notices please contact CAL at Level 11, 66 Goulburn Street, Sydney NSW 2000, Tel: (02) 9394 7600, Fax: (02) 9394 7601
Email: info@copyright.com.au
Website: www.copyright.com.au

For product information and technology assistance,
in Australia call **1300 790 853**;
in New Zealand call **0800 449 725**

For permission to use material from this text or product, please email
aust.permissions@cengage.com

National Library of Australia Cataloguing-in-Publication Data
ISBN: 9780170424981

A catalogue record for this book is available from the National Library of Australia.

Cengage Learning Australia
Level 7, 80 Dorcas Street
South Melbourne, Victoria Australia 3205

Cengage Learning New Zealand
Unit 4B Rosedale Office Park
331 Rosedale Road, Albany, North Shore 0632, NZ

For learning solutions, visit **cengage.com.au**

Printed in China by 1010 Printing International Limited.
6 7 24

TAFE NSW

WATER SUPPLY

PLUMBING SKILLS Series

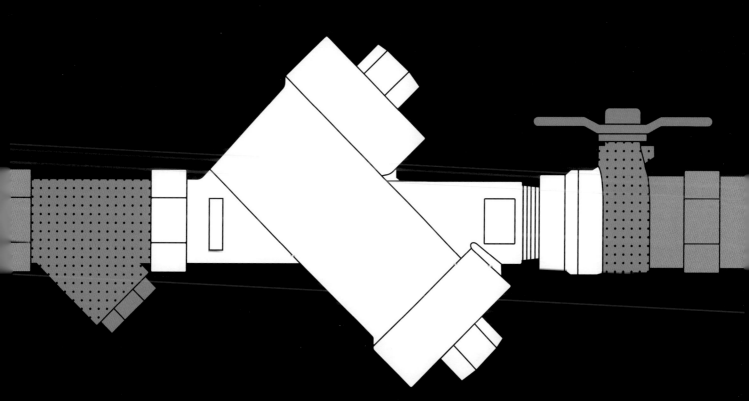

4e

THERE IS A STORY BEHIND THIS TEXTBOOK

It's more than just words on a page

We started by asking students like you, and your instructors, what you like about the textbooks you use and what would make them better. We listened to what you told us and now...

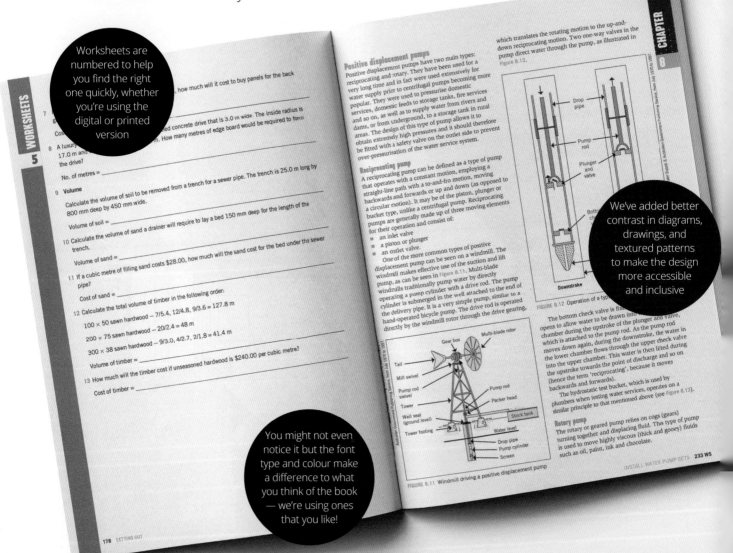

There are tabs throughout the text that make it easier for you to find what you're looking for

Worksheets are numbered to help you find the right one quickly, whether you're using the digital or printed version

We've added better contrast in diagrams, drawings, and textured patterns to make the design more accessible and inclusive

You might not even notice it but the font type and colour make a difference to what you think of the book — we're using ones that you like!

We'll keep listening so we can keep making even better textbooks. But now it's over to you. Check out the other great features of this textbook that are designed to help you learn in the 'Guide to the text' — you'll find it right after the table of contents.

This is a mock-up for illustrative purposes, pages not shown in this order in the actual textbook. Mock-up template from www.anthonyboyd.graphics

BRIEF CONTENTS

CONTENTS

Guide to the text

As you read this text you will find a number of features in every chapter that will enhance your study of plumbing and help you to understand how the theory is applied in the real world.

CHAPTER-OPENING FEATURES

The **Chapter overview** lists the topics that are covered in the chapter.

Identify the key concepts you will engage with through the **Learning objectives** at the start of each chapter.

FABRICATE AND INSTALL NON-FERROUS PRESSURE PIPING: COMMON MATERIALS USED IN THE SUPPLY OF WATER

1

Chapter overview

This chapter looks at the planning, identification, fabrication, jointing techniques and installation of non-ferrous pipes commonly used in the plumbing industry for the supply of water. By the end of the chapter, the skills to select and use tools and equipment to fabricate and install non-ferrous pressure piping for various applications should be achieved.

Different jointing and bending techniques for non-ferrous pipe and tubing to meet the requirements of manufacturers, industry and authorities are shown. Procedures on testing to comply with the requirements of manufacturers, industry standards and authorities for non-ferrous pressure pipe are referenced.

Learning objectives

Areas addressed in this chapter include:
- preparing for the efficient fabrication and installation of non-ferrous pressure piping
- identifying the installation requirements
- fabricating and installing pipework in accordance with authority, manufacturer and customer requirements
- ensuring the work area is clear, and that tools are well maintained and kept.

FEATURES WITHIN CHAPTERS

Engage actively with the learning by completing the practical activities in the **NEW Learning task** boxes.

From experience boxes explain the responsibilities of employees, including the skills they need to acquire and the real-life challenges they may face at work to enhance their employability skills on the job site.

Know your code icons highlight where Plumbing Standards are addressed to strengthen knowledge and hone research skills.

LEARNING TASK 2.1

NEW

1 If a house property service exceeds 30m, what diameter should it be?
2 Where should the water meter be located?

FROM EXPERIENCE

When excavating near existing services that are highly valued and easily damaged, such as optic fibre communication cables, 'non-destructive excavation' is recommended.

 AS/NZS 3500.1 PLUMBING AND DRAINAGE: WATER SERVICES

ix WS

Caution boxes highlight important advice on safe work practices for plumbers by identifying safety issues and providing urgent safety reminders.

Green tip boxes highlight the applications of sustainable technology, materials or products relevant to plumbers and the plumbing industry.

How to boxes highlight a theoretical or practical task with step-by-step walkthroughs.

 The plumber is directly responsible for identifying and preventing cross-connection to avoid contamination of the community's drinking supply.

GREEN TIP

Be prepared to place silt and sedimentation barriers in place prior to excavation to protect the environment from pollution. Heavy fines could be incurred otherwise.

HOW TO

HOW TO MAKE A SILVER-SOLDERED JOINT

The procedure for jointing silver soldering is as follows:
1   Cut the tube square (tube cutters are a purpose-built tool) and remove the internal burr.
2   Clean the pipe and fitting surfaces if oxidised or dirty.
3   Apply flux to the outside of the pipe (only necessary when using silver-soldering brass fittings).
4   Insert the pipe into the fitting and twist the pipe to spread the flux (if required).
5   Support the joint so that it does not move during soldering.
6   If there are flammable/painted surfaces near the joint, protect them with a heat-resistant mat.
7   Use a neutral flame.
8   Heat the joint until it is a dull maroon colour.
9   Touch the end of the silver soldering stick against the joint. When the joint is hot enough the filler rod will melt and begin to flow. Keep the flame moving at all times; if the flame is held in the one

Learn how to complete mathematical calculations with **Example** boxes that provide worked equations.

EXAMPLE 1.1

HOW TO CALCULATE THE HEAT LENGTH FOR A 90° BEND IN A DN 15 COPPER TUBE WITH A BEND RADIUS OF 90MM

Formula: $HL = \dfrac{2 \times R \times 3.1416 \times A}{360}$

where

HL   = heat length of the bend in millimetres = ?
R   = radius of the bend in millimetres = 90
3.1416 = the constant pi
A   = the angle the pipe is to be bent through = 90
360   = the number of degrees in a circle

Workings: $HL = \dfrac{2 \times 90 \times 3.1416 \times 90}{360}$

$= \dfrac{50893.9}{360}$

$= 141.372$

Answer: HL = 141mm

At the end of each chapter you will find several tools to help you to review, practise and extend your knowledge of the key learning objectives.

Review your understanding of the key chapter topics with the **NEW Summary**.

SUMMARY

- Non-ferrous materials are materials that do not contain iron.
- They are long-lasting because they are less corrosive than ferrous materials.
- Quality assurance helps to ensure a high standard of workmanship and processes, therefore reducing mistakes and problems.
- Always wear the appropriate personal protective equipment (PPE).
- Keep tools and equipment well maintained.

- Take the time to thoroughly read the plans and specifications before starting work.
- A site visit is vital prior to starting work to check for access and pipework location.
- Copper and plastic pressure piping are the most commonly used non-ferrous material.
- It is important to know the different jointing methods and the limitations of different materials.
- Ordering materials accurately reduces waste and saves time.

After you have worked through the chapter, reinforce the practical component of your training with the **NEW Get it right** section.

GET IT RIGHT

1 Which photo shows the correct use of the crimping tool?

2 Why is this method important?

3 What is the advantage of a battery-operated crimping tool compared to a manual crimping tool?

NEW

Worksheets give you the opportunity to test your knowledge and consolidate your understanding of the chapter competencies.

Worksheet icons indicate in the text when a student should complete an end-of-chapter worksheet.

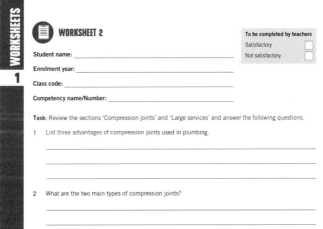

WORKSHEET 2

Student name: _____

Enrolment year: _____

Class code: _____

Competency name/Number: _____

To be completed by teachers

Satisfactory ☐

Not satisfactory ☐

Task: Review the sections 'Compression joints' and 'Large services' and answer the following questions.

1 List three advantages of compression joints used in plumbing.

_____

_____

_____

2 What are the two main types of compression joints?

_____

_____

COMPLETE WORKSHEET 2

Guide to the online resources

FOR THE INSTRUCTOR

Cengage is pleased to provide you with a selection of resources that will help you prepare your lectures and assessments. These teaching tools are accessible via cengage.com.au/instructors for Australia or cengage.co.nz/instructors for New Zealand.

MINDTAP

Premium online teaching and learning tools are available on the *MindTap* platform, the personalised eLearning solution.

MindTap is a flexible and easy-to-use platform that helps build student confidence and gives you a clear picture of their progress. We partner with you to ease the transition to digital – we're with you every step of the way.

The *Cengage Mobile App* puts your course directly into students' hands with course materials available on their smartphone or tablet. Students can read on the go, complete practice quizzes or participate in interactive real-time activities. *MindTap* is full of innovative resources to support critical thinking and help your students move from memorisation to mastery!

The *Series MindTap for Plumbing* is a premium purchasable eLearning tool. Contact your Cengage learning consultant to find out how *MindTap* can transform your course.

SOLUTIONS MANUAL

The **Solutions manual** provides detailed answers to every question in the text.

MAPPING GRID

The **Mapping grid** is a simple grid that shows how the content of this book relates to the units of competency needed to complete the Certificate III in Plumbing.

WEBLINKS

References from the text are provided as easy-to-access **Weblinks** and can be used in presentations or for further reading for students.

INSTRUCTORS' CHOICE RESOURCE PACK

This optional, purchasable pack of premium resources provides additional teaching support, saving time and adding more depth to your classes. These resources cover additional content with an exclusive selection of engaging features aligned to the text. Contact your Cengage learning consultant to find out more.

COGNERO® TEST BANK

A bank of questions has been developed in conjunction with the text for creating quizzes, tests and exams for your students. Create multiple test versions in an instant and deliver tests from your LMS, your classroom, or wherever you want, using Cognero. **Cognero test generator** is a flexible online system that allows you to import, edit and manipulate content from the text's test bank or elsewhere, including your own favourite test questions.

POWERPOINT™ PRESENTATIONS

Cengage **PowerPoint lecture slides** are a convenient way to add more depth to your lectures, covering additional content and offering an exclusive selection of engaging features aligned to the textbook, including teaching notes with mapping, activities, and tables, photos and artwork.

ARTWORK FROM THE TEXT

Add the **Digital files** of graphs, tables, pictures and flow charts into your learning management system, use them in student handouts, or copy them into your lecture presentations.

FOR THE STUDENT

MINDTAP

MindTap is the next-level online learning tool that helps you get better grades!
MindTap gives you the resources you need to study – all in one place and available when you need them. In the *MindTap Reader*, you can make notes, highlight text and even find a definition directly from the page.

If your instructor has chosen *MindTap* for your subject this semester, log in to *MindTap* to:
- get better grades
- save time and get organised
- connect with your instructor and peers
- study when and where you want, online and mobile
- complete assessment tasks as set by your instructor.

When your instructor creates a course using *MindTap*, they will let you know your course link so you can access the content. Please purchase *MindTap* only when directed by your instructor. Course length is set by your instructor.

FOREWORD

In Australia, the plumbing industry provides employment in a range of service areas, including water supply, fire and sprinkler systems, sanitary plumbing, drainage, gas installation, roof plumbing and mechanical services. The industry is one of the biggest employers of tradespeople in the country and they provide important services to consumers, business and other industries.

In order for plumbing enterprises to keep pace with global change and sustainable practices, the vocational and training (VET) sector must respond quickly and efficiently to meet industry needs. This text is designed to meet the requirements of the Construction, Plumbing and Services Training Package by providing information and activities that reflect the ever-changing skills required to undertake safe and effective activities in the water supply services area. The knowledge and skills derived from this text will provide learners with the tools for future learning and will prepare new and existing workers for a long and rewarding career in the industry.

I would like to thank Anthony Pingnam for his contribution of time and expertise in ensuring that the content of this publication is current and relevant to the building industry as it stands today and for the future of the industry.

Shayne Fagan
Head of Skills Team
Innovative Manufacturing, Robotic and Science
Western Sydney Region
TAFE New South Wales

PREFACE

Providing clean drinking water to our community is a massive responsibility for all involved. The plumber has a key part to play in installing water supply systems to residential, commercial, industrial and civil sites. Members of the plumbing industry take great pride in achieving the high standards set in Australia by remaining well informed on evolving work practices, new materials and new products.

This book is a collective effort by many plumbing teachers aiming to educate and inspire young apprentices in the industry. Its intention is to underpin their knowledge and develop problem-solving skills, combining safe work practices with professional skills and a thorough understanding of how and why procedures are done a certain way.

The text is aligned to the relevant training package competencies, with many references to AS/NZS 3500, which is also regularly referred to in the *Plumbing Code of Australia*. Due to some state or territory variations, it may be necessary for teachers to amend the information accordingly.

The passing of knowledge from elders is a time-honoured practice. The master and apprentice relationship is a tradition that goes back thousands of years and is still relevant in our society today. This book seeks to uphold that tradition in these modern times.

ABOUT THE REVISING AUTHOR

Anthony Pingnam, Grad. Dip. Adult Education, Cert. IV TAE, Cert. IV and Cert. III Plumbing Drainage and Gasfitting, teaches all streams in Plumbing, Drainage and Gasfitting at Certificate IV and Certificate III level. He has been teaching at TAFE NSW for 11 years. Prior to that, Anthony owned and operated a plumbing business for 25 years, undertaking all facets of plumbing in the commercial, industrial and residential sectors – including design, tendering and contract management from concept to completion – and employing and training many apprentices throughout that time. With this experience, becoming a teacher seemed a natural progression to enable him to impart his knowledge and to mentor apprentices.

ACKNOWLEDGEMENTS

This book would not be possible without the efforts from colleagues in the plumbing industry who have contributed their time and resources to write and update chapters in the previous editions. Thanks to Rob Young, John Humphrey, Shayne Fagan, Warren Breese, Gary Cook, Peter Smith, Shaun Kristen, Terry Bradshaw, Peter Towell and Steven Tuckwell for their efforts.

The authors and Cengage would like to thank the following reviewers for their advice and valuable feedback:

- Rob Gardiner – Box Hill Institute
- Russell Masterton – Box Hill Institute
- Grant Hastie – South Regional TAFE
- Tom Colley – Victoria Polytechnic
- Dean Carter – TAFE NSW

Additionally, we would like to extend our thanks to those who reviewed the previous editions of this text.

Finally, Anthony would like to thank his wife Marissa for her patience, understanding and photographical expertise.

UNIT CONVERSION TABLES

TABLE 1 Length units

Millimetres *mm*	Metres *m*	Inches *in*	Feet *ft*	Yards *yd*
1	0.001	0.03937	0.003281	0.001094
1000	1	39.37008	3.28084	1.093613
25.4	0.0254	1	0.083333	0.027778
304.8	0.3048	12	1	0.333333
914.4	0.9144	36	3	1

TABLE 2 Area units

Millimetre square *mm^2*	Metre square *m^2*	Inch square *in^2*	Yard square *yd^2*
1	0.000001	0.00155	0.000001
1000000	1	1550.003	1.19599
645.16	0.000645	1	0.000772
836127	0.836127	1296	1

TABLE 3 Volume units

Metre cube *m^3*	Litre *L*	Inch cube *in^3*	Foot cube *ft^3*
1	1000	61024	35
0.001	1	61	0.035
0.000016	0.016387	1	0.000579
0.028317	28.31685	1728	1

TABLE 4 Mass units

Grams *g*	Kilograms *kg*	Pounds *lb*	Ounces *oz*
1	0.001	0.002205	0.035273
1000	1	2.204586	35.27337
453.6	0.4536	1	16
28	0.02835	0.0625	1

TABLE 5 Volumetric liquid flow units

Litre/second L/sec	Litre/minute L/min	Metre cube/hour m^3/hr	Foot cube/minute ft^3/min	Foot cube/hour ft^3/hr
1	60	3.6	2.119093	127.1197
0.016666	1	0.06	0.035317	2.118577
0.277778	16.6667	1	0.588637	35.31102
0.4719	28.31513	1.69884	1	60
0.007867	0.472015	0.02832	0.01667	1
0.06309	3.785551	0.227124	0.133694	8.019983

TABLE 6 High pressure units

Bar bar	Pound/square inch psi	Kilopascal kPa	Megapascal mPa	Kilogram force/ centimetre square kgf/cm^2	Millimetre of mercury mm Hg	Atmospheres atm
1	14.50326	100	0.1	1.01968	750.0188	0.987167
0.06895	1	6.895	0.006895	0.070307	51.71379	0.068065
0.01	0.1450	1	0.001	0.01020	7.5002	0.00987
10	145.03	1000	1	10.197	7500.2	9.8717
0.9807	14.22335	98.07	0.09807	1	735.5434	0.968115
0.001333	0.019337	0.13333	0.000133	0.00136	1	0.001316
1.013	14.69181	101.3	0.1013	1.032936	759.769	1

TABLE 7 Temperature conversion formulas

Degree Celsius (°C)	(°F − 32) × 0.56
Degree Fahrenheit (°F)	(°C × 1.8) + 32

TABLE 8 Low pressure units

Metre of water mH_2O	Foot of water ftH_2O	Centimetre of mercury cmHg	Inches of mercury inHg	Inches of water inH_2O	Pascal Pa
1	3.280696	7.356339	2.896043	39.36572	9806
0.304813	1	2.242311	0.882753	11.9992	2989
0.135937	0.445969	1	0.39368	5.351265	1333
0.345299	1.13282	2.540135	1	13.59293	3386
0.025403	0.083339	0.186872	0.073568	1	249.1
0.000102	0.000335	0.00075	0.000295	0.004014	1

INTRODUCTION: PROPERTIES OF WATER

Chapter overview

This chapter looks at the general properties of water.

Learning objectives

Areas addressed in this chapter include:
- the origins and chemistry of water
- properties of water and its characteristics
- filtration and purification methods and the properties of 'hard' and 'soft' water, including sources of contamination and impurities
- drinking water (potable) supplies and protection measures
- hydrostatics and hydraulics.

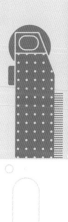

I.1 Background

A plumbing apprentice was asked by his TAFE instructor where he thought water originated from. The apprentice replied that it came from the tap. While that is partly true, a much greater depth of knowledge about water and its properties is required as a professional working in the water industry.

As plumbers we are entrusted with helping to maintain the general health of the people of Australia. We strive to achieve this by installing and maintaining a suitable water supply system, as well as by installing and maintaining an appropriate sanitary plumbing and drainage system to dispose of our waste products. In order to achieve this aim, an understanding of water systems and the plumber's role in installing and maintaining a safe and healthy water supply is necessary. Water is one of our most basic needs for survival. It is necessary for sustaining human, animal and plant life on Earth. Like many of nature's commodities, it is to be treated with respect and used economically.

In Australia the provision of safe drinking water (potable) is regulated by guidelines produced by the National Health and Medical Research Council (NHMRC) in collaboration with the Natural Resource Management Ministerial Council (NRMMC).

GREEN TIP

Water is precious and necessary for all living things. It certainly must not be wasted.

I.2 Where does water come from?

Water, from sources such as rivers, lakes and the ocean, is heated by the sun and converted into water vapour, which then rises back into the atmosphere as a gas. This process is called evaporation.

Plants take in water through their root systems and pass moisture back into the atmosphere via their leaves in a process called transpiration. Animals and humans pass moisture back into the atmosphere when they perspire (sweat).

Evaporation and transpiration form part of the natural water cycle (see Figure I.1). The cycle is completed when the water vapour is cooled and condenses to form clouds; when the condensation reaches dewpoint, the water vapour will form rain droplets and return to the surface as precipitation. The water may also fall in other forms; for example, as snow when the temperature is low enough.

See the references at the end of the chapter and investigate history and education links to find out more about where our drinking water comes from.

I.3 The chemistry of water

Water is a combination of two elements, hydrogen and oxygen, in the proportion of two to one. Its chemical symbol is H_2O, which in chemistry means that two atoms of hydrogen have combined with one atom of oxygen to form one molecule of water (see Figure I.2).

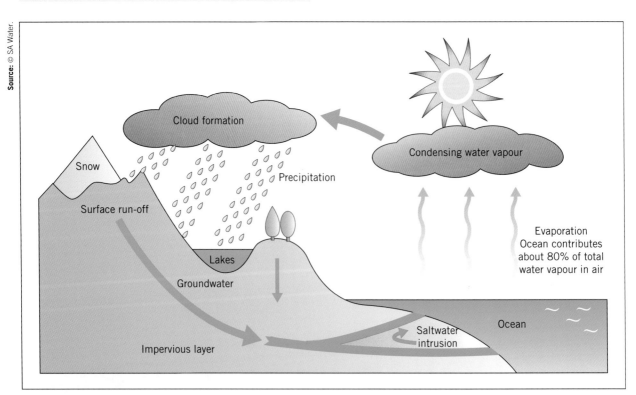

Source: © SA Water.

FIGURE I.1 The natural water cycle

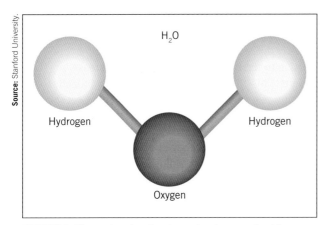

Source: Stanford University.

FIGURE I.2 The molecular structure of water as a liquid

The symbol represents only the composition of pure water such as is obtained through distillation; in this state it is a tasteless, odourless, transparent and colourless liquid, although when seen through considerable depth it can take on a bluish-green colouring.

States of water

Water may occur in three physical conditions or states:

1   In the solid state as ice. This condition starts at 0°C and continues at lower temperatures.
2   In the liquid state between the temperature range of 0°C and 100°C, where it is generally referred to as water (see Figure I.2).
3   In the gaseous state as steam at temperatures of 100°C and higher.

When it changes from one form to another, water is described as being in a changing state (see Figure I.3). Water changing from solid to liquid is said to be melting. When it changes from liquid to gas it is *evaporating*. Water changing from gas to liquid is called *condensation* (an example of this is the 'dew' that forms on the outside of a glass of cold drink). Deposition is when water changes from gas directly to solid form.

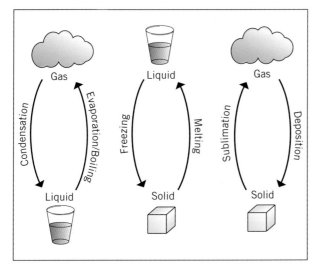

FIGURE I.3 The changing states of water

When water changes directly from solid to gas the process is called sublimation.

Depending on certain conditions, water can exist in both its solid and liquid states at 0°C.

Water hardness

As water evaporates, impurities are left behind. Pure water, in the strict chemical sense, is therefore unlikely to exist in nature. The purest water occurring in nature is probably rainwater that leaves the rain cloud over areas where there is no atmospheric pollution. This is unlikely to be the case over densely populated or highly industrialised areas.

In falling to earth from the cloud, water collects different types of impurities located in the atmosphere. Water has the capacity to absorb gases and to dissolve many solids. In its downward travel to the ground, as a result of collecting impurities, and depending on the type of ground over which it falls and travels across, the chemical composition of water is altered.

The terms 'soft water' and 'hard water' are used to indicate the capacity of water to dissolve soap and form suds. Hardness is caused by salts of magnesium and/or calcium. The 'softer' the water, the easier it is for suds to form from soap; the 'harder' the water, the more difficult it becomes to do so.

Hard water is not a serious health risk, but it can cause stomach upsets. It causes problems in the textile industry from the scum creating discolouring of fabrics. It also causes problems in boilers and heat exchangers where the scale caused by the salts coming out of solution may damage the boiler. The city of Adelaide has experienced these problems for many years due to the water supply having above-normal levels of calcium and magnesium salts. Rainwater that has fallen on granite or other volcanic rock types will usually be soft, clear and sparkling because such rocks have little effect on water quality. If it falls on boggy or peat-type soil, the water may remain relatively soft, but may become slightly acidic and take on a dirty-brown appearance. Although not pleasant in appearance, such water may still be potable.

On the other hand, water may fall on, or penetrate through, deposits of lime or magnesia. As these deposits dissolve in water, they make the water 'hard'. Such water tends to be a little more sparkling than in the other cases mentioned, and although its capacity to dissolve soap is much reduced, it may nevertheless be quite suitable for normal domestic use and for drinking.

There are two types of water hardness:

1   Temporary water hardness is caused by bicarbonates of calcium or magnesium.
2   Permanent water hardness is caused by calcium sulphates or magnesium sulphates.

Temporary hardness can be removed by simply boiling the water. When boiled, the carbonates are left at the bottom in the form of minute particles.

In certain areas plumbers may have to install water-softening devices to help provide water that is satisfactory for domestic usage. The typical water softener is a mechanical appliance that is plumbed into the home's water supply system. All water softeners use the same operating principle: they trade the minerals for something else, which is usually sodium. Permanent hardness can also be removed by a zeolite process, which works by replacing the magnesium and calcium salts, which cause hardness of water, with sodium salts.

FROM EXPERIENCE

Plumbers need to choose the correct water filter for the right application.

The pH scale

The term 'pH' stands for 'power of hydrogen' and is a measurement of the level of acidity or alkalinity of the water. The pH scale starts at 0 and ends at 14. A reading of 0 indicates a strong acidic content in the water, whereas a reading of 14 indicates a strong alkaline content. Swimming pools use this scale to balance the water levels for safe use. The midway point of 7 is known as neutral. Water should have a pH level of slightly higher than 7 to avoid acidity damage to piping systems (see Figure I.4).

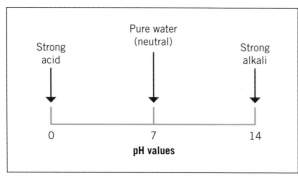

FIGURE I.4 pH values

LEARNING TASK I.1

1 Explain the transpiration process.
2 Which number on the pH scale represents neutral?

COMPLETE WORKSHEET 1

Characteristics of water

Water density varies depending on temperature. Water is most dense at 3.984°C (say 4°C) at sea level. At this temperature it weighs approximately 1000kg per cubic metre. Therefore, it is common to say that 1L of water has

a mass of 1kg. This is an important consideration when installing water services and storage tanks.

Any decrease or increase in temperature will be relative to the increase in volume and decrease in weight. This change in density as water approaches freezing point has wide-ranging implications for the planet we live on. Without this phenomenon, rivers, lakes and oceans would freeze solid, destroying all life within them. The buoyancy effect of ice on water creates an insulating mass of ice on the surface of the water that prevents the entire mass from freezing.

As a plumber you must be aware of, and make allowances for, this increase in volume when freezing occurs. The volume can increase by as much as 9% at atmospheric pressure and even more in a pressurised pipeline. If the increase in volume on freezing is not prevented, an increased pressure of up to 25 megapascals (MPa) may be generated in water pipes, which will in turn burst them in winter conditions.

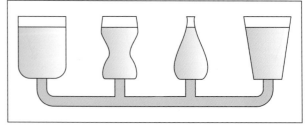
It is important to allow for expansion and contraction in pipework to avoid pipe bursts, especially in freezing conditions.

Water pressure

The branch of science that deals with fluids at rest is called hydrostatics, and that which deals with fluids in motion is called hydraulics. It is necessary for plumbers to know something about these two terms and how they affect their trade. For instance, a storage tank filled with water is subject to hydrostatic pressure and water supply reticulation system is subject to the laws of hydraulics.

Hydrostatics

The intensity of pressure exerted by water is directly proportional to the depth of the water and is measured in kilopascals (kPa). When water is static, the depth at any particular point is called the static head (head is another way of saying 'pressure'). Static head is measured in metres. If a number of containers of water are connected together by tubes below the surface of the water and the upper ends are not sealed, the water will stand at the same level in each, regardless of their size or shape. Since pressure at the same head is uniform, the pressure at the bottom of all the containers is the same, although they may not hold the same amount of water (see Figure I.5).

FIGURE I.5 Containers of varying shapes on the same level have the same outlet pressure and head level

A head of water of 1 metre has a pressure value of 9.80638 kPa. Mathematically, however, this is an awkward number to use for calculations. It is much more useful to round up and regard 1m head of water as having a value of approximately 9.81 kPa; or, if the head is in metres, we can use the formula:

$$P \text{ (in kPa)} = H \text{ (in metres)} \times 9.81$$

Every 1m of static water head corresponds to approximately 10kPa (actual pressure = 9.81kPa). Approximate conversions are:

10mm	=	0.1kPa
100mm	=	1kPa
1000mm	=	1m = 10kPa
10m	=	100kPa

For any given head, water transmits pressure equally in all directions. This is illustrated by a firefighter's hose, which first lies flat and then becomes cylindrical when filled with water and under pressure.

If a closed water container is supplied with water from a tank, the pressure on the top, sides and bottom of the container will differ only in proportion to the difference in head at each particular location. In Figure I.6 the pressure at take-off point 3 of the container is caused by head A. The pressure at take-off point 1 of the container is caused by heads A + B + C. The average pressure anywhere on the sides of the container can be calculated from the vertical height or head.

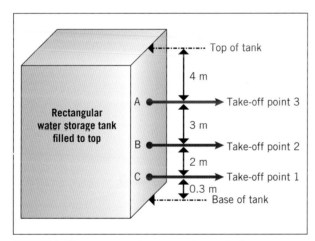

FIGURE I.6 Water pressure in a water storage tank

If a large-diameter pipe and a small-diameter pipe are fed with water at the same head, the pressure (kPa) in each will be the same, but the larger pipe will deliver more water because of its greater volume.

Hydraulics

A high static pressure does not necessarily mean that a large volume of water will be discharged from the end of a pipe connected to a source of supply. Friction between the moving water and the internal surface of the pipe is a major factor in the restriction of the volume of water delivered by a pipe. In general:

- The greater the velocity of water, the greater the friction and noise transmitted from the pipe.
- The smaller the diameter of the pipe, the greater the friction and noise transmitted from the pipe.
- The rougher the internal surface of the pipe, the greater the friction and noise transmitted from the pipe.

FROM EXPERIENCE

It is important to consider pressure, pipe size and smoothness when selecting piping materials.

LEARNING TASK I.2

1   How much pressure is generated by 1m head of water?
2   What would the water pressure be at an outlet if the height of the water supply above is 15m?

 COMPLETE WORKSHEET 2

I.4 Filtration and purification

Water supply authorities are required to provide a water supply that meets guidelines set down in the 2011 Australian Drinking Water Guidelines (ADWG). Typical procedures and processes used to satisfy these guidelines are outlined below.

Water from reservoirs and storage basins may contain organic matter and a variety of dissolved minerals. To make it fit for consumption, these may have to be removed by special treatment. Water purification can be achieved using some or all of the following processes:

- *Screening.* The water passes through a screen to remove large objects that may have entered the pipe from the dam.
- *Aeration.* In this process, air is passed through the water to eliminate soluble iron compounds and other impurities.
- *Flocculation.* A coagulant (such as ferrous chloride) is added to the water causing small suspended particles to aggregate or collect as a 'floc', which then becomes heavy and settles to the bottom. Mixing the coagulant with the water via an impeller speeds up this process.
- *Filtration.* The water is passed through one or more filters that retain solids and allow only clear water to pass through.
- *pH correction.* Lime is added to the water to adjust the pH level to between 7 and 7.5.
- *Softening.* This is the removal of magnesium and calcium salts from the water. These salts are the cause of water hardness.

- *Chlorination*. Chlorine is added to eliminate bacteria from the water that may be dangerous to health. Through this process, the water is sterilised.
- *Sterilisation*. Sterilisation of the water supply by the addition of chlorine kills and inhibits the growth of harmful organic bacteria.
- *Fluoridation*. Fluoride is added at the direction of the Australian Government Department of Health to aid in the prevention of tooth decay.

A typical flowchart of the water supply and filtration process is shown in **Figure I.7**.

Disease and contaminants

A number of organisms or substances that cause human disease can contaminate water supplies. The most serious of these are micro-organisms, which can have immediate and devastating effects on our health. Some chemical contaminants also cause human disease, the effects of which depend on the type and level of exposure. These dangers are why we have rules and regulations that place limits on how, where and why we install water supply systems.

Disease-causing organisms

Pathogenic (disease-causing) micro-organisms in drinking water pose the greatest potential threat to human health. Worldwide, more than three million people per year, many of them children under five years of age, die from waterborne and sanitation-related diseases. Most of these deaths occur in the developing world, where many communities have no access to clean or treated water or to adequate sanitation. While the potential threat remains, in most parts of Australia waterborne disease is controlled by good water management. However, in some parts of Australia water quality still remains a problem, especially in some rural and Indigenous communities.

Micro-organisms include bacteria, viruses and protozoa, only a few of which cause disease. However, micro-organisms in human and animal faeces are responsible for most waterborne diseases. In some parts of the world waterborne diseases such as dysentery, hepatitis, cholera and typhoid cause severe, and at times fatal, diarrhoea. *Cryptosporidium* and *Giardia* were brought to attention in

Source: © SA Water.

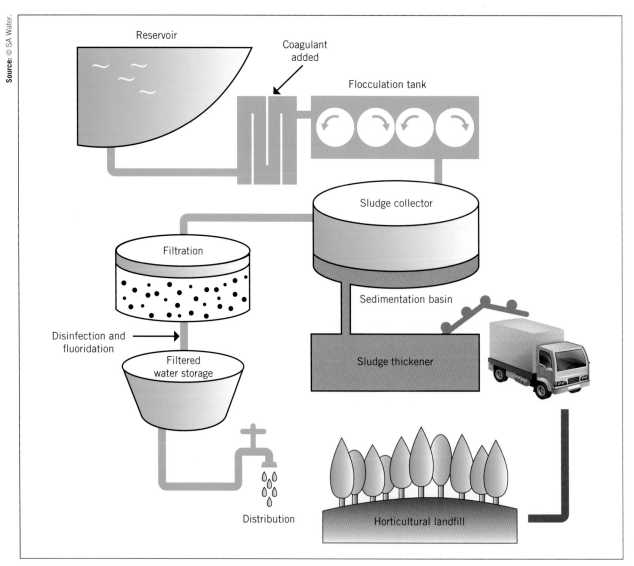

FIGURE I.7 Typical flowchart of the water supply and filtration process

Australia by the 1998 Sydney 'water crisis'. They are protozoans – parasites that consist of a single cell. *Cryptosporidium* and *Giardia* are a problem for the water supply industry because they are widespread in surface water, can survive for long periods and are difficult to treat.

Toxic substances

It is important to be aware of certain toxic substances that can occur in our waterways with harmful effects to our health.

Blue-green algae

Cyanobacteria, better known as blue-green algae, exists in rivers and dams if the water is stagnant, in warmer weather and if the ratio of nitrogen to phosphorous concentration is low.

It is a health hazard because of the toxins it releases. Some toxins result only in a skin rash, but others are more serious, causing liver and nerve damage. The toxins are released into the water and can remain even when the bacteria themselves have been removed. If contamination occurs, it may not be sufficient to boil the water. Boiling destroys the cells, but not all of the toxins. Therefore, special treatment is required to remove the toxins from water contaminated by blue-green algae.

Chemicals

Pathogenic micro-organisms have fairly immediate effects, but health effects from potentially harmful chemical and radioactive contaminants in drinking water become evident only after long exposure (typically many years). For example, low levels of arsenic in drinking water might increase the incidence of skin, lung or bladder cancer in a population that had been drinking the water for many years.

Chemicals of concern for drinking water include some naturally occurring chemicals such as nitrates, selenium and uranium; agricultural chemicals such as pesticides and fertilisers; and the chemical by-products formed when water is treated with a disinfectant (these disinfection by-products are discussed in detail in the section on water filtration and purification). However, the amount of these chemicals in our drinking water is generally very small – much lower than the levels that would be considered harmful to health. Indeed, we are exposed to higher levels of these chemicals in our environment and our food (although they are well below what is considered a harmful level).

Radioactive contaminants

The health effect most strongly associated with radioactive contaminants is cancer. Extremely low levels of radiation are a naturally occurring characteristic of water in our environment. Drinking water is likely to contribute only a very small proportion of a person's overall natural exposure to radiation.

GREEN TIP

Water authorities are responsible for maintaining an efficient water treatment process to prevent disease and contaminants entering the system.

LEARNING TASK **I.3**

1 Name three waterborne diseases.
2 Name the two parasites that were found in the water supply in Sydney, Australia, in 1998.

 COMPLETE WORKSHEET 3

SUMMARY

- The natural water cycle is made up of the combination of evaporation, transpiration, condensation and precipitation taking place over the natural water systems and land.
- The chemical symbol for water is H_2O – two atoms of hydrogen and one atom of oxygen.
- There are three states of water – liquid, solid and gas.
- Hard water is high in mineral content – calcium and magnesium.
- The pH scale measures the acidity and alkalinity of water from 0 (acid) to 14 (alkaline).
- One metre head of water has a pressure value of 9.81 kPa ($P = H \times 9.81$).
- Water purification (treatment) is a process of screening, aeration, flocculation, filtration, pH correction, chlorination, sterilisation and, in most cases, fluoridation.

REFERENCES

To find out more about your local authority, visit:
Icon Water (ACT): **https://www.iconwater.com.au/**
NHMRC: **https://www.nhmrc.gov.au/about-us/publications/**
australian-drinking-water-guidelines
Melbourne Water: **https://www.melbournewater.com.au**

PowerWater (NT): **https://www.powerwater.com.au**
QldWater: **https://www.qldwater.com.au**
SA Water: **https://www.sawater.com.au**
Sydney Water: **https://www.sydneywater.com.au**
Water Corporation (WA): **https://www.watercorporation.com.au**

WORKSHEET 1

To be completed by teachers

Satisfactory ☐

Not satisfactory ☐

Student name: _____

Enrolment year: _____

Class code: _____

Competency name/Number: _____

Task: Review the section 'I.2 Where does water come from?' and answer the following questions.

1 Explain the following terms in your own words:

a Evaporation

_____

_____

_____

_____

b Condensation

_____

_____

_____

_____

c Precipitation

_____

_____

_____

_____

2 What is the chemical symbol for water?

_____

3 Name the three different states (forms) of water. State the temperature range of these three forms.

a _____

b _____

c _____

4   The pH scale is a measurement of the amount of acid or alkaline in a substance. Neatly draw a diagram that indicates pH values.

5   State the ideal pH value for water.

_____

6   What are the two types of mineral salts found in hard water?

_____

_____

7   Temporary hardness can be removed by which method?

_____

8   Permanent hardness can be removed by which method?

_____

WORKSHEET 2

Student name: _____

Enrolment year: _____

Class code: _____

Competency name/Number: _____

To be completed by teachers

Satisfactory ☐

Not satisfactory ☐

Task: Review the section 'Characteristics of water' and answer the following questions.

1 At what temperature is water most dense?

_____

2 What is the mass (in kilograms) of 1000L of water?

_____

3 Explain in your own words why ice floats in water.

_____

_____

4 Describe in your own words the problems that can occur when water freezes within pipework systems.

_____

_____

5 What problems occur from high velocity within a pipe?

_____

_____

6 What can cause friction within a pipe?

_____

WORKSHEET 3

Student name: _____

Enrolment year: _____

Class code: _____

Competency name/Number: _____

To be completed by teachers

Satisfactory ☐

Not satisfactory ☐

Task: Review the section 'I.4 Filtration and purification' and answer the following questions.

1   Briefly describe in your own words each of the following water purification processes.

   a   Screening

   _____

   _____

   _____

   _____

   b   Aeration

   _____

   _____

   _____

   _____

   c   Flocculation

   _____

   _____

   _____

   _____

   d   Filtration

   _____

   _____

   _____

   _____

e pH correction

_____

_____

_____

_____

f Softening

_____

_____

_____

_____

g Chlorination

_____

_____

_____

_____

h Sterilisation

_____

_____

_____

_____

i Fluoridation

_____

_____

_____

_____

FABRICATE AND INSTALL NON-FERROUS PRESSURE PIPING: COMMON MATERIALS USED IN THE SUPPLY OF WATER

Chapter overview

This chapter looks at the planning, identification, fabrication, jointing techniques and installation of non-ferrous pipes commonly used in the plumbing industry for the supply of water. By the end of the chapter, the skills to select and use tools and equipment to fabricate and install non-ferrous pressure piping for various applications should be achieved.

Different jointing and bending techniques for non-ferrous pipe and tubing to meet the requirements of manufacturers, industry standards and authorities are shown. Procedures on testing to comply with the requirements of manufacturers, industry standards and authorities for non-ferrous pressure pipe are referenced.

Learning objectives

Areas addressed in this chapter include:
* preparing for the efficient fabrication and installation of non-ferrous pressure piping
* identifying the installation requirements
* fabricating and installing pipework in accordance with authority, manufacturer and customer requirements
* ensuring the work area is clear, and that tools are well maintained and kept.

1.1 Background

Non-ferrous materials are materials that do not contain iron. These types of materials are most commonly used in the plumbing industry to carry or store water as they are more resistant to corrosion (rust) than pipe materials that do contain iron, and therefore have a longer lifespan.

Tradespeople working in the plumbing services sector are required to identify and install a wide selection of pipe materials. Plans and specifications must be interpreted to allow the selection of the appropriate material for a given job. Tradespeople also must have the knowledge and skill to select and use the correct method/s for fabricating, joining, clipping and testing the particular pipe material using a variety of tools and techniques as per the manufacturer's specifications, industry standards and authority requirements.

1.2 Preparing for work

The plans and specifications for the job must be obtained to ensure the work carried out is compliant. Information from several sources is necessary to plan the fabrication and installation of non-ferrous pressure piping. This includes authorities' requirements, manufacturer's specifications, Australian standards, plans and specifications, and site inspection.

Work health and safety

It is important to know the work health and safety (WHS) requirements for each job. A risk assessment must be carried out identifying the hazards, accessing the risks and implementing control measures to ensure the safety of everyone.

Remember to wear the appropriate personal protective equipment (PPE). This can include:
- safety boots
- safety glasses
- ear plugs or ear muffs
- dust masks or respirators
- gloves
- high-vis workwear.

Quality assurance

Most companies have quality assurance procedures in place to promote a good standard of workmanship and efficient operations. This helps to prevent mistakes from happening.

The requirements of this policy can include:
- controls and procedures in the workplace
- use and maintenance of equipment
- interpreting specifications
- quality of materials
- handling procedures.

Organising and planning tasks

Fabricating and installing non-ferrous pressure piping for the supply of water can involve trench excavation, working at heights and working in confined spaces, so it is vital the correct procedures are followed. This involves planning and sequencing the order of tasks. Some preliminary tasks are:
- a site visit to assess access for deliveries, plant and equipment
- ensuring all fees and permits have been dealt with; for example, a road opening permit, footpath opening permit and water main drilling fee
- carrying out a 'Dial Before You Dig' enquiry
- being certain a thorough risk assessment has been done with a completed safe work method statement (SWMS).

It is important to notify other trades or the site supervisor of the tasks planned so there is minimal disruption and everyone can work together towards a common goal.

Tools and equipment

Tools and equipment must be maintained and kept in good working order. They should be cleaned after use and inspected for any damage. Electrical tools and leads must be regularly tagged by a qualified person to ensure that they are safe to use. Ladders and steps should be inspected for damage before using. All tools and equipment should be stored out of the weather.

Work area prepared

The work area must be risk assessed before commencing the job. Keeping the site clean and tidy will ensure a safer and hassle-free job. It will also save time in the long run due to fewer obstacles in the way.

LEARNING TASK 1.1

1 Where can information on the materials being used be obtained?
2 What is the purpose of quality assurance procedures?

1.3 Identifying installation requirements

Understanding plans and specifications is important to meet the job requirements for fabricating non-ferrous pressure piping. Plans and specifications will explain where the pipes and equipment are to be installed.

A site inspection prior to starting work is important, as it may identify a clash of services not shown as per plan. Checking the site will help to ensure the installation will not damage or interfere with surrounding structures, therefore saving time and money. Also access for deliveries, equipment and machinery can be determined.

When ordering materials, it is important to make an accurate 'take-off' of the materials from the plans and specifications to avoid excess waste of material and time. Be sure to check your order on delivery to make sure it is complete and not damaged.

Common uses of non-ferrous pressure piping

Non-ferrous piping materials commonly used in plumbing include:

- copper
- copper alloys (for example, brass)
- aluminium
- plastics (for example, polyethylene, polybutylene and cross-linked polyethylene).

Non-ferrous materials are used in plumbing for a wide variety of other uses, including:

- heated water
- cold water
- gas
- waste
- mechanical service pipes for heating and cooling buildings
- compressed air
- medical gases
- steam and condensate
- fuel oils
- irrigation.

When fabricating and installing pipework it is essential that the plumber knows the correct method/s for:

- joining pipes
- bending pipes
- supporting/bracketing pipes
- testing the installation for soundness.

FROM EXPERIENCE

Having the knowledge and skills to choose the correct installation method for a selected material earns customer confidence, creating future work opportunities.

Plumbing and gas-fitting standards, regulations and codes have sections that indicate where a material may or may not be used, and state the restrictions and/or limitations on the use of a given material. The Australian Standards and authorities' requirements work in with the plans and specifications for the job. The specifications will help to eliminate having to choose the appropriate materials and pipework fabrication for a specific task.

AS/NZS 3500.1 PLUMBING AND DRAINAGE: WATER SERVICES

LEARNING TASK 1.2

1 Why is a site visit important before starting work?
2 Name four services non-ferrous piping is used for.

1.4 Fabricating, installing and testing the pipe system

The non-ferrous pressure pipes and fittings mainly used in the supply of water are copper, brass and a range of polymers (plastics). Information on these materials and the jointing techniques are listed as follows.

Copper and copper alloys

The copper tube widely used in the plumbing industry is made to AS/NZS 1432. It is available in 6m hard drawn lengths for diameters up to 150mm and also annealed 18m coils for diameters from 8mm to 32mm. Copper tube is long-lasting, corrosion-resistant and has a smooth bore (it reduces friction) and it does not adversely affect the quality of the fluid flowing through it.

Blue-water phenomenon

Blue water is a phenomenon of copper tube that randomly affects large parts of the east coast of Australia, as well as New Zealand, the United States and Europe. It is characterised by the production of voluminous blue-green corrosion products in the tube bores and is primarily associated with cold, soft, unbuffered waters of high pH (alkalinity) and negligible levels of disinfectant residual.

It has been discovered that for corrosion to occur in Australia, the prevailing water chemistry must have certain characteristics. Tubes, of whatever surface condition, do not cause blue water or pitting to perforation in Adelaide, Brisbane or anywhere else where the source water is moderately 'hard', which means the water is high in calcium and magnesium. Minor surface pitting can be found in these waters, but rarely is a serious corrosion problem.

The extent of copper corrosion on the east coast of Australia, Perth, Tasmania, New Zealand and elsewhere encompasses a range of 'soft' water chemistries. All areas of contamination, however, seem to share significant characteristics:

- soft water, with total dissolved solids less than 300mg per litre
- low alkalinity (typically less than 30mg per litre)
- the ability to leach lime from cement-lined cast-iron pipes, leading to a relatively high pH at the end of the distribution system
- almost invariably cold water, or temperatures lower than 50°C if warm water is involved
- varying levels of chloride and sulphate, though corrosion is known to occur with both these analytes at less than 7mg per litre
- low residual disinfectant.

Although extensive research has been conducted on this subject, a precise and definitive diagnosis is yet to be determined. However, the evidence has shown that blue water can at least be controlled by a corrosion management program that provides buffering of the supply on either a localised or regional basis, together with maintenance of an adequate disinfectant residual.

Copper tube

Copper tube manufactured to AS 1432 is classified into four types: A, B, C or D. Each type refers to the wall thickness category of the tube, with Type A being the thickest and Type D the thinnest. Each type has colour-coded writing: Type A – green, Type B – blue, Type C – red and Type D – black. Copper tube is sized from the outside diameter of the tube (OD), unlike steel pipe that is measured from the inside diameter (ID). Copper tube can be easily annealed (softened) by heating, which makes bending and expanding tube ends easy.

Copper tube may also be purchased pre-insulated with a PVC polylag sleeve. This will reduce some heat loss, provide clearance for expansion and protect copper tube when laid in corrosive soils. It is also available with a purple PVC sleeve for use on recycled water services.

GREEN TIP

All copper tube offcuts should be scrapped to a metal recycler. It provides extra income while preventing waste.

Jointing copper

There are numerous methods used to join copper tubes. These include silver soldering (brazing), soft soldering, olive compression, croxed compression, flare compression, press-fit and push lock fittings. Refer to AS/NZS 3500.1 for jointing limitations, particularly soft solder, as there are new limitations in the latest standard stating that soft-soldered joints are not be used on new work but only on repairs to existing work.

AS/NZS 3500.1 PLUMBING AND DRAINAGE: WATER SERVICES

Silver-soldered/brazed joints

Silver soldering (also known as silver brazing, brazing or hard soldering) is a jointing process that relies on capillary attraction to draw solder into joints while intergranular penetration of the materials also takes place. This is where the heat allows the molten brazing rod to form a very strong bond with the material's surface. Silver soldering requires more heat than soft soldering and so generally oxygen/acetylene or oxygen/LPG equipment is used to heat the joints. For the capillary action to take place the surfaces must be clean and closely aligned.

There are three types of joints that can be silver soldered (see Figure 1.1):

- a tube end into a capillary fitting: (a) and (b)
- a tube end into a formed branch in another tube: (b)
- a tube end into an expanded tube: (c).

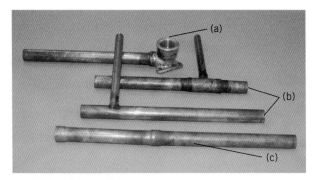

FIGURE 1.1 Types of joints

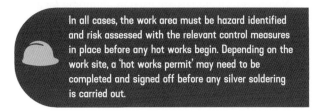

In all cases, the work area must be hazard identified and risk assessed with the relevant control measures in place before any hot works begin. Depending on the work site, a 'hot works permit' may need to be completed and signed off before any silver soldering is carried out.

To prepare for jointing, the outside surface of the tube and the inside surface of the fitting must be clean and free of oxides to ensure proper adhesion of the solder. New tubes and fittings are generally clean enough to make sound silver-soldered joints without further cleaning. If the fitting or tube-end surfaces are dirty, the surfaces should be cleaned with abrasive cloth or steel wool. All internal and external burrs must be removed to prevent friction loss.

No flux is required when silver soldering copper to copper joints. However, a general-purpose silver brazing flux must be used when silver soldering copper tube to a brass fitting.

The filler rod used contains copper, phosphorus and silver. Other additives may include tin, zinc and cadmium. The silver content of the rod varies with the type of work you are performing in accordance with Australian Standards and the specifications of a given project. The higher the silver content, the more expensive the rod will be. High-silver-content rods flow better for jointing purposes with additional strength. The silver content of a rod is indicated by the colour of the end of the rod (see Table 1.1).

TABLE 1.1 Silver-soldering filler rod specifications

Rod tip colour	Silver content	Temperature range (°C)
Yellow	2%	645–704
Silver	5%	645–740
Brown	15%	645–700
Gold	40%	660–780
Pink	50%	688–744

Source: © 2009 MM Kembla.

Capillary fittings

Capillary fittings may be made from copper or brass (see Figure 1.2). Capillary fittings are made for silver soldering

FIGURE 1.2 Samples of some of the plain capillary fittings available

and soft soldering. There are a large number of shapes and sizes of capillary fittings available. Therefore, plumbers must know how to describe each fitting when ordering them from plumbing suppliers or when communicating with workmates. Table 1.2 lists some of the terminology used when describing fittings.

Many plumbers still use an old numbering system to describe capillary fittings (see Figure 1.3). For example, a number 12 is an elbow and a number 24 is a tee. So a plumber who knows the number for a given fitting may ask for a '15mm no. 24' if they want a 15mm tee. If a

FIGURE 1.3 Capillary fittings
From the top:
20mm OD × 20mm OD brass socket or 20mm no. 1 DR
15mm OD × 15mm internally threaded brass connector or 15mm no. 2
15mm OD × 15mm externally threaded brass connector or 15mm no. 3

plumber does not know the number allocated to a fitting, they must know how to describe each fitting. Imperial measurements are still used occasionally when ordering fittings (20mm = ³/₄ inch; 15mm = ½ inch).

TABLE 1.2 Terminology used to describe fittings

Terminology	Description
External thread male iron (MI)	The externally threaded end of a fitting
Internal thread female iron (FI)	The internally threaded end of a fitting
Elbow bend	A fitting with a sharp 90° change of direction
Tee	A fitting allowing the connection of three pipe ends
OD	Outside diameter – used when describing the capillary end
BP/lugged	Used to describe fittings that have a backplate used to screw or fix the fitting securely against a wall, post, timber or other solid object
Dezincification-resistant (DR)	Indicates that the fitting is made from brass of a type (DR) that resists a form of corrosion called dezincification. Brass plumbing fittings must be made from DR brass. A longer radius fitting will allow better flow.

HOW TO

HOW TO DESCRIBE CAPILLARY FITTINGS

The method used to describe capillary fittings is as follows.

1 When describing the end of a fitting you should state the size and the type of end (OD, externally or internally threaded).
2 For a fitting with two ends, where one end is capillary (OD) and the other end threaded (externally or internally), ask for the capillary end first and then the

threaded end. If both ends are capillary, ask for the larger end first, if there is a difference in size (see Figure 1.4).
3 For fittings with three ends, the ends along the straight line are asked for first, stating the capillary (OD); or if the end sizes are different, state the larger capillary (OD) first, then the other end, followed by the branch (see Figure 1.5).

>>

FIGURE 1.4 Brass no. 1 and copper no. 1R (socket and reducing socket)

FIGURE 1.5 Capillary fittings with three ends
Clockwise from top left:
20mm × 15mm × 20mm brass tee or 20 × 15 brass no. 26
20mm × 20mm × 15mm brass tee or 20 × 15 brass no. 25
20mm × 15mm × 15mm brass tee or 20 × 15 brass no. 27
20mm copper tee or 20 no. 24

Expanded joint fittings

The second type of joint for silver soldering is an expanded joint. This is a joint made from the pipe without the need to supply a socket fitting. Expanded joint fittings must be silver soldered and have the advantage of only soldering one joint instead of two, as needed for a socket joint.

HOW TO

HOW TO MAKE EXPANDED JOINT FITTINGS

1   The end of a copper tube is annealed or softened (see Figure 1.6).
2   The tube end is then expanded using a tube expander (see Figures 1.7 and 1.8).
3   The end of another copper tube is cleaned then slipped into the expanded section (see Figure 1.9).
4   The joint is then silver soldered.

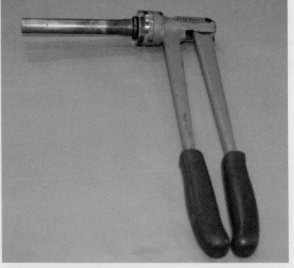

FIGURE 1.8 Tube expander handles closed together expanding tube end

FIGURE 1.6 Annealed copper tube end, ready to be expanded

FIGURE 1.7 Tube expander fitted into annealed tube end, ready to be expanded

FIGURE 1.9 Copper tube slipped into expanded tube end, ready for silver soldering

Branch-formed joints

The final type of joint for silver soldering is a branch-formed joint. This is a joint formed with a special tool that creates a branch connection. This replaces the need for a tee fitting and is very handy when retro fitting as only one soldered joint is needed instead of three for a tee fitting. It is also very cost effective.

Heat source

Oxygen/acetylene or oxygen/LPG equipment is generally used for silver soldering. LPG or MAPP gas may be used, but these heat sources take much more time to heat the joint to a temperature that will allow the silver-soldering filler rod to flow easily.

HOW TO

HOW TO MAKE BRANCH-FORMED JOINTS

To make branch-formed joints, a tee joint is made in a copper tube as follows:

1 The piece of copper tube is marked, lightly centre-punched and drilled where the branch is to be formed. Care should be taken to drill the correct size hole. Do not anneal the copper as the tube will deform when using the branch forming tool.

2 The hooked end of the branch former is inserted into the hole in the copper tube.

3 The bonnet of the branch former is held firmly against the copper tube while the branch is formed by using a ratchet to turn the hooked end, which pulls through the drilled hole forming the sides of the branch.

   Branch forming may not be used to form tees in the same size pipe (for example, DN 20 pipe with a DN 20 branch). The branch must be one pipe smaller (for example, DN 20 pipe with a DN 15 branch).

   Figure 1.10 shows the tools for making a branch, and Figure 1.11 shows a branch being formed.

4 A notching tool is used to punch tags off the branch end to contour with the main pipe to help prevent any protrusion.

5 A small dimple is made, using dimple pliers, in the pipe end being used as the branch to prevent the branch from being inserted too far into the joint (see Figure 1.12).

6 The branch pipe is now inserted into the formed branch ready to be silver soldered (see Figure 1.13). Note the position of dimples (on the high side) to prevent the pipe branch from penetrating into the main pipe.

FIGURE 1.11 Branch being formed

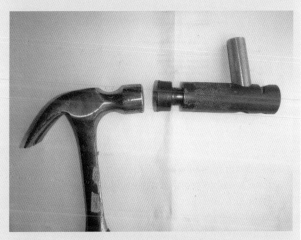

FIGURE 1.12 Notching tool in use

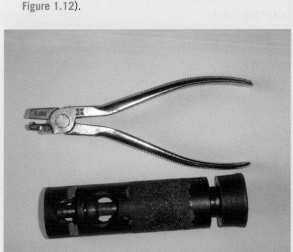

FIGURE 1.10 Notching tool and dimple pliers used in branch forming

FIGURE 1.13 Branch pipe inserted into formed branch, ready for silver soldering

Procedure for silver-soldered joints

The procedure to make a quality silver-soldered joint must be carried out in a step-by-step process as listed below.

HOW TO

HOW TO MAKE A SILVER-SOLDERED JOINT

The procedure for jointing silver soldering is as follows:

1   Cut the tube square (tube cutters are a purpose-built tool) and remove the internal burr.
2   Clean the pipe and fitting surfaces if oxidised or dirty.
3   Apply flux to the outside of the pipe (only necessary when using silver-soldering brass fittings).
4   Insert the pipe into the fitting and twist the pipe to spread the flux (if required).
5   Support the joint so that it does not move during soldering.
6   If there are flammable/painted surfaces near the joint, protect them with a heat-resistant mat.
7   Use a neutral flame.
8   Heat the joint until it is a dull maroon colour.
9   Touch the end of the silver soldering stick against the joint. When the joint is hot enough the filler rod will melt and begin to flow. Keep the flame moving at all times; if the flame is held in the one spot for too long you may melt the copper tube or damage the fitting.
10 Move the heat around the joint, melting the brazing rod into the joint.
11 Allow the joint to cool until the solder solidifies before moving it.
12 Wash off the flux with water and a damp rag (if necessary).

The advantages of silver soldering are as follows:
- There is visible proof of a sound joint.
- It is simple and quick.
- It is suitable for gas services.

The disadvantage is that it softens the copper tube near the joint.

 When silver soldering, tinted safety glasses should be worn for eye protection.

Soft-soldered capillary joints

This jointing method uses capillary action in which a liquid (in this case, molten solder) is drawn into the space between two closely aligned surfaces (the outside of the tube and the inside of the fitting). This method of jointing is usually used for jointing copper to brass fittings and copper to copper fittings (see Figure 1.14).

FIGURE 1.14 Soft solder, flux and a brass capillary fitting soldered onto a copper tube

Soft solder for use on water supply pipework is 'lead free'. Traditional 50/50 lead/tin solders are no longer used due to health problems associated with the use of lead. Lead-free soft solders are made from combinations of tin, silver, antimony and copper, with tin making up at least 95% of the solder.

To prepare for jointing, the outside surface of the tube and the inside surface of the fitting must be clean and free of oxides to ensure proper adhesion of the solder. The surfaces should be cleaned with abrasive cloth or steel wool before flux is applied.

Flux is used to:
- clean and remove oxides
- prevent new oxide from forming during soldering
- help the solder to flow.

A paste flux containing zinc chloride is applied to the surfaces being soldered.

Fittings may be made from copper or brass. There are two types of fittings: solder ring and plain (without solder ring) fittings.

Solder ring fittings

For a solder ring fitting, a ring of solder is placed into a groove around the fitting. This ring of solder has enough solder in it to make a sound joint. When liquefied by the heat, the solder ring flows into the capillary space around the fitting, making a sound joint (see Figure 1.15).

Plain fittings

Plain fittings (without solder ring fittings) have no solder ring (see Figure 1.16). Solder is added to the joint by the plumber.

Heat source

LPG, MAPP gas or air acetylene heating equipment are recommended for soft soldering. Oxygen and acetylene equipment must be used with care as it is easy to overheat the joint or those nearby and possibly cause leaks.

FIGURE 1.15 Solder ring fitting

FIGURE 1.16 20mm × 15mm copper elbow (no. 12R) and 20mm × 15mm × 20mm copper tee (no. 26)

FROM EXPERIENCE

When soldering near existing soft-soldered joints, wrap them with a wet rag or spray cool gel over them so they do not melt and leak.

Jointing procedure for soft-soldered capillary fittings

The procedure for jointing soft-soldered capillary fittings (see Figure 1.17) is similar to silver soldering, with a few exceptions.

FIGURE 1.17 Soft soldering a capillary joint

HOW TO

HOW TO MAKE A SOFT-SOLDERED JOINT

The tube and fittings must be thoroughly cleaned with steel wool or an abrasive cloth to a shiny finish for all joints.

1 Flux is required on all joints.
2 The joint must not be disturbed until it has cooled and all parts of a fitting must be soldered at the same time.
3 Not as much heat is required to perform the joint.

The advantage of a soft-soldered joint is it does not anneal (soften) the copper tube. Disadvantages include:

▪ It cannot be used on gas services.
▪ It cannot be used with annealed (coiled) tube.

COMPLETE WORKSHEET 1

Compression joints

In the plumbing industry compression joints are widely used and offer the advantages of:

▪ easy installation
▪ economy
▪ the ability to make sound joints in water-charged pipework
▪ a variety of sizes and styles and the ability to join a wide range of pipe materials
▪ the ability to disassemble pipework without damage.

There are different kinds of compression joints, and they can be divided into two main groups:

▪ manipulative joints, where the tube is manipulated (has its shape altered) as a part of the joint-making process (for example, by croxing or flaring the tube)
▪ non-manipulative joints, where nothing is done to the tube or pipe except to cut it to the correct length (for example, using a copper, brass or nylon olive).

Kinco nuts and olives (non-manipulative)

A brass nut compresses a nylon, copper or brass olive onto the pipe and into the joint (see Figure 1.18). This joint may also be croxed; otherwise it could fail under high pressure or high temperature if it is not installed correctly. Kinco nuts and olives (non-manipulative) are used to join copper tube to brass fittings and for the connection of water supply pipework to fixtures such as toilets, basins and kitchen sinks. These joints must meet the requirements as outlined in AS 3688–2005 Water supply – Metallic fittings and end connectors.

AS 3688–2005 WATER SUPPLY – METALLIC FITTINGS AND END CONNECTORS

Source: Alamy Stock Photo/Justin Kase z12z

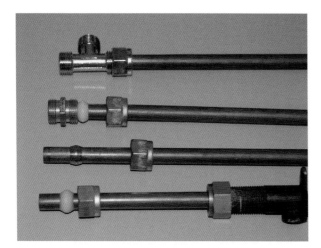
FIGURE 1.18 Kinco nuts and olives (non-manipulative)

Flare-type fittings (manipulative)

For flare-type fittings (manipulative), the joint is made by slipping a flare nut onto the tube first, then annealing the square end of the tube and flaring it with a flaring tool. The fitting can then be tightened onto the 60° angled end of the fitting.

Flare-type fittings are used to join copper tube to brass fittings or valves (see Figure 1.19). They are widely used on gas and water services. These joints must meet the requirements as outlined in AS 3688–2005 Water supply – Metallic fittings and end connectors.

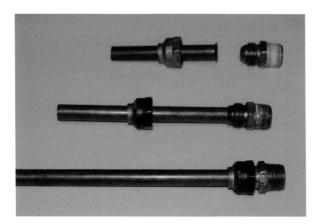
FIGURE 1.19 Flare-type fittings (manipulative)

A flaring tool is used to flare ends of copper tube (see Figure 1.20). Jointing compound may be used on water joints to enhance the seal but not on gas fittings.

Kinco nut and croxed joint (manipulative)

A Kinco nut and croxed joint (manipulative) is made by slipping a nut onto the tube first and then placing the croxing tool inside the square end of the tube (see Figure 1.21). The croxing tool (see Figure 1.22) is then used to form a ring around the tube that will stop the Kinco nut from coming off. Finally, hemp or a teflon tape grommet is wrapped around the crox to seal the joint between the crox and the male pipe thread being

FIGURE 1.20 Flaring tool

FIGURE 1.21 Making the joint
Top: Copper tube and nut showing formed crox
Centre: Hemp grommet wrapped around crox ready for assembly
Bottom: Assembled crox joint

FIGURE 1.22 Croxing tool

connected. This form of connection is generally used to join copper supply pipework to fixtures.

Push lock fittings (non-manipulative)

This method relies on a fitting being pushed over the tube end and the joint being made watertight by an O-ring.

The tube and fitting are held together by a barbed clip on the inside of the fitting that bites into the outside wall of the tube. Push lock fittings (non-manipulative) may be used to join copper, crossed-linked polyethylene, polyethylene and polybutylene tubes (see Figure 1.23).

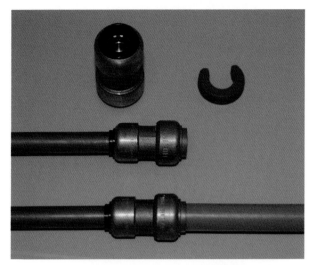

FIGURE 1.23 Push lock fittings (non-manipulative)

Both the inside and the outside of these tube ends must be cut straight, then de-burred and a witness mark applied to the tube end to indicate that the joint has been enlarged for its full depth. A witness mark is placed on the tube or pipe to check the depth that the pipe should go into the fitting. This helps to ensure that the joint is strong.

Prohibited locations of compression fittings

Kinco nuts and flare fittings must not be located:

- under concrete slabs
- where strain may be placed on the joint
- in wall cavities.

Kinco nuts and olives must not be used on gas installations.

Copper press-fit

Copper press-fit is a very popular way of jointing copper tube and fittings for gas and water services because no heat is required and it is a fast process. Fittings of up to 50mm have a rubber ring inserted into a groove. Larger fittings have a rubber ring and a stainless steel grip ring. The grip ring gives extra strength to the joint by stopping the pipe from pulling out of the fitting. To make the joint, the pipe is inserted into the fitting and the jaws of a battery-operated pressing tool are positioned over the fitting. The tool is then used to form the joint by crimping the fitting onto the pipe (see Figure 1.24).

Large services

Where copper is used on large-diameter services such as hydrants or to supply water to a multistorey property, other methods of mechanical jointing such as flange joints and roll-grooved joints are commonly used. Flanged jointing (see Figure 1.25) is commonly used on large

FIGURE 1.24 Copper press-fit tool

FIGURE 1.25 Copper adapter flanged fitting

copper tube applications, particularly when connecting a copper branch line to a cement-lined cast iron or cement-lined ductile iron main and/or for the connection of valves and ancillary items. Copper flanged adapters comprise a steel or brass flange and a copper socket that has a roll groove or expanded joint moulded into it. A rubber gasket is positioned between the two flanges, which when bolted together provides a watertight joint.

There is also a mechanical jointing method (see Figures 1.26 and 1.27) that incorporates the use of a grooved pipe, clamp and rubber/neoprene seal. This method is mostly used in above-ground installations.

Making a roll-grooved mechanical joint

When making the joint, the end of the pipe is either grooved using a specialised tool or has been pre-grooved by the manufacturer. When the pipe is cut into shorter lengths, it is usually roll-grooved by the installer.

Source: Victaulic Asia Pacific. Victaulic® is a registered trademark of Victaulic Company. All rights reserved.

FIGURE 1.26 Mechanical roll-grooved joint

Source: Victaulic Asia Pacific. Victaulic® is a registered trademark of Victaulic Company. All rights reserved.

FIGURE 1.27 Roll-groove fittings

HOW TO

HOW TO MAKE A ROLL-GROOVED MECHANICAL JOINT

To make the joint, the following procedure should be observed:

1 Roll-groove the ends of the pipes to be joined to the manufacturer's specifications.

2 Lubricate the inside edges of the rubber gasket with the lubricant supplied by the manufacturer.

3 Pull the rubber ring completely over the spigot end of one of the pipes to be joined until it is all the way on and the end of the pipe and the rubber gasket are flush.

4 Bring together the pipe ends to be joined so that they are just touching and in line with each other.

5 Feed the rubber ring back over the joint so that it is over both pipe ends and the centre of the joint is in the centre of the ring.

6 Apply a lubricant to the outside of the rubber gasket.

7 Place both pieces of the clamp over the joint and rubber gasket and make sure they are lined up with each other in the groove.

8 Place the nuts and bolts into the clamp and then tighten each alternately to the specified torque while ensuring the gasket is not pinched.

When making a roll-grooved joint in galvanised mild steel pipe, a similar procedure is followed, although a different roll-grooving tool is used. It should be noted that when making roll grooves for copper tubing, grooving tools intended for steel pipe must *not* be used or joint failure may occur.

A variety of roll-grooving machines are available from plumbing and mechanical services suppliers (see Figures 1.28 to 1.31). There are two distinctively different types: a manual tool with a handle, and a power-driven tool. There are several different brands and models within these categories. When purchasing a roll-grooving tool, or when using an existing machine, it is important that the correct roll-depth setting is used. Only roll sets specially made to form grooves to the Australian Standard for copper tube are suitable.

Source: : Victaulic Asia Pacific. Victaulic® is a registered trademark of Victaulic Company. All rights reserved.

FIGURE 1.28 Victaulic roll-grooving tool (hand use)

FIGURE 1.29 Manual roll-grooving tool

Source: Victaulic Asia Pacific. Victaulic® is a registered trademark of Victaulic Company. All rights reserved.

FIGURE 1.30 Roll-grooving attachment for a power-threading machine

Source: Ridge Tool Australia (RIDGID®).

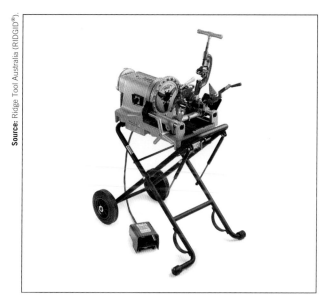

FIGURE 1.31 Power-threading/grooving machine

COMPLETE WORKSHEET 2

Plastic pipes (polymers)

Plastic pipes and fittings are the most common material being used by plumbers in Australia today due to reduced cost and installation time. When installing plastic pipes for hot and cold water services a number of restrictions apply. Plumbers should consult their codes and standards to ensure that they meet these requirements. Some general restrictions apply to plastic water supply pipes and fittings; they cannot be used:

■ for a meter assembly or a standpipe
■ within 1m of a water heater
■ in direct sunlight without further protection.

Other limitations are discussed in later chapters.

The types of plastic pipes available to plumbers include:

■ polyethylene (PE)
■ polybutylene (PB)
■ cross-linked polyethylene (PE-X)
■ unplasticised polyvinyl chloride (PVC-U)
■ composite cross-linked polyethylene/aluminium/cross-linked polyethylene (PE-X/AL/PE-X)
■ composite cross-linked polyethylene/aluminium/polyethylene (PE-X/AL/PE)
■ polypropylene (PP).

Similar to copper tube, plastic pipes are also colour-coded to identify the intended use. Purple is used for recycled water and green for rainwater. Pipe is generally available in 5m lengths and 25m, 50m and 100m coils.

AS/NZS 3500.1 PLUMBING AND DRAINAGE: WATER SERVICES

Jointing plastic pipes

The following types of joints are useful for various applications and different types of polymer products. Some require special tools and some are more cost effective than others. These are explained in more depth below.

Types of joints for plastic pipes include:

■ push to connect fittings
■ crimp ring fittings
■ heat fusion
■ compression sleeve.

Push to connect assembly

Push to connect fittings (see Figure 1.32) generally consist of the fitting body, an O-ring (which forms a seal between the pipe and the fitting) and a grab ring that secures the pipe in the fitting. This type of fitting is used on PE, PB and PE-X piping systems.

FIGURE 1.32 Assembled push to connect elbow

The pipe is cut to length and pushed into the fitting. An internal sleeve that prevents the pipe from distortion is built into most fittings. If it is not, a support sleeve is positioned inside the pipe end prior to pushing the pipe into the fitting. Most manufacturers place marks on the outside of the pipe to help indicate the depth that the pipe has to be pushed into the fitting. If there are no marks, a witness mark should be placed on the pipe prior to assembling the joint.

Care must be taken when handling pipe to be used with push to connect fittings to ensure that the outer surface of the pipe is free from damage such as scratches. The O-ring may not seal properly on the outside surface of the pipe if the pipe's surface is not smooth.

Crimp ring assembly

Crimp ring fittings generally involve pushing the pipe over the end of the fitting and inside a copper, stainless steel or soft metal ring. It is important to measure and make a witness mark on the pipe in order to check the pipe is fully engaged into the fitting before crimping (see Figure 1.33). A crimping tool is then used to compress the crimp ring and the pipe end onto the fitting (see Figure 1.34 and 1.35). Manual and battery-charged crimping tools are available. This type of fitting is used on PE, PB, PE-X and composite pipes such as PE-X/AL piping systems.

Source: Stilo45: Battery-operated crimping tool made by Intercable Tools GmbH.

FIGURE 1.34 Battery-operated crimping tool

Source: Reliance Worldwide Corporation (RWC).

FIGURE 1.33 Pipe, fitting and copper crimp ring

FIGURE 1.35 Manual crimping tool

Heat fusion

The heat fusion method of pipe jointing involves heating and melting the inside of a fitting and the outside of the pipe and pushing the pipe into the fitting to form a joint (see Figure 1.35). This method of pipe jointing is suitable for hot and cold water services, heating services, cooling services and industrial uses. The pipe material used is PP.

The pipe end is pushed into the socket side and the fitting socket is placed over the spigot side of the heat fusion welding machine until both have melted (see Figures 1.36 and 1.37). Manufacturers' instructions give tables indicating for how long the pipe and

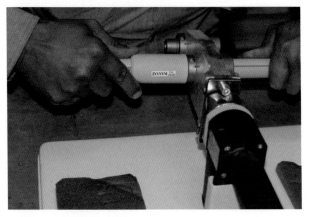

FIGURE 1.36 Heat fusion

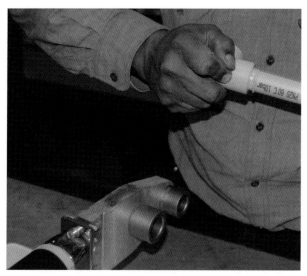

FIGURE 1.37 The pipe end and the fitting are pushed together and held together until the joint sets

fittings of a given size must be heated. In most cases, manufacturers also offer training for the use of their products.

Compression sleeve

The compression sleeve jointing method (see Figure 1.38) is suitable for use with PE-X and PE-X/AL/PE-X. This piping system is suitable for heating systems, gas, heated and cold water services. A brass compression ring is fitted and the tube is expanded using a similar tool to the one used for copper. The expanded tube end is then placed over the fitting. A compression tool is used to slide the compression ring over the expanded tube end and fitting (see Figures 1.39 and 1.40).

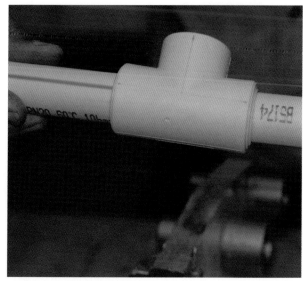

FIGURE 1.38 Completed heat fusion joint

FIGURE 1.39 Compression sleeve
This photograph shows the pipe being placed on the fitting, ready for completion of the joint.

FIGURE 1.40 Finished joint with the tool removed

PVC-U pipe and fittings (pressure pipe)

This pipe material is generally used for water services, irrigation systems and water mains. Generally, PVC-U pipe is joined by using a solvent cement; however, larger pipes used for water mains may be joined using rubber rings and gibault joints. When making a joint with solvent cement, the pipe is cut square, burrs are removed and the outside of the pipe end and the inside of the fitting socket are cleaned with priming fluid. They are then coated with PVC solvent cement of a type suitable for pressure fittings, assembled and held in place until the solvent sets.

Large services

Table 1.3 lists a number of plastics/polymers commonly used on large diameter services.

Jointing methods for PVC-U, PVC-M and PVC-O pipes

Polyvinyl chloride (PVC) pipes can be joined using the following methods:

- rubber ring jointing
- cold solvent cement welding (except PVC-O pipes, which should not be joined with solvent cement)
- flanged joints
- mechanical couplings.

TABLE 1.3 Materials suitable for use in the construction of water mains and their characteristics

Material	Characteristics
Unplasticised polyvinyl chloride (PVC-U)	• Lightweight • Does not corrode • Can be damaged easily (although with recent developments in plastics this disadvantage has been reduced) • May be installed in direct sunlight only under certain circumstances • Susceptible to becoming oval shaped (squashing) when high external soil/traffic loads are present
Polyvinyl chloride modified (PVC-M)	• Similar characteristics to PVC-U, although PVC-M is stronger, more flexible, lighter and possesses better flow characteristics than PVC-U pipe
Orientated polyvinyl chloride (PVC-O)	• Similar characteristics to PVC-U, although PVC-O is stronger, more flexible, lighter and possesses better flow characteristics than other PVC pipes • Solvent cement jointing is unsuitable

Polyethylene (PE)	• Similar characteristics to PVC-U, although PE possesses high impact strength, damage resistance, abrasion resistance, flexibility and high flow capacity; is available in longer lengths (which could be suitable for trench-less construction); and is weather resistant
	• Three grades of polyethylene pipe are available:
	  • MDPE (medium density polyethylene)
	  • HDPE (high density polyethylene)
	  • HPPE (high performance polyethylene)

Acrylonitrile-butadiene-styrene (ABS)	• Similar characteristics to PVC-U, although ABS has high impact strength and abrasion resistance, and is lightweight and weather resistant • Suitable for installations where the pipes are exposed to direct sunlight (although this may be subject to regulations, standards and local water utility requirements)
Glass-reinforced plastic (GRP)	• Lightweight • Available in long standard lengths of 6m, 12m and 18m • High strength and flexibility • High flow characteristics • Abrasion and corrosion resistant

Rubber ring jointing in PVC pipes

Elastomeric rubber ring jointing methods (see Figures 1.41 and 1.42) are similar to those employed in cement-lined cast iron/cement-lined ductile iron pipes.

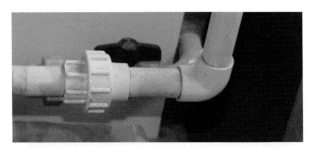

FIGURE 1.41 PVC union and elbow

FIGURE 1.42 Ring formed into a heart shape

HOW TO

HOW TO MAKE AN ELASTOMERIC RUBBER RING JOINT

When making the joint the following steps should be followed:

1 Ensure all jointing surfaces are completely dry before jointing.

2 Ensure the ring and the collars are free from dirt, dust and grit by wiping them out with a clean rag. Care must be taken during dusty or windy conditions to ensure the joint is kept clean and free of dust.

3 Prepare the rubber ring for insertion by flexing it into a heart shape.

>>

4   Insert the ring into the groove in the socket/collar. Ensure the ring is fitting snugly in the groove and not protruding into the pipe or socket.

5   Lower the pipe into the trench in accordance with proper handling procedures.

6   Ensure the pipe is aligned correctly and able to enter the socket straight and freely.

7   Lubricate the inside surface of the rubber ring with a thin film of the manufacturer's recommended lubricant. Only lubricant supplied by the manufacturer should be used. The lubricant can be applied using a glove, rag or brush.

8   With the same lubricant, lubricate the spigot of the pipe as far as the witness mark, paying particular attention to the chamfer.

9   Realign the pipe and enter the spigot carefully into the socket until it makes contact with the rubber ring.

10   Using the appropriate approved mechanical assistance while keeping the pipe straight along the centreline (axis) of both pipes being joined, push the spigot into the socket until the witness mark reaches the face of the socket while still remaining just visible.

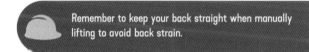

AS/NZS 3500.1 PLUMBING AND DRAINAGE: WATER SERVICES

Assembly of elastomeric joints

Assembly of elastomeric joints in PVC piping is usually a quick and easy process. However, as the joint is a compression-type joint and a fair amount of force is usually required, it is recommended that it should not be made without the use of some form of assisting mechanism. The joint will be more easily made, and will help satisfy WHS requirements for addressing strain hazards, by employing one of the following methods (depending on local conditions and the size of the pipe). Water services joined with rubber rings must have thrust blocks installed at bends more than 5°, tees, end caps, reducers, valves and inclines in excess of 1:5.

> Remember to keep your back straight when manually lifting to avoid back strain.

The crowbar/fulcrum technique

As illustrated in Figure 1.43, the joint can be pushed home using a crowbar or other suitable lever against the socket of the pipe. Note the timber block placed between the lever (as in Figure 1.44) and the pipe socket to protect it from damage. This leverage method should be used only on pipes no larger than 150mm.

The excavator method

This method is mainly used on large-diameter pipe. When an excavator or backhoe is available it can be used to complete the joint by pushing the spigot home (see Figure 1.44). To prevent damaging the pipe, a timber block is placed between the bucket and the collar. The joint is brought together in the usual way, ensuring it is properly aligned, and then the pipe is gently nudged forward with the machine's bucket until the joint is completed. Again, excessive force should not be used and care should be taken to ensure joints in the line that have already been made are not closed up past their

FIGURE 1.43 PVC-U pipe collar with rubber ring installed

FIGURE 1.44 The crowbar/fulcrum technique

witness marks, as this may affect allowances for movement and expansion.

The excavator/backhoe slewing method

Using the excavator/backhoe slewing method (see Figure 1.45), the pipe spigot is first positioned in the usual way, ensuring it is properly aligned and

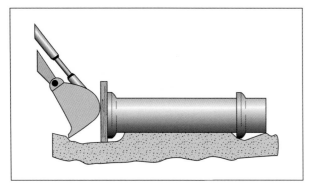

FIGURE 1.45 The trench excavator method

the excavator/backhoe's lifting hook, as shown in Figure 1.45. The bucket is then slowly slewed in the direction of the arrow, along the same axis as the pipe, drawing the pipe home into the collar and so completing the joint. Care should be exercised not to use excessive force or swing of the bucket as this may cause the pipe to come off its centreline, tilt or lift and damage the joint.

SOLVENT CEMENT JOINTING IN PVC PIPES

To avoid skin contact with solvents or cleaning fluid, gloves and glasses should be worn during this process. It is also essential that the solvent cement jointing process is carried out in a well-ventilated area.

ready for jointing. A sheathed webbing lifting sling is then placed around the collar end of the pipe and pulled tight. The other end of the sling is attached to

HOW TO

HOW TO MAKE A SOLVENT CEMENT JOINT IN PVC PIPE AND FITTINGS

The process for solvent cement jointing in PVC pipes is as follows:

1   Ensure that the end of the pipe is cut square and chamfered to the recommended tolerance, and that all burrs are removed. Chamfering of the spigot is of utmost importance as it promotes the even distribution of the solvent cement within the joint. A square end will push the glue off the collar and out of the joint.

2   Wipe the surface of the pipe end (spigot) and inside the collar with a clean dry cloth.

3   Inspect the pipe and collar/socket for any imperfections such as scratches and gouges. If the scratches or gouges exceed 10% of the pipe wall thickness, then the pipe end should be discarded or cut off until a clean undamaged spigot is achieved.

4   Dry fit the joint. Do not force the pipe into the socket. If the pipe will not go into the socket, or bind up before reaching the full depth of the socket, the tolerances of the pipe or collar may not be correct. If this is the case, the joint must be abandoned as the finished product may not result in a successful full-strength watertight joint.

5   Place a witness mark on the pipe spigot at a distance equal to the socket depth less 10mm. Do not scratch the witness mark into the surface of the pipe. Felt-tipped pens only should be used.

6   Wipe over both the spigot and the socket surfaces with a clean dry cloth for a second time to remove any dust immediately before making the joint.

7   Using a swab (usually supplied) or a dedicated clean rag moistened with a cleaning fluid recommended by the manufacturer (usually methyl ethyl ketone, or MEK), thoroughly wipe the spigot up to, but not beyond, the witness mark.

8   Ensuring the solvent cement is properly stirred or shaken before use, evenly apply solvent cement to the spigot and the socket with a clean dedicated brush. Do not use the same brush or swab as the one used for the cleaning fluid. Do not apply solvent cement onto the pipe beyond the first witness mark.

9   Once both socket and spigot are freshly and evenly coated with solvent, and while ensuring the axis of both the pipe and the socket are in line, immediately push the pipe, in a smooth even motion, into the socket to a depth where the witness mark on the spigot reaches the face of the socket. It is essential that the joint is completed as quickly as possible, while the solvent cement is still wet on the spigot and inside the collar. Do not attempt to make the joint if the solvent cement has dried out.

10   Wipe any excess solvent cement from around the socket and, where possible, from inside the joint. Ensure joints are supported and do not use cleaning fluid to clean up excess solvent cement.

11   Do not disturb joints for at least five minutes after jointing. Pipe testing should not be carried out above working pressures until 24 hours after jointing.

12   To ensure the integrity of the solvent joint, the joint and the pipe must be kept as close to normal room temperature as possible. In hot conditions, shading of the jointing work area for a minimum of one hour can reduce the temperature to a more desirable level and make for an easier jointing process. In hot or wet conditions, a cover over the jointing work area will prevent direct sunlight or precipitation from compromising the jointing process.

Source: Vinidex Pty Ltd, Polyethylene Design Manual ©.

FROM EXPERIENCE

Applying solvent cement to the socket first reduces the risk of the spigot coated with cement coming into contact with dirt.

Different drying and curing times apply depending on the pipe size, ambient temperature and type of glue, so it is important to read the manufacturer's instructions.

Tip: Always screw the lid on the primer and solvent cement immediately after each joint to prevent spillage.

> Working in confined spaces with solvent cement can cause physical harm from toxic vapours, so suitable PPE is important. Refer to the product's Safety Data Sheet (SDS).

Jointing methods for PE pipe

Joining lengths of PE can be achieved by the following techniques.

Mechanical compression couplings

Mechanical compression couplings (see **Figure 1.46**) are used for pipe sizes up to DN 110. Mechanical joints are generally specified for temporary services and where welding is impractical. Mechanical assembly requires the use of fittings, generally working on the compression principle. Mechanical compression fittings can also be used for joining PE to pipes or fittings of a different material. Mechanical jointing fittings are generally self-restraining, thus eliminating the need for thrust blocks and restraints.

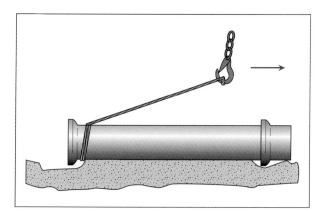

FIGURE 1.46 The excavator/backhoe slewing method

Electro fusion

Electro fusion (or socket fusion) joints (see **Figure 1.47**) are normally used for pipe sizes DN 90 and above. Fittings are available in the size range DN 16 to DN 710. Manufacturers offer training in the use of their products, and larger sizes may be available.

Electro fusion welding is when the cleaned/shaved ends of pipe are inserted into a socket that is heated to

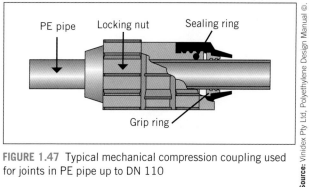

FIGURE 1.47 Typical mechanical compression coupling used for joints in PE pipe up to DN 110

create a bond between the socket and the pipe. Resistive heating wires are embedded in the socket. These resistive heating wires are connected to a control box and an electrical current is then passed through the wires. This applied heat melts the interior of the socket and the exterior of the pipe to form the welded joint.

Butt fusion

Butt fusion joints (see **Figure 1.48**) are normally used for pipe sizes DN 90 and above and are often chosen as the preferred jointing method for larger water mains. However, butt fusion welding is the primary technique used for large pipe sizes such as 250mm diameter and above, and is usually the only jointing method used for pipe diameters greater than 500mm.

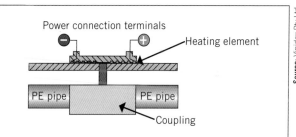

FIGURE 1.48 Cross-section of a typical electro fusion joint

In butt fusion welding, also known as hotplate welding, a heated plate is placed between the squared and cleaned/shaved ends of the pipes being joined. The plate heats the pipe ends and is quickly withdrawn, and the molten pipe ends are then pushed together. The manufacturer's recommended pressure is used when the pipes are brought together so that molecular segments can diffuse across the contact surfaces, producing what is known as an interment contact.

Flanged joints (PE flanged stubs)

Flanged PE stubs to join to PE pipe by electro fusion or butt fusion are available in sizes DN 90 to DN 450 (see **Figure 1.50**). Flanged PE stubs to join PE pipe by mechanical compression are available in sizes up to DN 160. The flange requires a corrosion-resistant backing plate of Class 316 stainless steel for below-ground applications or hot-dipped galvanised steel for above-ground installations. The nuts and bolts used to make

Source: Vinidex Pty Ltd, Polyethylene Design Manual ©.

FIGURE 1.49 Butt welder machine

Source: Vinidex Pty Ltd, Polyethylene Design Manual ©.

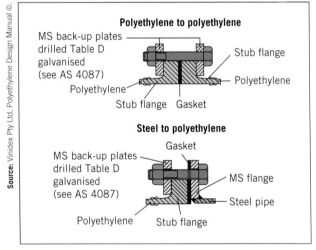

Polyethylene to polyethylene

MS back-up plates
drilled Table D
galvanised
(see AS 4087)

Stub flange

Polyethylene

Polyethylene

Stub flange   Gasket

Steel to polyethylene

Gasket

MS back-up plates
drilled Table D
galvanised
(see AS 4087)

MS flange

Steel pipe

Polyethylene   Stub flange

FIGURE 1.50 Cross-section of a typical PE flanged joint

the flanged joint should be the same material as the backing plate. Class 316 stainless steel nuts and bolts should be identified by an A4 marking.

Flange gaskets are to comply with WSA 109. Hydrant installations require full-face flanges, with bolting details complying to AS/NZS 4087.

AS/NZS 4087 METALLIC FLANGES FOR WATER WORK PURPOSES

Making an electro fusion joint in PE pipe

For the electro fusion jointing method to be effective it is essential to pay particular attention to preparation of the jointing surfaces. Of equal importance is the attention that should be paid to the removal of the oxidised surface along the pipe spigot for a distance equal to, if not slightly more than, the depth of the socket, as well as ensuring both the spigot and the insides of the socket jointing surfaces are clean. The pipe is prepared for jointing by removing a layer of material to a specified depth, thus removing the oxide and allowing fusion of a clean, oxide-free pipe spigot.

An electro fusion joint is made as follows:

1   To allow for efficient jointing, ensure there is enough space to permit safe access to the work area. If the joint is to be completed in situ, such as in a trench, there should be a free space with a minimum clearance of 150mm around the joining area. More clearance may be necessary when joining larger pipes.

2   Because the temperature of the jointing process is so critical, cover the open end of the pipe at the opposite end from the fitting joint to make sure airflow through the pipeline does not happen during the heating and cooling cycles of the jointing process.

3   Ensure the pipe ends being joined are cut square to the axis of the pipe and remove any burrs and swarf.

4   So as to remove any traces of dirt, mud and/or other contamination, wipe the pipe surface being prepared for the jointing process with an isopropanol wipe (an alcohol-based cleaning solution). Methylated spirits, acetone, methyl ethyl ketone (MEK) and other solvents are *not* suitable for preparing the joining surface. Each wipe should be used only once.

5   Ensure the joint area is completely dry and the alcohol left by the wipe has completely evaporated before proceeding with the jointing process.

6   Check the pipe for roundness (ovality). If necessary, use an appropriate re-rounding tool to remove any ovality and ensure the pipe is round and within specified tolerances.

7   While still in its bag, place the socket of the fitting alongside the pipe end; place a witness mark on the pipe at half the fitting depth plus about 20mm. This will ensure that the scraped area on the pipe can be visually checked once the joint is completed.

8   Ensure the pipe clamps are of the correct size for the pipes being joined.

9   Using an appropriate preparation tool, remove the entire surface along the pipe from the spigot up to the witness mark. Do not use tools such as metal files, rasps or emery paper as they are not suitable end preparation tools.

10 Insert the pipe ends into the fitting so that they are in contact with the centre stop. If there is no centre stop, set up an accurate witness mark on the pipe spigot, as previously explained. Ensure the pipe ends are correctly aligned along their axis in order to avoid any stress during the jointing process. This is of particular importance when using coiled pipes.

11 Place the pipes into the correct-size pipe clamps or other suitable support mechanism to ensure the pipe joint cannot move. Also ensure the fitting is satisfactorily supported to prevent it sagging during the electro fusion procedure (see Figure 1.51).

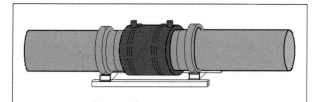

FIGURE 1.51 Pipes inserted into the socket fitting and clamped ready for welding

Source: Plastics Industry Pipe Association of Australia, Industry Guidelines, POP 001, *Electrofusion jointing of PE pipe and fittings for pressure applications*, Issue 7.3, 2015, http://www.pipa.com.au.

12 The required jointing time is usually indicated on the fitting or specified on the data sheet generally supplied with the fitting. Check that the correct time is shown on the control box display (see Figure 1.52); if not, enter the appropriate jointing time into the control box timer. (Automatic control boxes are available that alleviate the need to enter fusion times. They use a barcode system to set the appropriate times for each size and type of fitting.)

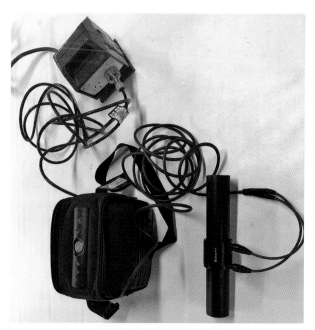

FIGURE 1.52 Electro fusion welder

13 Activate the electro fusion machine by pressing the start button. Observe the machine display to ensure the heat cycle is progressing.

14 If the electro fusion cycle terminates before the specified time has elapsed, check for faults as indicated by the control box warning lights or display. Check for a possible cause of the breakdown, such as inadequate fuel in the generator or power supply failure. If the cycle is terminated early, do not attempt a second fusion cycle until the entire joint has cooled down to less than 45°C. Some manufacturers may recommend replacement of the fitting rather than attempting a second electro fusion weld.

15 The completed joint should be left in the clamps for cooling. The time needed will be specified on the fitting or its data sheet, or shown in the display of the automatic control box.

Source: Plastics Industry Pipe Association of Australia, Industry Guidelines POP 001, *Electrofusion jointing of PE pipes and fittings for pressure applications*, Issue 6, 2005, http://www.pipa.com.au.

 COMPLETE WORKSHEET 3

Bending tube

There are a number of methods available to plumbers to create changes of direction in tubes. These methods include:

- elbows
- bends
- lever-type benders
- bending springs
- sand bending
- supported bends for plastic tube
- long-radius curves.

Time and money can often be saved by bending the pipe yourself rather than buying and installing elbows and bends. Bending the pipe yourself will also result in fewer joints and so less chance of leaks. Sometimes it may just look better to bend the tube, such as when using gold or chrome tube. Perhaps the building design may require you to make bends of a radius that fits the building better. And at other times you may not have fittings readily available and so have to do without.

To ensure good flow, it is important when bending tube that the pipe remains round throughout the length of the bend. Bending machines, springs and sand are all used to ensure that pipes do not lose their round shape when bent. They all support the side walls of the pipe being bent, preventing them from closing in or kinking. An example of the sides of a pipe closing in is when a garden hose is bent sharply. Two opposite sides of the pipe squeeze in, and the two sides at 90° to these move out. It is possible to reduce the flow out of the end of the hose or, if you bend the hose 180° back onto itself, you can stop the flow completely.

To ensure good flow it is essential that a plumber uses bends that are free from defects.

Lever-type benders

The most common method used by plumbers to make bends in copper tube up to and including DN 20 is with a tube bender (see Figure 1.53). The bend is made by fitting the copper tube into the bender and, by pulling the levers, bending the tube to the angle required (up to 180°). The sides of the tube are held round throughout the bend by the tube bender. Benders may also be used on composite pipe. The radius of the bend when using tube benders is not adjustable.

FIGURE 1.53 Bend in copper tubing made with lever-type benders

Bending springs

Support can be given to tubes during bending by using internal or external bending springs (see Figure 1.54). They can be used on copper tube and composite pipes.

FIGURE 1.54 Bending copper tube using an external spring

External springs

For copper tubes, the tube where the bend is required must be annealed (softened) and cooled down. Slide the spring over the annealed tube and bend the tube until it is bent a little past the required angle. Then bend the tube back to the required angle. (The overbending and bending back of the tube makes it easier to slide the spring off the tube after the tube has been bent.) Slide the spring off the tube.

Internal springs

These are suitable for making bends near the ends of the tube only. For copper tube, the area to be bent must be annealed. This is not required for composite tube. Slide the spring into the tube. Again, overbend the tube and bend it back to the required angle, then slide the spring out.

FROM EXPERIENCE

Internal springs are handy to use when pulling a bend on copper tube tails protruding from a wall instead of soldering a fitting.

Sand bending

Sand bending uses sand packed tightly inside the pipe (usually copper tube or steel pipe) to keep the pipe round during the bending process (see Figure 1.55). This method would be used to custom-make bends to a given radius for reasons such as aesthetics (looks).

 Dry sand is used as any moisture would turn to steam while heating. The steam would expand and blow the plugged end, which could possibly burn a person.

The first thing that has to be done when making a sand bend is to select a bend radius. This may be as small as three to five times the pipe diameter for copper tubes up to DN 20. Larger tubes will require a larger bend radius.

When sand bending copper tube, the section of the tube to be bent is the only section that is heated. The length of the section of tube to be bent has to be calculated. It is called the heat length. This is done by selecting the bend radius and using the formula:

$$HL = \frac{2 \times R \times 3.1416 \times A}{360}$$

where

HL	= heat length of the bend in mm
R	= radius of the bend in mm
3.1416	= the constant pi
A	= the angle the pipe is to be bent through
360	= the number of degrees in a circle

Supported bends for plastic tubes

Most plastic tube manufacturers make a pipe-bending guide for use with their pipes. These guides are used to make relatively short-radius bends in plastic pipes (see Figure 1.56).

Long-radius curves

All tubes can be bent without support using long-radius curves (see Figures 1.57 and 1.58). The problem with these is that they are not aesthetically pleasing, so they are commonly used in unseen areas, such as within walls. This is particularly so with plastic tubes where

EXAMPLE 1.1

HOW TO CALCULATE THE HEAT LENGTH FOR A 90° BEND IN A DN 15 COPPER TUBE WITH A BEND RADIUS OF 90MM

Formula: $\quad HL = \dfrac{2 \times R \times 3.1416 \times A}{360}$

where

HL $\quad$ = heat length of the bend in millimetres = ?

R $\qquad$ = radius of the bend in millimetres = 90

3.1416 = the constant pi

A $\qquad$ = the angle the pipe is to be bent through = 90

360 $\qquad$ = the number of degrees in a circle

Workings: $\quad HL = \dfrac{2 \times 90 \times 3.1416 \times 90}{360}$

$$= \dfrac{50893.9}{360}$$

$$= 141.372$$

Answer: HL = 141mm

FIGURE 1.55 Sand bending copper

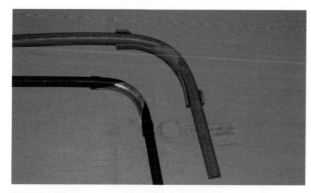

FIGURE 1.56 Metal and plastic support guides used with plastic pipes

the cost of the extra tube used is small in comparison to the cost of a fitting and the time taken to install the fitting (including the added possibility of a leak).

Clipping and fixing

Appropriate clips and fixings that are recommended by the manufacturer should be used. This will ensure the

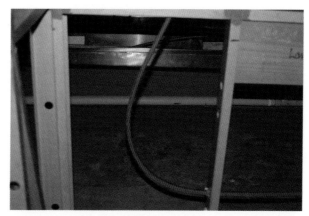

FIGURE 1.57 Long curve within a wall frame using plastic tube

FIGURE 1.58 Alternative using an elbow

clips are compatible with the pipe material, preventing any corrosive effect and therefore being longer lasting.

Refer to AS/NZS 3500.1 for the recommended vertical and horizontal spacing of clips for different pipe materials.

AS/NZS 3500.1 PLUMBING AND DRAINAGE: WATER SERVICES

Testing

Once the pipe system has been installed, it must be tested to make sure it doesn't leak. The testing must be carried out to regulatory authority requirements, standards and codes of practice. The test results must be recorded in a format required by the regulatory authority.

A test bucket is required for cold-water and hot-water installations. A test bucket is a piston pump pressurising the water service as the pump handle is manually moved up and down. Alternatively, an electric pump can be used for larger services.

More information on testing requirements is covered in Chapters 2 and 3.

AS/NZS 3500.1 PLUMBING AND DRAINAGE: WATER SERVICES

LEARNING TASK `1.3`

1   Name three jointing methods for copper pipe.
2   Name three types of polymer pipe.
3   Name three jointing methods for polyethylene pipe.
4   What is the advantage of a long-radius bend?

1.5 Cleaning up

Clearing the work area and disposing of the materials at the end of the job is just as important as setting the job up. This ensures a good reputation with clients.

- Any excess material should be recycled or reused wherever possible to prevent waste and to save costs.
- Cleaning tools and equipment as they are packed away makes the job easier for when they are used next time.

- Be sure to sign off the required documents upon completion of the work. The old saying 'The job's not finished till the paperwork is done' is as relevant today as ever!

LEARNING TASK `1.4`

1   What type of documents need to be submitted upon completion of work?
2   Why is recycling excess material important?

 COMPLETE WORKSHEET 4

SUMMARY

- Non-ferrous materials are materials that do not contain iron.
- They are long-lasting because they are less corrosive than ferrous materials.
- Quality assurance helps to ensure a high standard of workmanship and processes, therefore reducing mistakes and problems.
- Always wear the appropriate personal protective equipment (PPE).
- Keep tools and equipment well maintained.

- Take the time to thoroughly read the plans and specifications before starting work.
- A site visit is vital prior to starting work to check for access and pipework location.
- Copper and plastic pressure piping are the most commonly used non-ferrous material.
- It is important to know the different jointing methods and the limitations of different materials.
- Ordering materials accurately reduces waste and saves time.

REFERENCES

ABB: **https://www.baldor.com/brands/baldor-reliance**
Crane Enfield Metals: **http://www.cranecopper.com.au**

Iplex Pipelines: **http://www.iplex.com.au**
Reliance Valves: **https://www.rmc.com.au**

GET IT RIGHT

1   Which photo shows the correct use of the crimping tool?

2   Why is this method important?

3   What is the advantage of a battery-operated crimping tool compared to a manual crimping tool?

WORKSHEET 1

Student name: _____

To be completed by teachers

Satisfactory ☐

Not satisfactory ☐

Enrolment year: _____

Class code: _____

Competency name/Number: _____

Task: Review the text from section '1.1 Background' to the heading 'Compression joints' and answer the following questions.

1　Why is non-ferrous pressure piping less corrosive than ferrous pressure piping?

_____

_____

_____

_____

2　State four common uses for non-ferrous pipes.

_____

_____

_____

_____

3　Describe how copper tubes are classified.

_____

_____

4　State three reasons for insulating copper tube.

_____

_____

_____

5　Complete the following statement in relation to capillary action. 'Capillary action is where a _____ is drawn into the _____ between two closely _____ surfaces.'

6   List three methods that may be used to join copper tube.

_____

_____

_____

7   List the three joint types commonly used when silver soldering copper tube.

_____

_____

_____

8   Is the following statement true or false? 'No flux is required when silver soldering a joint between a copper tube and a copper capillary fitting.' Circle the correct answer.

T   F

9   How is the silver content of a silver-soldering filler rod indicated?

_____

10   What heat source is commonly used by plumbers when silver soldering?

_____

_____

11   Complete the table below.

Terminology	Description
MI	
	A fitting with a sharp 90° change of direction
DR	
Tee	
	Outside diameter
FI	

12   Is the following statement true or false? 'When ordering a capillary tee, you should ask for the branch first and then the ends along the straight line.' Circle the correct answer.

T   F

13 Identify each of the fittings in the following picture showing capillary fittings with three ends.

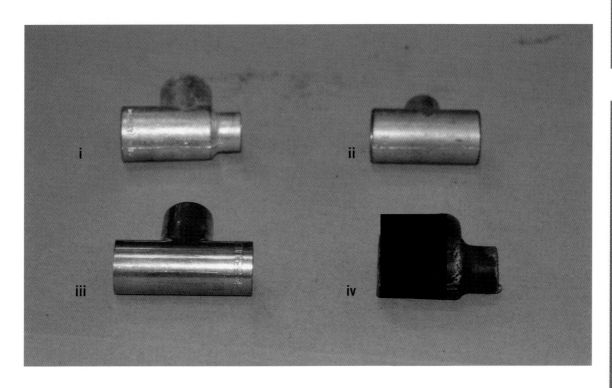

_____

_____

_____

_____

14 According to AS/NZS 3500.1, what are the limitations for soldered joints?

_____

15 What is the heat source used for soft soldering?

_____

WORKSHEET 2

Student name: _____

To be completed by teachers

Satisfactory ☐

Not satisfactory ☐

Enrolment year: _____

Class code: _____

Competency name/Number: _____

Task: Review the sections 'Compression joints' and 'Large services' and answer the following questions.

1   List three advantages of compression joints used in plumbing.

_____

_____

_____

2   What are the two main types of compression joints?

_____

_____

3   What are Kinco nuts and olives commonly used for?

_____

_____

_____

4   State two uses for flare-type fittings.

_____

_____

5   Is the following statement true or false? 'Push lock fittings require the ends of the tubes to have their shape altered prior to assembly.' Circle the correct answer.

    T   F

6   Is the following statement true or false? 'Flanged jointing is commonly used on large copper tube installations.' Circle the correct answer.

    T   F

7   List three locations where compression fittings must not be used.

_____

_____

_____

8 Explain in your own words the differences between manipulative and non-manipulative joints.

_____

_____

_____

_____

9 Is the following statement true or false? 'Copper tube must be annealed prior to joining using the press-fit system.' Circle the correct answer.

T   F

10 Is the following statement true or false? 'Roll-grooved joints are commonly used on large diameter services such as fire services.' Circle the correct answer.

T   F

11 What type of joint is shown in the picture below?

Source: *Copper Tube Handbook, Copper Development Association Inc., McLean, Virginia.*

_____

WORKSHEET 3

Student name: _____

Enrolment year: _____

Class code: _____

Competency name/Number: _____

To be completed by teachers

Satisfactory ☐

Not satisfactory ☐

Task: Review the sections from 'Plastic pipes (polymers)' to 'Bending tube' and answer the following questions.

1 Is the following statement true or false? 'When using push to connect type fittings on a polybutylene water service, the pipe end must be heated prior to assembly.' Circle the correct answer.

   T   F

2 Explain the term 'witness mark' and describe its purpose.

   _____

   _____

   _____

3 Describe the heat fusion method of jointing pipes.

   _____

   _____

   _____

4 Is the following statement true or false? 'A special tool is required when using the compression sleeve method for jointing plastic pipes.' Circle the correct answer.

   T   F

5 Is the following statement true or false? 'A battery-operated crimping tool is more suitable to operate in a confined space than a manual crimping tool.' Circle the correct answer.

   T   F

6 State three locations where polymer pipes are not permitted for use.

   _____

   _____

   _____

7   List five types of polymer pipes used by plumbers.

_____

_____

_____

_____

_____

8   Is the following statement true or false? 'Water services joined with rubber rings must have thrust blocks installed at bends more than 5°, tees, end caps, reducers, valves and inclines in excess of 1:5.' Circle the correct answer.

    T   F

9   Which type of plastic pipe cannot be solvent cement-welded?

_____

WORKSHEET 4

Student name: _____	**To be completed by teachers**
	Satisfactory ☐
	Not satisfactory ☐

Enrolment year: _____

Class code: _____

Competency name/Number: _____

Task: Review the section 'Bending tube' and answer the following questions.

1 List seven methods used to create changes of direction in pipework.

_____

_____

_____

_____

_____

_____

_____

2 What is the test pressure and time for water services as per AS/NZS 3500.1?

_____

_____

3 What is the horizontal spacing for clips on 20mm copper tube as per AS/NZS 3500.1?

_____

4 Is the following statement true or false? 'Short-radius bends create more friction loss than long-radius bends.' Circle the correct answer.

T   F

5 What is the minimum bend radius suitable for bending 20mm copper tube?

_____

6 Why must dry sand be used when sand bending?

_____

_____

7 Internal bending springs are only suitable to make bends where?

_____

_____

8   What safety provision must be adhered to prior to removing the sand from a heat bend?

_____

_____

9   What role does a template have in a heat bend?

_____

_____

10   What is the formula used to calculate the heat length for a sand bend?

_____

_____

_____

11   Why is it important to leave the workplace clean and tidy at the end of the day?

_____

_____

_____

📋 **WORKSHEET RECORDING TOOL**

Learner name		Phone no.	
Assessor name		Phone no.	
Assessment site			
Assessment date/s		Time/s	
Unit code & title			
Assessment type			

Outcomes

Worksheet no. to be completed by the learner	Method of assessment WQ – Written questions PW – Practical/workplace tasks TP – Third-party reports SC – Scenarios RP – Role plays CS – Case studies RW – Report writing PF – Portfolio	Satisfactory response	
		Yes ✓	No ✗
Worksheet 1	WQ	☐	☐
Worksheet 2	WQ	☐	☐
Worksheet 3	WQ	☐	☐
Worksheet 4	WQ	☐	☐

Assessor feedback to the learner

Feedback method (Tick one ✓):	☐ Verbal ☐ Written (if so, attach) ☐ LMS (electronic)

Indicate reasonable adjustment/assessor intervention/inclusive practice (if there is not enough space, a separate document must be attached and signed by the assessor).

Outcome (Tick one ✓):	☐ Competent (C) ☐ Not Yet Competent (NYC)

Assessor declaration: I declare that I have conducted a fair, valid, reliable and flexible assessment with this learner, and I have provided appropriate feedback.

Assessor name:	
Assessor signature:	
Date:	

Learner feedback to the assessor

Feedback method (Tick one ✓):	☐ Verbal	☐ Written (if so, attach)	☐ LMS (electronic)

Learner may choose to provide information to the RTO separately.

		Tick one:	
Learner assessment acknowledgement:		**Yes** ✓	**No** ✗
The assessment instructions were clearly explained to me.		☐	☐
The assessment process was fair and equitable.		☐	☐
The outcomes of assessment have been discussed with me.		☐	☐
The overall judgement about my competency performance was fair and accurate.		☐	☐
I was given adequate feedback about my performance after the assessment.		☐	☐

Learner declaration:
I hereby certify that this assessment is my own work, based on my personal study and/or research. I have acknowledged all material and sources used in the presentation of this assessment, whether they are books, articles, reports, internet searches or any other document or personal communication.
I also certify that the assessment has not previously been submitted for assessment in any other subject or at any other time in the same subject and that I have not copied in part or whole or otherwise plagiarised the work of other students and/or other persons.

Learner name:	
Learner signature:	
Date:	

2 INSTALL WATER SERVICES: MAIN TO METER

Chapter overview

This chapter addresses the skills and knowledge required to install a water service from the utility's water main, up to and including installation of a water meter (property service), using approved materials and techniques. Information is provided for the installation of recycled water services. (Commissioning of water services is covered in Chapter 6.)

Learning objectives

Areas addressed in this chapter include:
* planning and preparation
* setting out and installing water services from the main to the meter
* materials selection
* testing the property service and clean-up.

2.1 Background

Water for human consumption, irrigation and firefighting services, as well as recycled water in more recent years, is delivered to most large communities through a network of private and public water mains systems. A water main is a pipeline that conveys drinking or recycled water throughout the community and is owned and maintained by the water supply utility or the local council.

This chapter outlines the installation of a property service (main to meter). The installation of a property service involves all approved pipework and fittings used for the supply of water to a property from the water main, up to and including the meter assembly, or to the stop tap if there is no meter. A single connection to the main and a single property service for each water service type (drinking and recycled water systems) must be provided for each property.

AS/NZS 3500.0 GLOSSARY

2.2 Installing a property service (main to meter)

In some new estates a separate drinking (potable) and non-drinking (if available) water property service is installed for each property (see Figure 2.1) These services must only be connected to the water utility's appropriate water main. The minimum size of a residential property service is usually DN 20 (DE 25 for polyethylene services); however, the utility may allow a DN 25 service when the total length of the water service exceeds 30m. The local authority's website will provide details and procedures for a main tapping (see, for example, Figures 2.2, 2.3 and 2.4).

GREEN TIP

Drinking water is becoming more precious and scarcer, so using a recycled water supply is an excellent way to help address this problem.

The water utility may approve a larger-diameter property service to accommodate local requirements in remote areas. These services are called trunk service

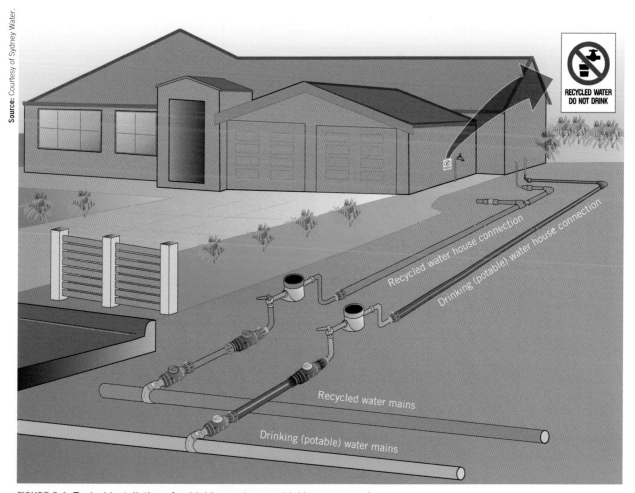

Source: Courtesy of Sydney Water.

RECYCLED WATER DO NOT DRINK

Recycled water house connection

Drinking (potable) water house connection

Recycled water mains

Drinking (potable) water mains

FIGURE 2.1 Typical installation of a drinking and a non-drinking water service

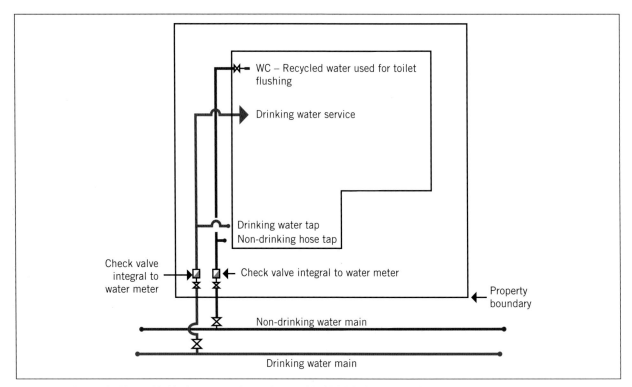

FIGURE 2.2 A recycled (non-drinking) water service and a potable (drinking) water service

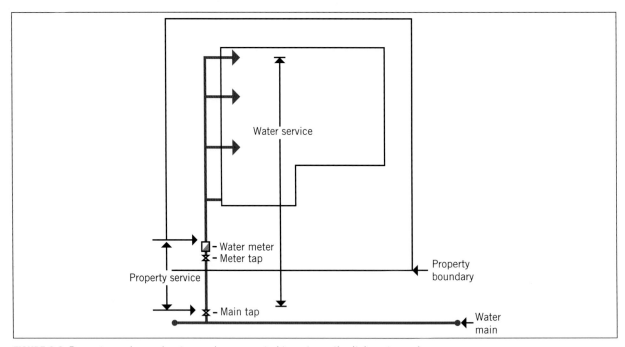

FIGURE 2.3 Property service and water service connected to water authority's water main

lines, trunk service or joint service. Individual properties then connect to this service in the same way that a connection is made to the utility's water main.

The water utility may require a split service to feed two properties where the water main is located on the other side of the road (a long service).

FROM EXPERIENCE

A recycled water main is always located closest to the property boundary and is purple in colour.

LEARNING TASK 2.1

1 If a house property service exceeds 30m, what diameter should it be?
2 Where should the water meter be located?

Single residential only application

(Office use only)   Case no. ☐☐☐☐☐☐☐☐   P.S.P. no. ☐☐☐☐☐☐☐☐

Particulars of property

Lot no. _____   Street no. _____   Street name _____

Suburb _____   P/code ☐☐☐☐   Melways Ref. _____

An A4 scaled site plan is required to be submitted with the application detailing the building outline, and dimensions from the property foundations to the property boundaries and preferred sewer locations. **Note: If the application is not completed in full and/or a plan is not accompanying the application,** it may not be processed.

Particulars of owner

Name _____   Phone _____   Fax _____

Particulars of applicant

By lodging this application, the applicant warrants they are either the owner or occupier of the property, or they are authorised by the owner or occupier to make this application on their behalf.

Name _____   Phone _____   Fax _____

Address _____   P/code ☐☐☐☐

Signature _____   Date _____

Particulars of licensed plumber (Mandatory)

Name _____   Phone _____   Licence no. _____

Detail of work / fees

New connections (Must include potable water, recycled water and sewer when available) **Note:** Standard Connection is for 20mm only.

* ☐ Single dwelling sewer connection with dry tapping 20mm (Applies where recycled water is not available)   Total $496.00

* ☐ Single dwelling sewer connection with wet tapping 20mm   Total $458.00

New connections which include class 'A' recycled water

* ☐ Single dwelling sewer connection with dry tapping 20mm and class 'A' recycled water   Total $1,270.40
  Dry tapping connection (If available) 20mm

Water tanks

* ☐ Does the development include below ground water tank interconnected with the drinking water supply?   Yes ☐ No ☐

If 'yes', please specify _____

Dry tapping Is the site available for dry tapping installation?   Yes ☐ No ☐

Note: Dry Tapping location must be accessible and clear of obstructions, if not a re-booking fee may apply.
Note: 20mm Dry Tappings are installed (Includes Potable and Class 'A' Recycled Water Connections). This includes the meter assembly and garden hose bib tap. Where Class 'A' Recycled Water is utilised, mandatory inspections are required. Inspection fees total $323.40 which is included in the fees. *GST of $29.40 is included in this charge.
Note: Contribution fees may apply to some backlog (Septic to Sewer) connections.

Other - sewer only

* ☐ Sewer only (Demolishing & rebuilding - retain water meter) $45.00   * ☐ House extensions (Recycled) $45.00
* ☐ Septic to sewer connection $45.00   ☐ Drain alteration / cut & seal $45.00
* ☐ House extensions $45.00   Details of work _____

Water only Note: All application fees and charges are GST free. Vacant land contribution fees may be applicable.

* ☐ Dry tapping (20mm only) - sewer unavailable $451.00   ☐ Wet tapping (20mm only) $413.00
* ☐ Plugging & re-tapping (Wet area & 20mm only) $463.00   ☐ Plugging only (20mm only) $125.00

Method of payment

Cash ☐   Cheque ☐   Credit card ☐

Card type   Mastercard ☐   Visa ☐   Card holder's name _____

Card no. ☐☐☐☐ ☐☐☐☐ ☐☐☐☐ ☐☐☐☐   Expiry date ____ / ____ / _____

Signature _____

Please notify me by

Mail ☐   Fax ☐   Email ☐   Email address _____

FIGURE 2.4 An example of a plumbing application form (although most applications are now completed online)

The service pipe then splits into individual services, feeding each property. A short service is where a property service is connected to a water main located on the same side of the street as the property. In this case, a split service may not be required.

HOW TO

HOW TO INSTALL A PROPERTY SERVICE AT THE TIME OF THE DEVELOPMENT OF A SUBDIVISION AND TERMINATE

1. A service connection ball valve (see Figures 2.5 and 2.6) is located within the property below ground level, to which a licensed plumber will connect once the property development has started, when they install the meter riser. The property service connection ball valves are located below ground with a water service marker tape brought to ground level by the developer at the time of installation. Within some local water utility areas these valves may be covered by a protective plastic cover.

2. Not all water utilities require a service connection ball valve to be fitted. It is important to check with the local water authority to find what the procedure is in the local area.

3. An isolation (ball) valve is installed at the meter and laid on its side below finished ground level.

4. An isolation (ball) valve (meter tap) is installed at the meter at the required height above finished ground level. Some local water utilities may require it to be tagged and pinned to prevent unauthorised use (see Figure 2.7). When working in a unfamiliar area, find out the local water authority requirements for a compliant installation.

FIGURE 2.6 Property service connection ball valve and meter riser

FIGURE 2.5 Property service connection ball valve marking tap

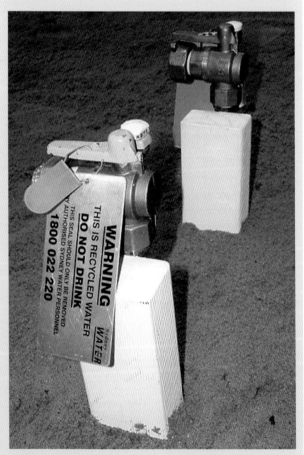

FIGURE 2.7 Meter isolation (ball) valve terminated at the required meter location

2.3 Planning and preparation

Being prepared and planning well will help to avoid costly problems and help the job to run more efficiently. The flowchart in Figure 2.8 illustrates the process for determining the planning for the installation of a property service.

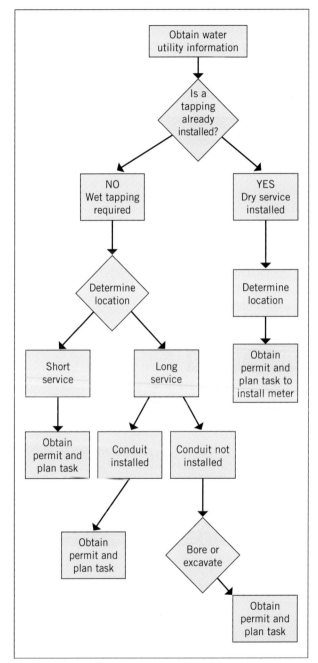

FIGURE 2.8 Installation flowchart

FROM EXPERIENCE

Planning ahead and being well organised results in time and cost efficiency.

HOW TO

HOW TO CARRY OUT PRELIMINARY WORK BEFORE THE PROPERTY SERVICE INSTALLATION BEGINS

The owner or licensed plumber must do the following:

1 Pay the appropriate fees and obtain a road opening permit from the local council. Familiarise yourself with the local council and water authority applications and requirements.
2 Obtain a copy of the plans, which must be approved by both the local council and the water utility.
3 Obtain appropriate consent forms (plumbing permit) and pay the required fees to the water utility.
4 Understand the local council or water utility's requirements as they may differ depending on the area.
5 Carry out a 'Dial Before You Dig' enquiry.

Excavation within public property

Before carrying out any excavation within public property (road or footpath), it is essential that a check be made to identify and locate all underground services. This is particularly important when using mechanical equipment. It is advisable to contact Dial Before You Dig (http://www.1100.com.au or telephone 1100; see Figure 2.9) to arrange the location information for services to be sent to you from the utilities, such as:

- electricity cables
- stormwater mains
- sewer mains
- water mains
- communication (telephone and television) cables
- gas mains.

FIGURE 2.9 Dial Before You Dig information

Source: Copyright © Dial Before You Dig 2007.

Due to the danger to life and amenities, *no work should proceed* until this process has been thoroughly completed, the services identified and the area marked (see Figure 2.10).

Source: iStockphoto/Hailshadow.

FIGURE 2.10 Footpath marked by a locating contractor in readiness for work

FROM EXPERIENCE

When excavating near existing services that are highly valued and easily damaged, such as optic fibre communication cables, 'non-destructive excavation' is recommended.

 In the planning and preparation stage, it is important that a risk analysis is carried out addressing the excavation method, whether trench support is required and the type of barricades needed.

Traffic control and road opening permits

When working in an area where there is any type of traffic movement, properly trained personnel must be employed to arrange traffic management plans and control the flow of traffic in and around the area. Not only should traffic be controlled in this manner but this may also be a WHS requirement in your state or territory. Failure to implement proper traffic control could be an infringement of the relevant WHS Act applicable to your workplace and result in prosecution or a fine. This is usually part of the company's quality assurance policy to ensure a safe work procedure.

When excavation works are to be carried out in a public area, all applications and permits should be obtained from the relevant authority, such as the local council or road and traffic authority. This usually requires the payment of a restoration fee. This may also affect the tender price, as each authority uses a different scale when calculating this fee. The rates can differ quite substantially in different areas. The costs are calculated on the type, load capacity and size of the road. For example, a main highway with a concrete surface would be very expensive to restore. Local emergency services, such as the police, fire brigade and ambulance services, may also need to be notified, especially if roads are to be closed, as this may affect the emergency routes to hospitals, fires and so on.

The size of the property service and meter

The diameter of a property service is sized to meet the demand and flow rate requirements of the user or users. This will be covered within the Certificate IV unit – Plan, size and lay out hot and cold water services and systems. Property service connections up to DN 65 are connected to the water utility's main via a single tapping or twin tappings (drillings). Larger services are connected by a tee and valve inserted in the main (see Figures 2.11, 2.12 and 2.13).

 Remember to check the local authority's guidelines prior to commencing any works.

Each state and territory may have supply tables that indicate:
- the drilling size requirements
- the size of the water service
- piping material type
- the total length of the water service for drilling size
- the size of the drilling
- the number of drillings for the property service size
- the size of the water meter.

LEARNING TASK 2.2

1 Where does the information for underground services come from?
2 What type of excavation is advisable to expose optic fibre cables?
3 How are water services larger than 65mm connected to the water main?

COMPLETE WORKSHEET 1

2.4 Selecting property service materials

Each material has its advantages and disadvantages. Being able to recognise the limitations of each type of material will help ensure the longevity and serviceability of the installation. While the materials listed in AS/NZS 3500.1 and covered in Chapter 1 of this book are approved for use in a property service, the

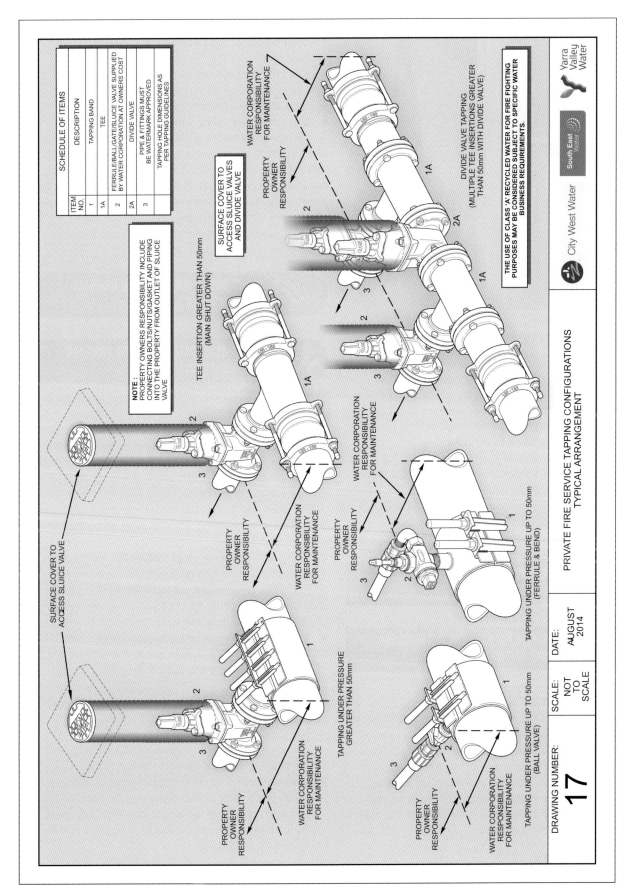

SCHEDULE OF ITEMS	
ITEM NO.	DESCRIPTION
1	TAPPING BAND
1A	TEE
2	FERRULE/BALL/GATE/SLUICE VALVE SUPPLIED BY WATER CORPORATION AT OWNERS COST
2A	DIVIDE VALVE
3	PIPE & FITTINGS MUST BE WATERMARK APPROVED TAPPING HOLE DIMENSIONS AS PER TAPPING GUIDELINES

NOTE:
PROPERTY OWNERS RESPONSIBILITY INCLUDE CONNECTING BOLTS/NUTS/GASKET AND PIPING INTO THE PROPERTY FROM OUTLET OF SLUICE VALVE

TEE INSERTION GREATER THAN 50mm (MAIN SHUT DOWN)

SURFACE COVER TO ACCESS SLUICE VALVES AND DIVIDE VALVE

WATER CORPORATION RESPONSIBILITY FOR MAINTENANCE

PROPERTY OWNER RESPONSIBILITY

DIVIDE VALVE TAPPING (MULTIPLE TEE INSERTIONS GREATER THAN 50mm WITH DIVIDE VALVE)

THE USE OF CLASS 'A' RECYCLED WATER FOR FIRE FIGHTING PURPOSES MAY BE CONSIDERED SUBJECT TO SPECIFIC WATER BUSINESS REQUIREMENTS.

SURFACE COVER TO ACCESS SLUICE VALVE

PROPERTY OWNER RESPONSIBILITY

WATER CORPORATION RESPONSIBILITY FOR MAINTENANCE

PROPERTY OWNER RESPONSIBILITY

WATER CORPORATION RESPONSIBILITY FOR MAINTENANCE

TAPPING UNDER PRESSURE UP TO 50mm (FERRULE & BEND)

SURFACE COVER TO ACCESS SLUICE VALVE

PROPERTY OWNER RESPONSIBILITY

WATER CORPORATION RESPONSIBILITY FOR MAINTENANCE

TAPPING UNDER PRESSURE GREATER THAN 50mm

PROPERTY OWNER RESPONSIBILITY

WATER CORPORATION RESPONSIBILITY FOR MAINTENANCE

TAPPING UNDER PRESSURE UP TO 50mm (BALL VALVE)

City West Water   South East Water   Yarra Valley Water

PRIVATE FIRE SERVICE TAPPING CONFIGURATIONS TYPICAL ARRANGEMENT

DRAWING NUMBER:	DATE:	SCALE:
17	AUGUST 2014	NOT TO SCALE

FIGURE 2.11 Installation of a tee and valve

Source: South East Water. This diagram is current at time of printing. It appears for training purposes only and should not be reproduced without the express permission of the owner, South East Water.

FIGURE 2.12 Tee and valve main connection

FIGURE 2.13 Main tap connection

local water utility may restrict the type of materials permitted to be used. For example, within Sydney Water's (NSW) area of operation, only copper (type A and B) and polyethylene (PE) material may be used. PE pipe and fittings must be manufactured to AS/NZS 4130 and rated to a minimum of PN 12.5.

Pipe equivalency tables are listed in AS/NZS 3500.1 showing the appropriate nominal diameter (DN), based on internal diameter for different pipe materials.

Learners should check with their supervisor/ instructor for the sizing requirements in their local area.

AS/NZS 3500.1 PLUMBING AND DRAINAGE: WATER SERVICES

AS/NZS 4130 POLYETHYLENE (PE) PIPES FOR PRESSURE APPLICATION

Copper (type A and B)

Copper is required if pressures are in excess of 1200kPa. Copper may be used on most property services and is essential to use for risers to the meter assembly. Some water utilities prefer the use of PE, as it provides insulation and also corrosion protection for metallic water mains. Type A and B copper tubes are the only types of copper tube allowed for use between the main and meter (property service).

Polyethylene (PE)

PE has the following classification:
- allowable size range DN 20, 25, 32, 40, 50 and 63
- maximum pressure 120m head (1200kPa) for type PN 12.5.

Pipe identification for PE pipes is as follows:
- drinking water (non-dual water areas): black with blue stripes.

The following colours cannot be used outside of dual water areas:
- drinking water (dual water areas): blue
- recycled water: black with purple stripes.

PE pipe must not be used where it is subject to direct sunlight, as part of the water meter assembly or vertical riser, or as specified in AS/NZS 3500.1 (Water services) and AS/NZS 4130 (Polyethylene (PE) pipes for pressure application). It must be a single length of pipe and be free of joints or fittings between the main isolation valve and the service connection valve or meter riser.

AS/NZS 3500.1 PLUMBING AND DRAINAGE: WATER SERVICES

AS/NZS 4130 POLYETHYLENE (PE) PIPES FOR PRESSURE APPLICATION

Only copper pipes may be used as part of the meter riser; where copper pipes are used as part of a non-drinking water service they must be sheathed with purple PE. The non-drinking water pipe must have a 75mm purple identification tape installed on top of the pipe, running longitudinally, and fastened to the pipe at not more than 3m intervals.

All pipes and fittings must display a Standards mark WaterMark as required by AS/NZS 5200, as applicable.

Bedding and backfill requirements

The bedding and backfill requirements for a property service are covered within AS/NZS 3500.1. Learners should refer to AS/NZS 3500.1 to gain a more in-depth understanding. Some of the main points are as follows:
- Water services must be surrounded with no less than 75mm of compacted sand or fine-grained soil. Avoid contact with any hard-edged object such as rocks.
- Water services must be backfilled within the public property as required by the local council requirements. Learners should contact their local council to find out what the backfill requirements are within public property in the local area.
- Unless specified to the contrary by the local water utility or council, copper pipes may be installed in soil excavated from the trench in which they are to be installed, provided the soil is compatible and free from rock and rubble.

AS/NZS 3500.1 PLUMBING AND DRAINAGE: WATER SERVICES

Corrosive areas

Corrosive areas are those that contain compounds consisting of magnesium oxychloride (magnesite) or its equivalent, coal wash, ash, sodium chloride (salt), ammonia or materials that may produce ammonia. A benign soil is typically considered to be a sandy, free-draining soil of low salt content that is non-corrosive. Anything else should be treated as a potentially aggressive/corrosive area. If copper tube is being used, it will need to be protected by insulating the pipe with a plastic sleeve (lagging) if installed in a corrosive area.

Depth of cover

Any water service located below ground on public and private property must have a minimum cover as indicated below, unless otherwise stipulated by the local water utility or council:

- 75mm under concrete slabs and footings
- 300mm not subject to vehicular loading (excluding fire services)
- 600mm for fire services not subject to vehicular loading
- 450mm subject to vehicular loading but not under a carriageway
- 600mm subject to vehicular loading under a sealed carriageway
- 750mm subject to vehicular loading under an unsealed carriageway.

AS/NZS 3500.1 PLUMBING AND DRAINAGE: WATER SERVICES

2.5 Identifying installation requirements

Site inspection

A site inspection allows you to visualise any areas relevant to the task that may not be apparent from the plans and specifications, such as:

- access
- traffic conditions
- site conditions
- water main location
- suitable position of the meter (as required by the local water utility)
- obstructions that may impede the installation
- hazard identifications
- confirmation of the information obtained from the plans and specifications.

A risk analysis on the initial site inspection will identify all the apparent hazards. The risks can be assessed so the appropriate control measures can be planned and put in place.

Job sequencing

Once all the required information is obtained, the job is planned by sequencing it into logical steps (see Figure 2.14). This sequencing is an ideal opportunity to carry out a risk assessment and prepare a safe work method statement (SWMS) or job safety analysis (JSA) for the task. Recording and reviewing these procedures will help to refine and enhance your company's quality assurance procedures. The sequencing and planning will also identify what safety equipment, personal protective equipment and tools are needed for the task.

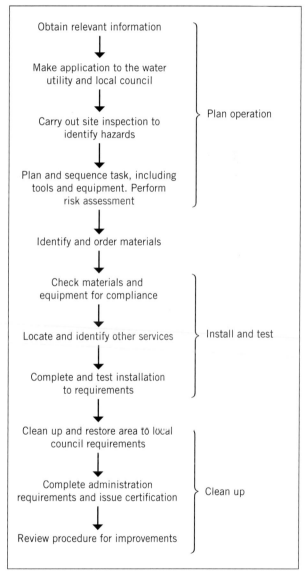

FIGURE 2.14 Installation sequence

Personal protective equipment

Personal protective equipment (PPE) may include, but is not limited to, any of the in-depth list of PPE that can be found in Chapter 6: Handle and store plumbing materials in *Basic Plumbing Skills*.

Tools and equipment

Tools and equipment may include, but are not limited to, any of the in-depth list of tools that can be found in Chapter 7: Use plumbing hand and power tools in *Basic Plumbing Skills*.

Before starting work

Before commencing any installation work, a final check of the information gathered at the planning stage should be made. Some companies will have a checklist for this procedure as part of their quality assurance program. The checklist should include, but is not limited to, the following questions:

1 Are you familiar with the local authorities' regulations and requirements?
2 Do you have a copy of the approved plans and specifications?
3 Do you have the appropriate permits and have all the fees been paid?
4 Have the materials ordered been checked for compliance and quality?
5 Are all the tools and equipment required available and in serviceable order?
6 Has all PPE been checked for correct operation and condition?
7 Have you made a final site inspection to ensure that there have been no alterations or obstructions since the planning inspection?

This is also your last chance to check the calculations used to determine pipe size and material compliance to standards, codes and regulations.

AS/NZS 3500.1 PLUMBING AND DRAINAGE: WATER SERVICES

FROM EXPERIENCE

Showing initiative by selecting the correct tools and equipment for the job is a valued skill.

Prepare the work site

To protect the workers and the public, the work site must be clearly defined and secured to prevent unauthorised access. Special consideration should be given to vehicular traffic.

Planning the work around vehicular traffic requires specific requirements such as traffic control, signage, barricades and road plates.

Determine the water main location

Before either arranging or making a tapping in the main, licensed plumbers with a main drilling accreditation must find a hydrant or valve cover in either the footpath or the road. It is important to ensure that it is the correct main, as some streets may have more than one water main laid in them. A plate affixed to a wall, fence or telegraph pole identifies the location of the main (see Figures 2.15 and 2.16). This plate is marked in a series of different abbreviations that will help you determine the location (see Table 2.1). It also has stamped into it the size of the main and the distance from the plate to the main. Other local water utilities may use a cat's eye (reflector) affixed to the centre of the road to identify a hydrant location (see Figures 2.17 and 2.18).

FIGURE 2.15 Hydrant indicator placed on a power pole

FIGURE 2.16 Hydrant valve indicators placed on a power pole

TABLE 2.1 Hydrant/service valve identifying abbreviations

Abbreviation	Meaning
HR	Hydrant in road
HP	Hydrant in footpath
AV	Area valve
DV	Dividing valve
SV (or SVP)	Service valve in footpath
SV (or SVR)	Service valve in road

To obtain the line of the main you need to repeat the above step up or down the road. The line between the two hydrants or valves is the line of the utility's water main.

GREEN TIP

Be prepared to place silt and sedimentation barriers in place prior to excavation to protect the environment from pollution. Heavy fines could be incurred otherwise.

Source: Ray-O-Lite Aust Pty Ltd.

FIGURE 2.17 Drinking water hydrant indicator

Source: Ray-O-Lite Aust Pty Ltd.

FIGURE 2.18 Non-drinking water hydrant indicator

Excavate for tapping and connection of the service

Mark out the excavation and be sure that the meter is directly opposite and at a right angle (90°) to the main. This is important so the main tap can be easily located in the future. Where the excavation crosses any identified other services they must be located by hand (excavation) and identified to prevent costly damage, delays and possible fatality.

The following requirements must be met:

1 Any excavation for connection to water mains must be in accordance with the water utility's requirements (for an example, refer to Figure 2.13). The excavation for drilling must be a minimum of 1m × 1m in size for a single drilling, with a clearance below the main of a minimum of 150mm.
2 The excavation must be maintained in a dewatered condition.
3 The excavation from the water main into the property can be an open trench if it is close by. However, if the water main is in or on the other side of the road, an under-road boring machine should be considered.

4 Drilling of the main or installation of the tee and valve must be carried out as required by the water utility (for an example, refer to Figure 2.12).
5 If a protective sleeve around a cast iron/ductile iron main is damaged, it must be repaired to the water utility's specifications by the licensee.

Determine the water meter location

Water meters that are used for billing purposes (see Figure 2.19) must be installed:

- within the property
- as near as practicable to the street alignment
- immediately downstream of the meter isolating valve
- directly opposite the connection and at a right angle (90°) to the main
- at the front of the property within easements.

FIGURE 2.19 Water meter installation

Where the property service is required to be offset within a private easement or right-of-way, a below-ground isolating valve must be provided downstream of the offset:

- as required by the relevant water utility in other locations
- horizontally
- so that it is protected from damage – if located in frost-sensitive areas, meters must be protected against damage caused by freezing of the water. This can be achieved by installing the water meter underground in an accessible box.

It must be readily accessible for reading, maintenance or removal, and be clear of obstacles; and meters DN 50 or larger must be supported independently of the piping.

FROM EXPERIENCE

Learners are reminded to check the local authority's guidelines prior to commencing any works. See Figures 2.20, 2.21 and 2.22 for water meter variances.

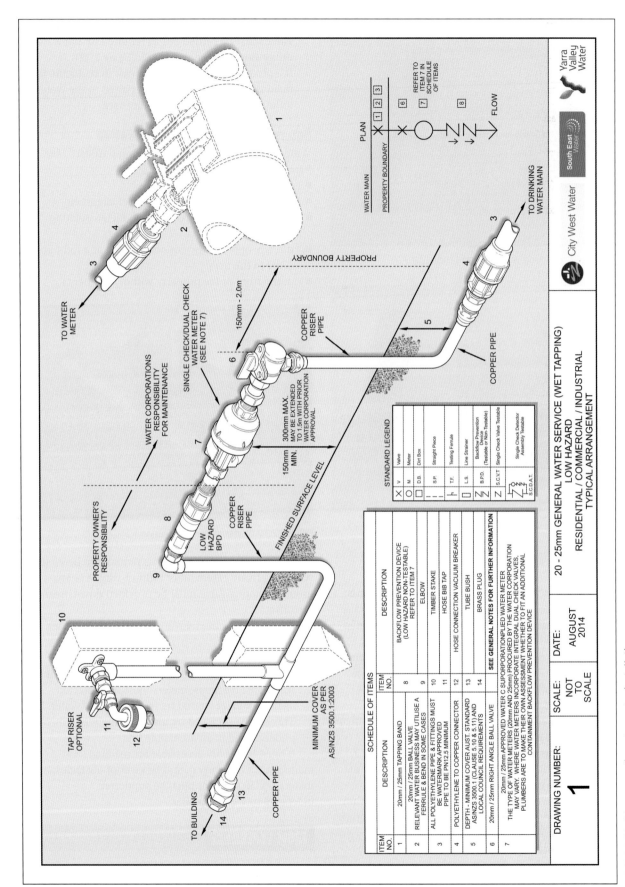

FIGURE 2.20 Diagrammatic water meter installation

Source: South East Water. This diagram is current at time of printing. It appears for training purposes only and should not be reproduced without the express permission of the owner, South East Water.

Source: South East Water. This diagram is current at time of printing. It appears for training purposes only and should not be reproduced without the express permission of the owner, South East Water.

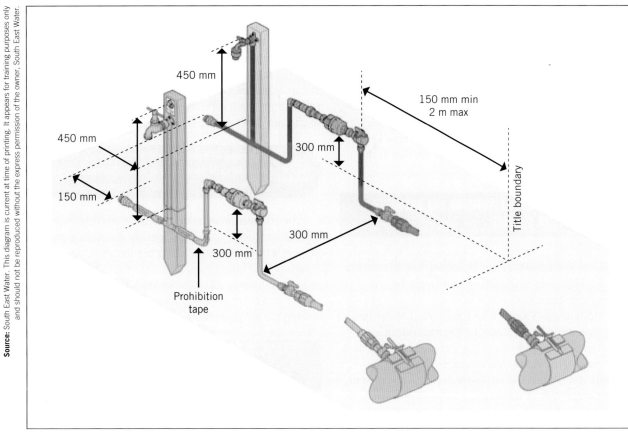

FIGURE 2.21 Drinking and recycled water meter assembly

Source: South East Water. This diagram is current at time of printing. It appears for training purposes only and should not be reproduced without the express permission of the owner, South East Water.

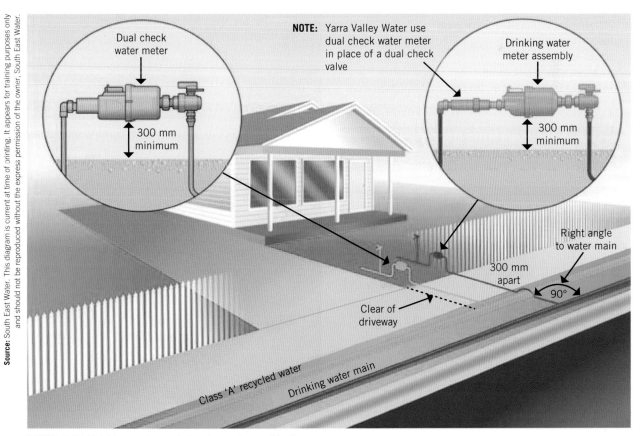

FIGURE 2.22 Drinking and recycled water meter positioning

Some water utilities may have other specific requirements, such as:

- a minimum clearance of 300mm is to be left under the meter (though water utilities in some areas may only require 150mm clearance under the meter)
- a minimum clearance of 300mm is required when installing containment backflow prevention devices (at the meter)
- water meters must be installed within 1m of the front boundary and within 1m and parallel to the side boundary.

 AS/NZS 3500.1 PLUMBING AND DRAINAGE: WATER SERVICES

Recycled water meter installation requirements

Non-drinking (recycled) water meters must be:

- purple in colour
- a minimum of 300mm from the drinking water meter
- installed with a minimum of 150mm clearance between the underside of the meter and the finished ground level. Additional clearance (150mm) is required if a testable backflow device is installed on the outlet side of the meter.

Check with the local water utility to ascertain the local installation requirements and whether special inspections are required in their recycled water area. Generally, in recycled water areas, mandatory inspections are required at three stages:

1 meter to dwelling
2 rough-in
3 commissioning and fit-off.

Inspections for meter to dwelling and rough-in may be combined.

 AS/NZS 3500.1 WATER SERVICES

 COMPLETE WORKSHEET 2

2.6 Installing water services

Tapping (drilling) the main

The tapping (drilling) can be made as either a 'dry tapping' or a 'wet tapping'.

- Dry tappings are usually made during the development stage of a subdivision prior to pressurisation of the main before the roads are sealed. The property service is terminated as described earlier in this chapter.
- Wet tappings are carried out on pressurised water mains as directed by the water utility.

As a general rule, drinking water mains are installed in either cast iron/ductile iron or from PE materials; however, other materials may have been used in the past. For example, while asbestos cement (AC) mains are no longer installed, there are many still in existence. When tapping AC mains, special precautions must be observed as older mains are likely to contain asbestos.

 The proper PPE and hazardous materials handling must be used when working with asbestos. Refer to the Safety Data Sheet (SDS) for Asbestos.

The connection of a main isolation (ferrule) valve to a cast iron/ductile iron main is made via drilling and tapping a thread in the wall of the pipe; or more currently by a tapping saddle and then drilling the main, as shown in Figure 2.23. The connection of a property service to a PE main is made via a tapping saddle and drilling the main, as shown in Figure 2.24.

FIGURE 2.23 Connection to a ductile iron drinking water main

FIGURE 2.24 Connection to a PE recycled water main

HOW TO INSTALL A TAPPING SADDLE

1 Prepare the area of pipe to be covered by tapping saddle. Be sure it is clear of any joints or fittings.
2 Position the bottom half of the tapping saddle making sure it is away from scored, pitted or damaged areas, as this will not provide a good seating area.
3 Fit the top of the tapping saddle around the pipe so that the bolts pass through the bolt holes in the top of the tapping saddle.

4 With one hand underneath the bottom half of the tapping saddle pressing upwards, locate the bolt holes and screw on the nuts supplied with the other hand.
5 Finger-tighten the nuts so that the gap between the two halves of the tapping saddle is equal on both sides.
6 Tighten the nuts to 20Nm (15ft/lb).

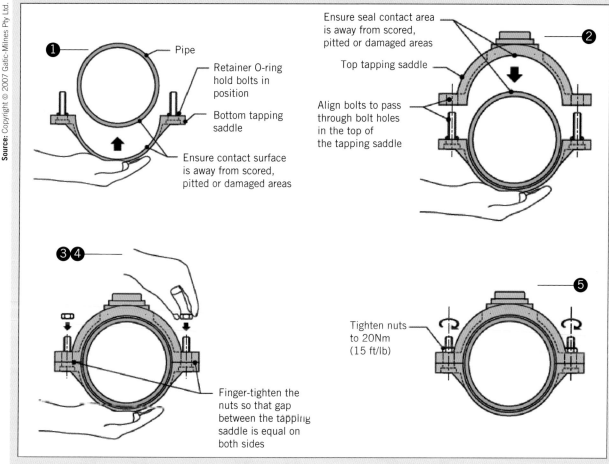

FIGURE 2.25 Installing a tapping saddle

Only licensed plumbers with the required qualifications and approved by the water utility are allowed to carry out either wet or dry tappings. Tapping a water main already under pressure allows services to be connected at a later date without interrupting the water supply for existing consumers.

Additional requirements for lip seals

When installing tapping saddles with lip seals, make sure the lip seal is in the correct position. The lip seal should be placed in the sealing groove with the sharp inner side of the lip seal facing out towards the pipe. A small moulding line running around the lip seal also indicates this should be facing down towards the pipe.

If the lip seal is placed in the incorrect position, the joint may leak.

Seal installation

An O-ring is installed as shown in Figure 2.26.

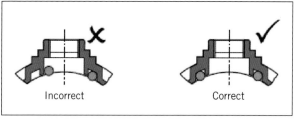

FIGURE 2.26 O-ring installation

Source: Copyright © 2007 Gatic-Milnes Pty Ltd.

HOW TO DRILL TAPPING AND FIT MAIN COCK

1 Fit an appropriate tapping band to the pipe and screw in the selected under-pressure tapping ferrule.
2 Unscrew the bonnet assembly, remove the jumper valve and fully open the poly ferrule plug. (A ball valve may be used in lieu of the under-pressure tapping ferrule where acceptable.)
3 Screw the tapping machine into the top of the ferrule, using the correct adapter as required.
4 Rotate the feed nut until the drill tip makes contact with the pipe.
5 Apply pressure to the drill bit via the feed nut, turning the ratchet at the same time to cut the hole.
6 When drilling is completed, reverse the feed nut to retract the drill bit until clear of the poly ferrule plug.
7 Allow water escaping from the ferrule to flush the system, then close the poly ferrule plug. (A hose may be connected to the ferrule to divert the waste water.)
8 Close the poly ferrule plug, remove the tapping machine and replace the jumper valve and bonnet.
9 Connect the service pipe and open the poly ferrule plug to charge the service line.

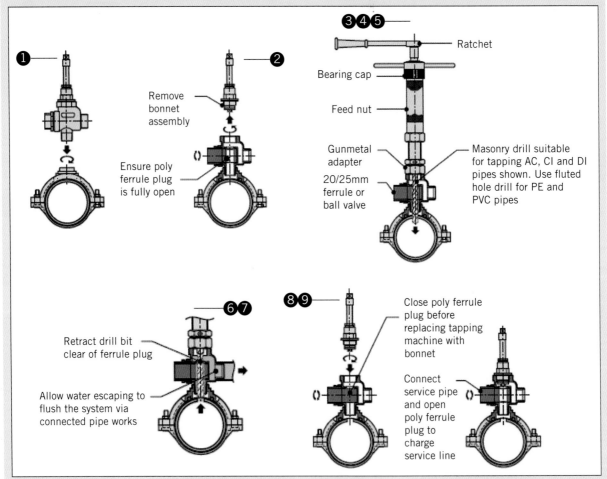

Source: Copyright © 2007 Gatic-Milnes Pty Ltd.

FIGURE 2.27 Tapping sequence

Standard hand-operated under-pressure tapping machines allow tappings of 20mm and 25mm to be made in all pipe types without the need for electrical power in remote locations or on new construction sites.

Important operating instructions

Regardless of the machine used, always follow these important operating instructions:

- Operate by hand only; added leverage will damage the machine and the drill bit.
- Minimise sideways movement during operation; rocking of the machine will damage the drill bit.
- Always limit the feed rate to ensure a smooth cutting action; the feed nut may need to be restrained to prevent the drill tip from jamming.
- Do not use the upper handles to retract the feed nut; this may loosen the bearing cap, which, if removed under pressure, will cause the following results:
  - dislodgement of the circlip
  - damage to the bearing
  - injury to the operator.

- Keep the bearing cap screwed on tightly.
- Never place your head directly over a live tapping as it can blow off if bumped or damaged in the installation/excavation process.

Tapping small-diameter copper mains

The following procedure is the usual method used for connecting a property service to a copper trunk main:

1 Tapping to a small-diameter copper main is made by inserting a brass tee with a female iron (FI) centre outlet in the main, using a compression coupling or by silver soldering. Under no circumstances are soft-soldered joints to be used.

2 The FI outlet and the ferrule should be kept in a vertical position to allow the service pipe to be connected as early as possible.

Insulating bush

All metallic water services *must* be insulated from the main by the use of an insulating bush either:

1 fitted between the main tap and the property service, or

2 incorporated into the tapping saddle, into which is screwed the main tap.

The insulator is designed to prevent corrosion between the dissimilar metals (cast iron/ductile iron and copper) and to help prevent stray electrical currents passing from the property services to the main, thus reducing the life of the main and possibly causing it to become live, should an electrical fault exist.

Installing a long service

If the water main is on the other side of the street (long service) to the property, there are three alternative methods of installing the service:

1 Use the conduit supplied by the developer when the road was laid. Markers are usually located in the concrete kerb.

2 Arrange for an under-road boring machine to drill a small hole under the road through which the service can be pushed.

3 Use a water drill to drill a small hole under the road through which the service can be pushed.

Proximity to other services

A non-drinking water service must be separated from drinking water services as follows:

- Above-ground installations of non-drinking water services must not be installed within 100mm of any parallel drinking water service except when installed in pipe duct or structurally separated.
- Below-ground installations of non-drinking water services must not be installed within 300mm of any parallel drinking water supply.

Some local water utilities may require the non-drinking water service to be installed on the left-hand side (facing the property).

AS/NZS 3500.1 PLUMBING AND DRAINAGE: WATER SERVICES

Electrical safety

When carrying out any work on a metallic drinking or non-drinking water service connected to the water utility main, consideration needs to be given to electrical safety precautions and earthing to protect against potential electrical shock. Plumbers must always check for stray current before working on any metallic service. Further information may be found in AS/NZS 3500.5 and SafeWork NSW's website (under 'Electrical hazards for plumbers' at https://www.safework.nsw.gov.au). See also http://www.plumbingconnection.com.au/bonding-straps.

 Bonding straps must be used when disconnecting a meter or when cutting a metallic water service to prevent an electrical shock (see Figure 2.28). Plumbers have died from electrocution when not using bonding straps.

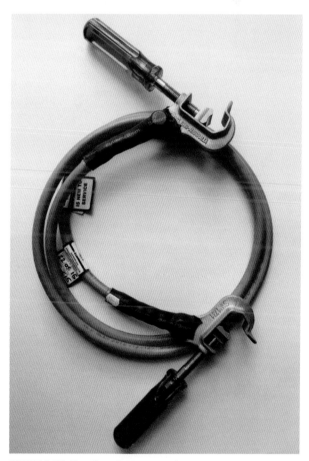

FIGURE 2.28 Bonding strap connection

Temporary interconnection between the drinking and non-drinking water services

While both the drinking and non-drinking water services need to be charged with water during construction, any interconnection between the drinking

FIGURE 2.29 Temporary interconnection between drinking and non-drinking water services

and non-drinking water services *is permitted at the meter assembly only* (see **Figure 2.29**). A temporary bypass is to be installed with a shut-off valve between the drinking and non-drinking water services at the water meter assembly only; a connection is not permitted to be made to the non-drinking water meter isolation valve. All meter assemblies and temporary bypasses may need to be inspected by a water utilities inspector.

Example of a property water service

Figure 2.30 shows an example of water services for a Class 2 building.

This example is based on:

- a minimum head of 30m
- the highest fixture being 13m above the main
- a 5m minimum head being required at any fixture outlet.

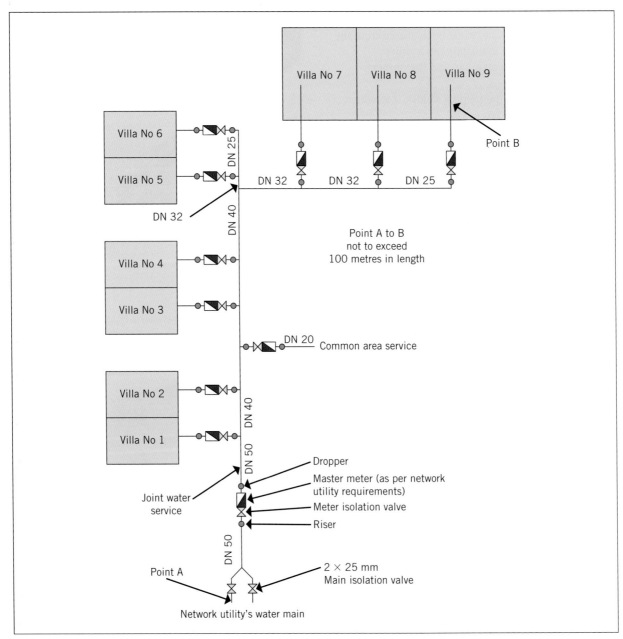

FIGURE 2.30 Water services for a Class 2 building

The size of the service may change if any of the design requirements indicated above were to change. (The sizing of a water service is not part of this unit's requirements.)

Example of combined fire and domestic water services

The general layout for combined fire and domestic water services is shown in **Figure 2.31**. This is commonly known as a trident service because the one property service is feeding three services – the domestic water service, the fire sprinkler service and the fire hydrant service.

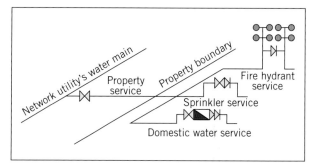

FIGURE 2.31 Combined fire and domestic water services

LEARNING TASK 2.3

1 Who is authorised to carry out main tappings?
2 What do the letters 'HP' on signed fixed to telegraph pole mean?
3 Describe a 'long service'.
4 How is stray current tested?

2.7 Testing the service and cleaning up

Before backfilling any property service, it must be tested to the requirements of AS/NZS 3500.1, such as, but not limited to, the following:

- The pipework must be flushed to remove any foreign matter until the water runs completely clear.
- The service must be subjected to a hydrostatic pressure test of 1500kPa for a minimum of 30 minutes.
- The service must show no signs of leakage during the test.

When carrying out a test it is important to ensure that all pipework is isolated from any fixture or appliance that may be damaged by the test.

AS/NZS 3500.1 PLUMBING AND DRAINAGE: WATER SERVICES

Remember that the job is not complete until the work area is cleaned up and the rubbish removed. Be sure to reduce waste by storing any material left over that can be reused.

Clean all tools and equipment when packing up and keep them lubricated and maintained.

Document information in the diary. This would include job address, description of work done, the names of people involved in the job, and the time and the dates spent on that job. Some companies would have a generic worksheet that would be completed. This would form part of the company quality assurance policy.

LEARNING TASK 2.4

1 Why is it important to flush pipes before testing?
2 Why must tapware and fixtures be disconnected when testing the water service?
3 How are tools lubricated?

COMPLETE WORKSHEET 3

SUMMARY

- The service from the water main up to and including the water meter is called the 'property service'.
- Make sure all permits and fees are completed before starting work on the site.
- Make sure a thorough search for underground services is carried out.
- The property service is to be installed at 90° to the water main.
- Copper or polyethylene are generally the piping materials used for property services.

- Water meters must be installed within 1m from the front property boundary.
- Extra inspections are usually required for recycled water service installations.
- Licensed plumbers must be accredited to install main tappings.
- Always check for stray electrical current and use bonding straps when disconnecting a service.
- Test the water service to 1500kPa for 30 minutes as per AS/NZS3500.1.

REFERENCES

South East Water: **http://www.southeastwater.com.au**

GET IT RIGHT

1   Which water meter is incorrectly installed?

2   Why is this an issue?

3   What is the correct procedure?

WORKSHEET 1

Student name: _____

Enrolment year: _____

Class code: _____

Competency name/Number: _____

To be completed by teachers

Satisfactory ☐

Not satisfactory ☐

Task: Review the sections '2.2 Installing a property service (main to meter)' and '2.3 Planning and preparation' and answer the following questions.

1 In your own words, define a water main.

_____

_____

_____

2 Describe a dual water system.

_____

_____

_____

3 What is the minimum size of a property service?

_____

4 What colour is used to identify a recycled water service?

_____

5 In your own words, describe three tasks the licensed plumber must complete before installing a property service

_____

_____

_____

6 Dial Before You Dig is a service provided to help locate various public utilities services. List four of these utilities services.

_____

_____

_____

_____

7 The water utility may require a DN 25 service if the total length of service exceeds how many metres?

_____

8 Up to what diameter is a single or twin tapping used?

_____

9 How would a 100mm water service connect to a water main?

_____

10 Is the potable (drinking) water main or the recycled water main closer to the property boundary?

_____

WORKSHEET 2

To be completed by teachers

Satisfactory ☐

Not satisfactory ☐

Student name: _____

Enrolment year: _____

Class code: _____

Competency name/Number: _____

Task: Review the section '2.4 Selecting property service materials' and answer the following questions.

1   Give four reasons why it is recommended to carry out a site inspection before starting work on a water service.

_____

_____

_____

_____

2   What is the recommended minimum cover for a water service located below ground in public or private property as per AS/NZS 3500.1?

   a   Under a concrete slab or footing: _____ mm

   b   Not subject to vehicular loading (excluding fire services): _____ mm

   c   Fire services not subject to vehicular loading: _____ mm

   d   Subject to vehicular loading but no carriageway: _____ mm

   e   Subject to vehicular loading under a sealed carriageway: _____ mm

   f   Subject to vehicular loading under an unsealed carriageway: _____ mm

3   What is the recommended minimum support for a water service as per AS/NZS 3500.1?

_____

4   Which classes of copper are allowed to be used for property services?

_____

5   Name two limitations using polyethylene pipe.

_____

6   Identify the following symbols in relation to hydrant and service valves for water mains:

HR _____

HP _____

SVP _____

SVR _____

7   List three requirements of a non-drinking supply property service.

_____

_____

_____

8   Copper must be used on a service that exceeds _____ kPa.

9   What colours are used to identify PE pipe used for drinking water in a non-dual area?

_____

10 Where should a water meter be located in relation to the water main and why?

_____

_____

11 What is the minimum clearance below a water meter?

_____

12 What support requirements are needed for water meters 50mm and larger?

_____

_____

13 What is the minimum height a water meter can be installed with a backflow prevention device?

_____

14 At what three stages must a recycled water installation be inspected?

_____

_____

_____

WORKSHEET 3

Student name: _____

Enrolment year: _____

Class code: _____

Competency name/Number: _____

Task: Review the section '2.4 Selecting property service materials' and '2.7 Testing the service and cleaning up' and answer the following questions.

1   Define the term 'dry tapping' in relation to drilling a water main.

_____

_____

2   What are the minimum dimensions of an excavation for a main tapping?

_____

3   What provides the seal between a tapping band and a PVC water main?

_____

_____

4   What is the purpose of an insulating bush on a cast iron main tapping?

_____

_____

5   Underground drinking water services should have _____ mm clearance from a non-drinking water service.

6   An above-ground recycled water service must not be installed within _____ mm of a potable water service.

7   What is a long service?

_____

_____

8   What is the minimum pressure and time a water service is tested for, as per AS/NZS 3500.1?

_____

9   When should bonding straps be used, and why?

_____

_____

_____

10   Why is it important to check the local authority's requirements before commencing work?

_____

_____

_____

WORKSHEET RECORDING TOOL

Learner name		Phone no.	
Assessor name		Phone no.	
Assessment site			
Assessment date/s		Time/s	
Unit code & title			
Assessment type			

Outcomes

Worksheet no. to be completed by the learner	Method of assessment WQ – Written questions PW – Practical/workplace tasks TP – Third-party reports SC – Scenarios RP – Role plays CS – Case studies RW – Report writing PF – Portfolio	Satisfactory response	
		Yes ✓	No ✗
Worksheet 1	WQ	☐	☐
Worksheet 2	WQ	☐	☐
Worksheet 3	WQ	☐	☐

Assessor feedback to the learner

Feedback method (Tick one ✓):	☐ Verbal ☐ Written (if so, attach) ☐ LMS (electronic)

Indicate reasonable adjustment/assessor intervention/inclusive practice (if there is not enough space, a separate document must be attached and signed by the assessor).

Outcome (Tick one ✓):	☐ Competent (C) ☐ Not Yet Competent (NYC)

Assessor declaration: I declare that I have conducted a fair, valid, reliable and flexible assessment with this learner, and I have provided appropriate feedback.

Assessor name:	
Assessor signature:	
Date:	

Learner feedback to the assessor

Feedback method (Tick one ✓):	☐ Verbal	☐ Written (if so, attach)	☐ LMS (electronic)

Learner may choose to provide information to the RTO separately.

	Tick one:	
Learner assessment acknowledgement:	**Yes** ✓	**No** ✗
The assessment instructions were clearly explained to me.	☐	☐
The assessment process was fair and equitable.	☐	☐
The outcomes of assessment have been discussed with me.	☐	☐
The overall judgement about my competency performance was fair and accurate.	☐	☐
I was given adequate feedback about my performance after the assessment.	☐	☐

Learner declaration:
I hereby certify that this assessment is my own work, based on my personal study and/or research. I have acknowledged all material and sources used in the presentation of this assessment, whether they are books, articles, reports, internet searches or any other document or personal communication.
I also certify that the assessment has not previously been submitted for assessment in any other subject or at any other time in the same subject and that I have not copied in part or whole or otherwise plagiarised the work of other students and/or other persons.

Learner name:	
Learner signature:	
Date:	

SET OUT AND INSTALL A WATER SERVICE: METER TO POINTS OF DISCHARGE

3

Chapter overview

This chapter focuses on the installation requirements of a water service from the outlet bend of a meter up to the connection points of fixtures, appliances and outlets, including drinking and non-drinking water services and hot water services.

Learning objectives

Areas addressed in this chapter include:
- planning and preparation
- material selection
- setting out and installing water services from the meter to the point of discharge
- testing the house service and clean-up.

3.1 Background

Once the water supply has been connected to the property from the authority's main, it is metered and then connected to the points of discharge/use. Appropriate material selection and the correct setting out of the installation are important as much of this work will be concealed, so will be hard to access at a later time if mistakes are made at this point. Correct planning and preparation at this stage is essential before the walls are sheeted or rendered.

3.2 Planning and preparation

As discussed in Chapter 2, it is important to carry out your pre-planning work:

1 Obtain the required permits from the water utility and any other authority that has control over the work that may have specific local requirements.
2 Check the plans for any specific installation requirements such as fixture set-outs and the type of tapware.
3 Check AS/NZS 3500 and Plumbing Code of Australia (PCA) for installation requirements.
4 Select the material type to be used (copper or polymer).
5 Carry out a site inspection.
6 Ensure quality assurance and work health and safety.
7 Arrange any special installation requirements such as backfill material and depth of cover.
8 Organise the appropriate tools and equipment including personal protective equipment (PPE).

Design considerations

It is important to plan the installation well so the job runs smoothly and is kept to a high standard. Designing the most effective way to supply water to fixtures and outlets by choosing the appropriate material, fittings and fixings will help achieve a professional job.

Drinking and non-drinking services

When designing drinking and non-drinking water services you need to consider:

■ the most direct route from the water meters to the fixtures, appliances and outlets
■ the location and type of the hot water system – it should be installed central to and as close as possible to the most frequently used outlet, which is usually the kitchen sink. It is a good rule to locate the hot water heater so *no more than 2L of cold water is drawn off from a hot tap* before hot water is present. If this cannot be achieved then consideration should be given to the installation of a flow and return hot water system, or the installation of more than one hot water heater, and to the type and location of temperature control devices.

■ running the hot and cold water services close to each other; if installed horizontally, the hot water service should be installed above the cold water pipe
■ ensuring hot water branches are as short as is practicable.

> **GREEN TIP**
>
> It is important to plan the hot water system location well, so less energy and cold water draw-off is achieved with lower running costs.

Non-drinking water usage

Non-drinking (recycled) water may only be used for:
■ toilet flushing
■ garden watering/irrigation
■ washing cars
■ filling ornamental ponds
■ firefighting (check with the local authority if it may be used in your local area)
■ construction and industrial purposes (only with special approval of the water utility)
■ clothes washing (check with your supervisor/ instructor if it may be used in your local area). Non-drinking (recycled) water must *not* be used for:
■ drinking
■ cooking or other kitchen purposes
■ personal washing, such as bathing
■ evaporative coolers (although some state/territory health authorities have given approval for highly treated recycled water to be used for these purposes in some areas)
■ household cleaning
■ internal wash-down outlets (taps)
■ swimming pools
■ recreation involving water contact (such as children playing under sprinklers)
■ the irrigation of fruit trees and crops that are eaten raw or unprocessed.

AS/NZS 3500.1 PLUMBING AND DRAINAGE: WATER SERVICES

> **GREEN TIP**
>
> Using recycled water is very important to help conserve precious drinking water, which will become even more scarce in the future.

Thermal insulation and frost protection

If a water service is to be installed in an area subject to regularly low temperatures (below 0°C), the service must be protected to prevent the water from freezing. This can be achieved by installing in-ground pipe to a minimum depth of 300mm and providing the pipe with a protective waterproof insulation and a trace heating system.

A water service located on a metal roof must be designed to prevent it from coming into contact with the roof material because dissimilar metals in contact will cause electrolysis. Consideration needs to be given to installing pipes in other locations. Where this is not possible, pipes need to be insulated with the required minimum thick thermal insulation for that area and provided with a waterproof sheathing.

FROM EXPERIENCE

Showing an awareness of the environment that you are working in by using correct applications and materials increases customer confidence.

Insulation of hot water pipes

Thermal insulation of hot water pipes must meet the minimum thermal insulation R-value for that area. This information may be found in AS/NZS 3500.4.

Thermal insulation must be provided for a domestic hot water service as follows:

- on the cold water inlet pipe between the isolating valve and the hot water heater
- for a minimum of the first 500mm of pipe from the outlet of the hot water heater or 150mm down the first vertical leg of a heat trap (if fitted)
- on multiple installations, on the hot water manifold to a point 500mm past the last hot water branch.

AS/NZS 3500.4 PLUMBING AND DRAINAGE: HEATED WATER SERVICES

LEARNING TASK 3.1

1 Name two design considerations for a water service.
2 Where is pipe insulation required on hot water services?

3.3 Selecting the materials

Unless otherwise stated in the plans or specifications, the types of materials that may be used are covered within AS/NZS 3500 and include:

- copper
- polybutylene (PB)
- cross-linked polyethylene (PE-X)
- polypropylene (PP)
- polyethylene (PE)
- unplasticised polyvinyl chloride (PVC-U)
- modified polyvinyl chloride (PVC-M; used on large water services)
- oriented polyvinyl chloride (PVC-O; used on large water services).

When selecting material you need to consider the limitations, such as:

- what the pipework is going to be used for
- water quality and its temperature
- the type of ground (for considerations such as corrosion and leaching)
- the possibility of chemical attack from the environment
- the material compatibility of material and products
- frost protection
- the water pressure within the water utility's supply system
- any special material installation requirements, such as:
  - some polymer pipes and fittings may not be installed in direct sunlight
  - polymer pipes may not be used between the cold water isolation valve and the hot water heater
  - polymer pipes may not be used as part of a temperature pressure relief valve
  - polymer pipes and fittings may not be used within 1m of the outlet of a hot water heater
  - polymer pipes and fittings may be installed on the outlet side of a temperature control valve.

Installation requirements for materials should always be checked for the minimum requirements, which are stated in AS/NZ 3500.1. When selecting a material other than copper, it is important to refer to the Australian equivalent pipe size guide in AS/NZS 3500.1 (see Table 3.1 for an example).

TABLE 3.1 Example from the Australian equivalent pipe size guide

Material size (DN)	Acceptable equivalent size according to material type				
	Copper	PVC	PB	PE-X	PP
20	20 (15mm internal bore)	20 (15mm internal bore)	22	25 (20.5mm internal bore)	25 (20.5mm internal bore)

Apart from the installation requirements of AS/NZS 3500.1 and AS/NZS 3500.4, the installation of each material is also covered by other standards such as AS 1432 (copper tubes for plumbing, gasfitting and drainage application), AS 2492 (cross-linked polyethylene (PE-X) pipe for hot and cold water applications) and AS/NZS 2642 (polybutylene pipe systems).

AS/NZS 3500.1 PLUMBING AND DRAINAGE: WATER SERVICES

AS/NZS 3500.4 PLUMBING AND DRAINAGE: HEATED WATER SERVICES

COMPLETE WORKSHEET 1

LEARNING TASK 3.2

1   Name four types of piping material used on hot and cold water services.
2   Name three limitations to consider when choosing water service material.

3.4 Installing the service

The hot and cold water piping system installation usually takes place at 'rough-in' stage. This must be planned ahead in the construction timetable so that different trades can be notified and booked in advance for efficient job progression.

Rough-in

'Rough-in' is the term used for the stage at which the pipework is installed inside walls, ceilings and under floors before the wall sheeting or rendering and ceiling sheeting takes place.

It is important to have clear communication with the builder/client and to be aware of the construction stage. It is far easier to install the pipework before any walls are rendered or sheeted. This helps to prevent clashes with other trades. The installation of the water service from the meter to the building is called the front run and can also be considered part of the rough-in.

The installation of the pipework at the relevant time will help you to:

■ avoid damaging:
  – brickwork
  – wall sheeting
  – ceilings
  – floors
  – other services
■ provide access for the correct supporting of pipework
■ provide clear access for the installation, saving time.

It is important to remember the special installation requirements when installing services underground, such as:

■ a minimum of 300mm separation between any drinking and non-drinking water service
■ a minimum of 100mm separation between any water service not greater than DN 65 and any electrical supply (provided the electrical cable has an orange marking tape along its length)
■ a minimum of 300mm separation between any water service greater than DN 65 and any electrical supply (provided the electrical cable has an orange marking tape along its length)
■ a minimum of 600mm separation between any water service and any electrical supply where the electrical supply has no marking tape or protection.

Depth of cover

Any water service located below ground within private property must have a minimum cover (see Figure 5.5, 'Typical installation in a trench', in AS/NZS 3500.1:2003), as indicated below:

■ 450mm minimum when subject to vehicular traffic
■ 75mm when located under a concrete slab or house
■ 300mm in all other locations not subjected to vehicular load.

AS/NZS 3500.1 PLUMBING AND DRAINAGE: WATER SERVICES

GREEN TIP

Remember to contact Dial Before You Dig (http://www.1100.com.au or telephone 1100) to locate existing services before excavating.

Before installation

It is important to consult with the builder/client to establish the finished wall. This will have a direct impact on the depth that the fittings are recessed into the wall and the type of tapware to be installed. It will also determine the type of connection point required.

Termination methods

When installing hot and cold water outlets horizontally (such as recess sets or washing machines) the hot water outlet is to be on the left and the cold water outlet on the right (see Figures 3.1, 3.2 and 3.3). If they are installed vertically, the cold water outlet must be installed in the lower location and the hot water outlet in the upper location. These set-out methods are used at rough-in stage for the following fixtures:

■ showers
■ baths
■ basins
■ sinks
■ dishwasher isolation valves
■ washing machine isolation valves.

Setting out

It is very important to have all the correct information at rough-in stage, such as the type of tapware to set out for. It is common practice to install single lever mixer taps in the wall for showers and baths, and they must be installed at rough-in stage. Also the type of shower rose chosen will determine where the outlet will be positioned, such as a fixed arm shower rose or a hand shower on a sliding rail. There is no advantage in proceeding with the rough-in unless you are absolutely sure it is correct, as this may result in the use of the

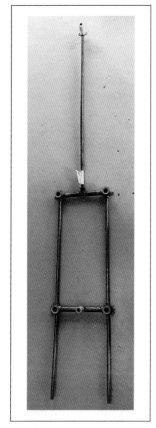

FIGURE 3.1 Combined shower/bath recess tee

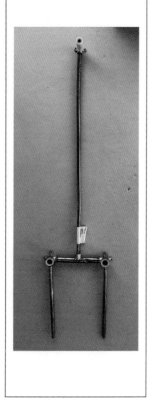

FIGURE 3.2 Shower recess tee

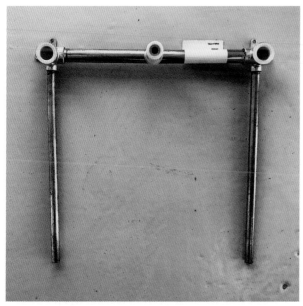

FIGURE 3.3 Bath recess tee

screwed directly onto the spindles or they are held in position by a spring. If the recess bodies are installed too deep, the tap handles may not screw on. Alternatively, if they are not installed deep enough, the flange may not rest closely on the tiles or wall and there will be an unsightly gap. The depth requirements may also be different depending on whether the spindles are half-turn ceramic disc or just ordinary jumper valve sets. Each set of tapware has its own requirements in regard to the depth of recess bodies (generally, the noggin is set back 35mm from the face of the stud). Therefore, setting the recess at an inappropriate depth at the rough-in stage may result in having to chop out tiles to move them deeper into the wall, or purchase expensive recess adapters or extended spindles to change them to the proper depth so that the dress fittings can be fitted correctly.

FROM EXPERIENCE

Clear communication with the builder/client will help avoid costly mistakes at the rough-in stage.

Installation of noggins

When setting out the installation, you may need to install support noggins for lugged/back plate elbows and recess sets (see Figures 3.4 and 3.5). If they are to be supported and fixed to noggins, it is important that the noggins are set back to the required depth for the type of tapware, as stated above. Care should also be taken to ensure that the noggins are installed with equal setback on both sides and that they are level and plumb. Also noggins must be installed to support the fixing of wall hung basins, wall hung vanity units, cisterns, soap

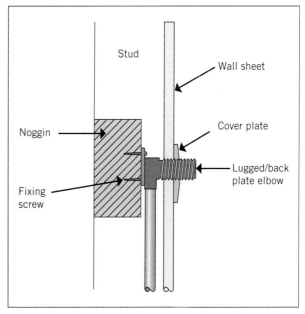

FIGURE 3.4 Lugged/back plate elbow fixed to noggin

wrong type of fitting, or incorrectly installing tapware or valves.

As an example, when installing recess bodies for tap sets, the depth in the wall at which they are positioned may depend on where the wall finishes and the dress flanges that are to be fitted. This adjustable depth is dependent on whether the dress flanges are

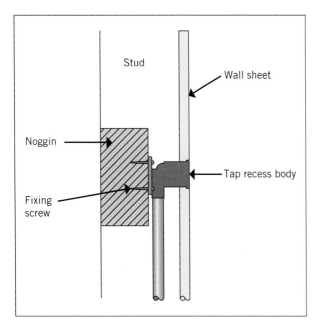

FIGURE 3.5 Tap recess body fixed to noggin

holders, toilet roll holders and shower screens. Poorly fixed noggins and lugged fittings can lead to fittings and fixtures becoming loose with use.

Over time, timber noggins tend to dry and shrink, so if nails are used for fixing they may become loose. It is better to use screws as the fixing method for clips, hangers, lugged/back plate elbows and recess sets. Where possible, install the fitting in the centre of the noggin; if screws or nails are used on the edge of noggins they may split the timber.

Pipe supports

Australian Standards require that all pipework installed has some form of support to hold the pipes in place. There are a number of other factors that may influence the plumber's choice of pipe support method for a given installation. These factors include:

- the ability of the pipe to hold itself between supports (for example, 15mm copper tube may sit straight between two timbers placed 1.5m apart, whereas 15mm plastic tube in the same situation would sag between the timbers – it is supported at a maximum spacing of 600mm) as per AS/NZS3500
- the weight of the pipe and its contents
- compatibility (for example, steel saddles or brackets would be unsuitable for use with copper tube without an insulator being placed between the two different metals)
- expansion and contraction
- appearance
- corrosion resistance – supports used outside need to be more corrosion-resistant than those inside (out of the weather), and areas near the coast may also require more corrosion-resistant materials
- the method used to fix the supports to the building.

Copper pipes

- DN 15 and 20 copper pipes must be supported at a maximum distance between supports of 1.5m vertically and horizontally.
- DN 25 copper pipes must be supported at a maximum distance between supports of 2m vertically and horizontally.

Saddles and clips (see **Figures 3.6** and **3.7**) used for supporting pipes need to be of a compatible material or coated with an inert, non-corrosive coating. Screws should be used instead of nails when fixing saddles and clips to timber.

Source: ANTA Learner Guide for BCPDR3004A Install water main pipe systems. This file is licensed under the Creative Commons Attribution-Share Alike 3.0 Unported licence.

FIGURE 3.6 Saddles and stand-off clips for copper pipes

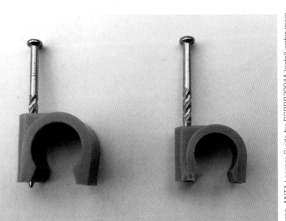

Source: ANTA Learner Guide for BCPDR3004A Install water main pipe systems. This file is licensed under the Creative Commons Attribution-Share Alike 3.0 Unported licence.

FIGURE 3.7 Polymer saddles

Polymer pipes

- DN 15 pipes need to be supported at a maximum distance between supports of 600mm when installed horizontally and 1200mm when installed vertically.
- DN 20 pipes need to be supported at a maximum distance between supports of 700mm when installed horizontally and 1400mm when installed vertically.
- DN 25 pipes need to be supported at a maximum distance between supports of 750mm when installed horizontally and 1500mm when installed vertically.

Polymer pipes conveying hot water may require a lesser distance between supports to prevent the pipe from sagging.

Saddles

Saddles (see Figure 3.8) hold the pipes tightly against the building structure to which they are fixed. Saddles may be made from copper, stainless steel, galvanised mild steel, plastic, or plastic and nylon-coated mild steel.

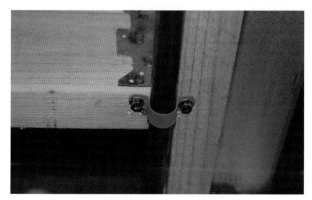

FIGURE 3.8 Copper tube fixed to timber with a nylon-coated saddle

Stand-off brackets

Stand-off brackets (see Figure 3.9) hold the pipes a set distance clear of the building structure and may be made from stainless steel, galvanised mild steel, plastic, or plastic and nylon-coated mild steel. They are used so the pipe is not in contact with the structure and so the pipe can clear other services.

FIGURE 3.9 Stand-off bracket used to support copper tube

Hanging brackets

Hanging brackets (see Figure 3.10) are used to support pipes hanging below the building structure. They may be made from galvanised mild steel or painted mild steel and nylon-coated mild steel.

Bracket systems

There are a number of bracketing systems on the market. They may incorporate saddles, stand-off brackets and hangers in the one system. One bracket

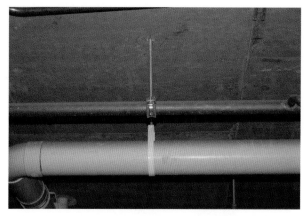

FIGURE 3.10 Copper tube supported by a threaded rod hanger

may hold a number of pipes of various sizes and material types (see Figure 3.11).

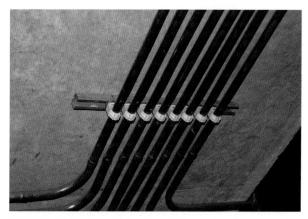

FIGURE 3.11 Copper pipework supported using channel and plastic brackets

Installing pipes in timber wall frames

When installing water service pipes and fittings in or fixed to timber wall frames, you may be required to drill or notch the timber stud (see Figure 3.12). It is good practice to install horizontal pipe runs at a height between 300mm and 450mm from floor level. Wall sheets are not usually fixed in that area so nails and screws into pipes can be avoided. It is also easy to locate pipework if alterations are made at a later stage.

- A hole of a maximum of 25mm diameter may be drilled through timber studs for pipes to pass through.
- A notch of a maximum of 20mm deep may be cut in the depth of a timber stud to install pipes.
- A notch of a maximum of 25mm deep can be cut into a timber stud for a bath installation.
- A notch of a maximum of 10mm wide may be cut in the breadth of a timber stud to install pipes.
- Holes and notches may not be closer than three times the hole or notch width.

Remember that it is better to use screws as the fixing method in timber studs. It is also best to use an insulator between any pipe, timber and clip; this will

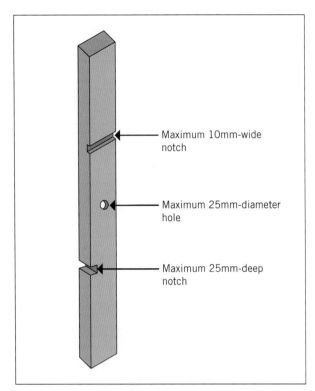

FIGURE 3.12 Notching timber wall studs

allow for expansion and contraction, so preventing a creaking noise.

Drilled holes in timber studs

Where pipes pass through wooden studs, the hole drilled should be larger than the pipe to allow a clear space for movement of both the finished wall and the pipe. The space around the pipe may be filled with silicon that, when set, will allow movement and keep the pipe secure. A side-fix clip may also be used to hold the pipe.

Holes in metal studs

Metal wall frames come with pre-drilled holes. The manufacturer also supplies polymer insulation grommets of different sizes. These are designed to hold the pipe firm, ensuring there is no direct contact between the pipe and the frame but still allowing expansion and contraction. It also protects polymer from the sharp edge of the metal frame and stops electronic corrosion of metal pipes (see Figure 3.13). If there is no pre-drilled hole, a hole should be drilled or punched in the metal frame, ensuring that the size of the hole is accurate (see Figure 3.14). Under no circumstances should the hole be hacked into the metal frame, as shown in Figure 3.15.

Installing pipes in masonry walls

When installing water service pipes and fittings in a brick or masonry wall, it is important to lag all pipework

FIGURE 3.13 Plastic tube passing through a metal stud with side-fix clip in place

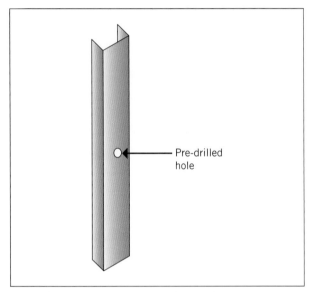

FIGURE 3.14 Metal stud with pre-drilled hole

FIGURE 3.15 Incorrectly installed water pipe in a metal stud

with a plastic sheath to allow for expansion and contraction because the pipework is encased in cement render (see Figure 3.16).

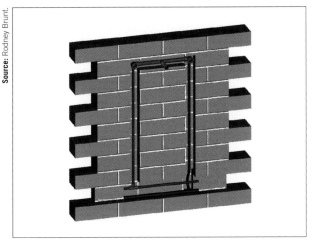

FIGURE 3.16 Installation of a recess set in a brick wall

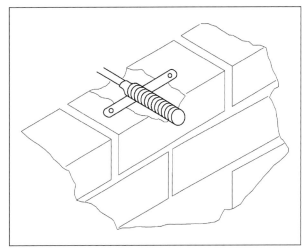

FIGURE 3.17 Winged connector

When carrying out any work using cement or concrete always use the correct PPE and avoid skin contact.

HOW TO

INSTALL HOT AND COLD WATER IN A MASONRY WALL

1 Mark out where the pipes and outlets are to be located on the brick wall with crayon or spray paint.
2 Wear appropriate PPE: goggles, respirator, ear muffs and safety boots.
3 Cut the chase (trench) into the wall 25mm deep with a diamond blade saw (water-cooled double blade).
4 Chop the chase out to a consistent depth.
5 Install the hot and cold water piping and terminate at the required fixtures.
6 Make sure all pipework is lagged to allow for expansion and contraction.
7 Use a spirit level to check all outlets are level and plumb.
8 Cap off and test pipework and fittings to the required test pressure and time.
9 Cement render pipes and fittings so all chases are finished flush with the wall.

Winged connectors (see Figure 3.17) may be cemented into the brickwork; they are designed to be held in place by their brass wings. They are sometimes preferred as connection points for hot, cold and tempered water connection for hot water heaters, and for external standpipes (hose taps) on brick walls.

Tap heights

The heights and fittings shown in Table 3.2 should be used only as a guide; you will need to check with the builder/client if they have other height requirements and tap specifications.

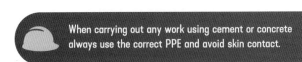

COMPLETE WORKSHEET 2

LEARNING TASK 3.3

1 Describe the term 'rough-in'.
2 What does the term 'front run' mean?
3 What is the maximum size hole that can be drilled in a timber stud?
4 What is a requirement for hot and cold water pipes chased in a brick or masonry wall?

3.5 Sizing a water service

A water service must be sized to meet the flow rate and fixture requirements of a particular building, and this information is covered in depth in the Certificate IV unit – 'Plan, size and lay out hot and cold water services and systems'. Below is some basic information about flow rates and sizing of water services.

The flow of water through pipes

While static pressure is a guide to the volume of water that might be expected to flow through a piping system, other factors determine the volume that will actually flow and create pressure loss. The pipe sizing tables contained in AS/NZS 3500 satisfy the requirements of flow and friction loss. Correct installation is still required to ensure that the piping system performs as required. The most important factors are:

1 *The diameter of the pipe.* The larger the bore of the pipe, the more water it can deliver.

CHAPTER 3

Source: Rodney Brunt.

TABLE 3.2 Guide for tap and outlet heights

Fixture	Outlet height	Centres	Termination fitting
Sink (hod/wall-mounted tap set)	1000mm	Centre of sink	Recess tee
Sink (bench-mounted tap set)	600mm	200mm offset from centre	Lugged/back plate elbow (see **Figure 3.18**)
Laundry trough/tubs (wall-mounted tap set)	1000mm	Centre of trough	Recess tee
Laundry trough/tubs (bench-mounted tap set)	500mm	200mm offset from centre	Lugged/back plate elbow (see **Figure 3.18**)
Clothes washing machine	1350mm	200mm offset from centre (may be installed vertically or horizontally)	Lugged/back plate elbow (see **Figure 3.18**)
Dishwashing machine	300mm		Lugged/back plate elbow for each connection point; usually only cold water connection
Basin	600mm	200mm offset from centre	Lugged/back plate elbow (see **Figure 3.18**)
Shower (taps)	950mm	200mm offset from centre	Recess bodies (may be installed vertically or horizontally)
Shower rose	1850mm		Lugged/back plate elbow (see **Figure 3.18**)
Bath	600mm or to suit tiles	May be installed at the plug end or the centre of the side wall	Recess set
Water closet (close coupled)	150mm	150mm right of centre	Lugged/back plate elbow (see **Figure 3.18**)
Water closet (mid-level)	450mm	150mm right of centre	Lugged/back plate elbow (see **Figure 3.18**)

Source: Reece Group.

FIGURE 3.18 Lugged/back plate elbow (19 BP)

2   *The length of the pipe.* The longer the pipe, the greater the friction that occurs between the water and the internal surface of the pipe; therefore a greater pressure loss occurs.

3   *The condition of the pipe.* Different pipes have different internal bores, which cause more friction and eddies through the rough surface and restrict the water flow.

4   *Changes in direction.* Sharp bends can create eddies, which in turn cause friction and restriction of the water flow. Frictional loss varies with the angle of the bend and the radius of curvature. The sharper the angle and the shorter the radius, the greater frictional losses will be. A radius of five times the pipe bore on the centre line has the same amount of friction loss as the same straight length of pipe.

5   *The quality of jointing.* Burred pipe ends, excess sealing compound and gaps between pipe ends restrict the flow of water.

6   *Passage through valves.* When water passes through a valve, it is restricted and is usually forced to change direction once or twice, depending on the design of the valve.

7   *Enlargement or contraction of size.* To gain maximum flow, enlargements and contractions of pipe diameter should be made as gradual as possible.

Water is a virtually non-compressible liquid. This characteristic can have a direct impact on the water supply piping and may result in an effect called water hammer (a distinct loud hammering sound that can occur in a poorly designed or poorly installed piping system; this is discussed in Chapter 4). The kinetic energy contained in the moving water (which has a mass of 1 kilogram per litre) can create a shock wave that transmits as a violent vibration into the surrounding structures due to water's inability to absorb the impact of this change in velocity when the flow of water is suddenly halted.

To help alleviate this problem and overcome the noises transmitted by the transfer of water through the piping system to the point of use, AS/NZS 3500 stipulates maximum vertical and horizontal support spacings and places maximum limits of 500kPa pressure in a building and a maximum velocity of 3m per second.

AS/NZS 3500.1 PLUMBING AND DRAINAGE: WATER SERVICES

Flow rates

Each tap, valve and outlet is required to meet a minimum flow (discharge) rate; for example:
- water closet (cistern), basin (standard outlet), shower, sink (aerated outlet): 6L per minute
- bath: 18L per minute
- laundry tub: 7L per minute.
- washing machine/dishwasher: 12L per minute
- 20mm hose tap: 18L per minute
- 15mm hose tap: 12L per minute.

AS/NZS 3500.1 PLUMBING AND DRAINAGE: WATER SERVICES

AS/NZS 3500.4 PLUMBING AND DRAINAGE: HEATED WATER SERVICES

Minimum size cold water branches

The following restrictions apply to cold water branch lines.
- A branch with a minimum internal diameter of 12.5mm (DN 18 copper) must not exceed 6m in length and may supply water to one fixture or outlet and in addition:
  - a make-up tank supplying a gravity-fed hot water system
  - a flushing device.
- A branch with a minimum internal diameter of 10mm (DN 15 copper) must not exceed 3m and may supply a combination bath and shower unit, or laundry trough and washing machine, or a kitchen and dishwasher.

- A DN 10 branch must not exceed 1m and may be used to connect hot and cold water to a mixing tap. It must be installed in a reasonably straight run.

Supply to a hose tap (stand pipe) has a minimum pipe size of DN 15 and must be installed a minimum of 450mm above finished ground level.

Minimum size hot water branches

For a storage water heater in excess of 170kPa, the minimum sizes are:
- a DN 18 service from the heater to the first branch
- a DN 15 branch to a kitchen sink or basin
- a DN 15 branch to a sink and laundry
- a DN 15 branch to a bathroom and one other room
- a DN 15 branch picking up all fixtures within a bathroom.

The pipe sizes indicated above and the others outlined in AS/NZS 3500.4 are a guide only – a larger pipe diameter may be required to meet the minimum flow rates, as indicated above.

AS/NZS 3500.1 PLUMBING AND DRAINAGE: WATER SERVICES

AS/NZS 3500.4 PLUMBING AND DRAINAGE: HEATED WATER SERVICES

Other considerations

As per AS/NZS 3500.1, there are minimum requirements for separation between different services and colour identification.

Proximity to other above-ground services

A drinking water service requires:
- a minimum of 100mm separation from any parallel non-drinking water service
- a minimum of 25mm from any gas service or electrical cable, wire or conduit.

Pipe identification

Pipe identification for PE is as follows:
- drinking water – black
- recycled water – purple.

Any copper tube used as part of a non-drinking water service must be sheathed with purple PE (see Figures 3.19 and 3.20).

LEARNING TASK 3.4

1 What can cause pressure loss in a water service?
2 What can cause 'water hammer'?
3 What is the maximum length of a DN15 cold water copper branch?
4 What is the minimum size pipe to connect to a hot water heater outlet?

Source: Reliance Worldwide Australia.

FIGURE 3.19 Cross-linked PE pipe – the pipe for recycled water is purple and is clearly marked 'recycled/reclaimed'

Source: Courtesy Crane Copper Tub.

FIGURE 3.20 Copper pipe with purple PE sheathing for recycled water

3.6 Testing the service

All water services need to be tested before being covered or concealed from view. Prior to any testing, the services need to be flushed to remove all air and any foreign matter. The testing needs to be carried out with fixtures and appliances isolated (not subject to the pressure test). If the service can be charged (filled) with water prior to applying pressure it will save considerable time.

AS/NZS 3500.1 requires all water services to hold a test pressure of 1500kPa (15 bar) for a minimum of 30 minutes. A test bucket (see **Figure 3.21**) is the most common method used for testing a water service.

AS/NZS 3500.1 PLUMBING AND DRAINAGE: WATER SERVICES

Source: ANTA Learner Guide for BCPDR3004A Install Water Main Pipe Systems. This file is licensed under the Creative Commons Attribution-Share Alike CC BY 3.0 Unported licence.

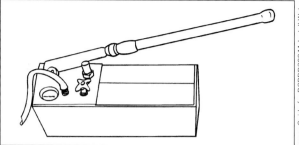

FIGURE 3.21 Test bucket

A temporary interconnection is generally made between the hot and cold services by removing the jumper valves (tap washers) in a recess set (bath or shower) and sealing the male thread with an approved cap. The test bucket (full of water) is connected to a convenient point within the system and pressure applied until the required pressure of 1500kPa is reached (see **Figure 3.22**). As required by the standard, the system is left pressurised for a minimum of 30 minutes. The system is visually checked for leaks; if none are found and the test gauge on the test bucket indicates it is holding at 1500kPa, this system may be backfilled or covered. The test bucket can be removed and the male thread sealed with an approved plug.

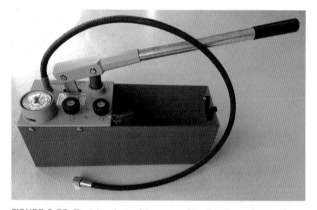

FIGURE 3.22 Test bucket with gauge fitted to hose for testing water pipework

If there is a non-drinking water service it needs to be tested separately in a similar manner.

A check should be made with the local water utility if an inspection is required of the hot and cold water services and of any non-drinking water services.

It is recommended that the water service be charged with water during construction; should the service be damaged it will leak, requiring it to be reported and fixed. If the construction is going to take some time, the service should be regularly flushed with clean water.

LEARNING TASK 3.5

1 Why is it important to flush pipework before testing?
2 What piece of equipment is used to pump the water service to test pressure?

3.7 Alternative water supply: rainwater

Authorities are now encouraging property owners to install rainwater storage tanks (for new residential developments it is compulsory), with some state and territory governments offering a rebate for their installation based on the size and end usage. (See HB 230 and AS/NZS 3500 for 'Above-ground rainwater tank installation with charged (wet) system for all downpipes – cross-section', 'Above-ground tank with pit pump installation – uncharged (dry) system' and 'Flexible rainwater storage device – bladder under floor construction'). A rainwater collection system can provide water for a number of uses, including:

- toilet/urinal flushing
- clothes washing machines
- hot water systems
- garden irrigation
- car washing and similar outdoor use
- filling ornamental ponds
- filling swimming pools and spas
- fire-fighting (subject to the requirements of AS 2419.1, AS 2118 and AS 2441).

Properly maintained rainwater tanks can provide good-quality drinking water. If rainwater tanks are to supply water for cooking or drinking, the property owner should have in place an adequate system of filtering, cleaning and maintenance. It is important to be aware of potential risks associated with microbial and chemical contamination. Rainwater tanks in urban areas can be contaminated with airborne contaminants from heavy traffic, smelters and heavy industry. Rainwater tanks can also be contaminated from roof or plumbing materials or with bacteria from bird or animal droppings. Water from rainwater tanks is not recommended for drinking or cooking unless it has been properly filtered, in line with local authority guidelines.

A first-flush device (see Figure 9.1, 'Example of pre-storage filters for rainwater tanks', in HB 230–2008) should be installed on any rainwater collection system where the water is to be used for drinking or cooking or be connected to an appliance. The first-flush devices should be cleaned (emptied) after rainfall. Tanks must be installed with mosquito proofing on the inlet and overflow. The mosquito proofing installed on the inlet must be above the spill level of the tank. A rainwater tank overflow must discharge to a legal point of discharge and be fitted with a suitable non-return valve.

 Installing an adequate filtering system that removes parasites and bacteria will help to mitigate health risks.

If a fixture or outlet is connected to a water supply from a rainwater tank, the tank will need to have a top-up system connected to the utility's water supply (see Figure 15.6, 'Above-ground rainwater tank installation with mains water top-up and rainwater supplied to appliances in the household', in HB 230–2008). A drinking water service interconnected with a rainwater service requires an approved backflow prevention device installed at the point of interconnection, and a further device installed for property containment on the outlet of the meter (discussed in Chapter 10). Rainwater piping is coloured green to prevent cross-connection with drinking supplies.

GREEN TIP

Rainwater harvesting is now compulsory for most new developments to supply water for flushing toilets and an outside hose tap for washing and irrigation.

Other points to bear in mind when using rainwater tank supply include the following:

- All rainwater tanks and outlets are required to be identified with an approved sign (see the current version of AS 1319) (see also Figure 3.23).
- The catchment area and storage capacity need to be considered (toilet flushing quickly depletes stored water).
- Pressure requirements need to be checked; for example, will a pump be required to meet minimum pressure standards?

Source: © 421 Environmental Products.

FIGURE 3.23 Rainwater identification sign

LEARNING TASK 3.6

1 What can rainwater be used for?
2 What is a first-flush device?

 It is required to label all rainwater outlets with an approved sign stating 'rainwater'. Learners can refer to HB 230–2008.?

3.8 Cleaning up

After the rough-in stage is complete, the work area should be left clean and tidy by sweeping and removing all debris and rubbish. Any leftover material should be stored and reused on another job to prevent waste and to save costs. All packaging waste should be placed in its respective bin. Paper and cardboard waste should be placed in the recycling bin.

All tools and equipment should be cleaned, lubricated and stored safely for reuse. Discard and replace any cutting tools or blades that are blunt so they are ready to use next time.

Complete all documentation required and be sure all mandatory inspections have been completed.

 COMPLETE WORKSHEET 3

SUMMARY

- Gather all information from the latest plans and specifications and pay all fees.
- Confirm with client the type of tapware and fixtures.
- Mark out rough-in accurately from all the information gathered.
- Choose the appropriate piping material and check specifications.
- Choose the most direct path for the hot and cold water service, reducing the amount of bends for less friction loss.
- Consider the most convenient location for the hot water heater.
- Use appropriate guidelines when installing a recycled water service and check for extra inspections.
- Notch and drill appropriate sizes in timber stud walls.
- Correctly size pipework (refer to AS/NZS3500.1).
- Be sure pipework is appropriately clipped and additional noggins to fix fixtures are installed.
- Flush pipework and test water service to 1500kPa for 30 minutes.
- Clean and tidy work area and pack away all tools and equipment.

REFERENCES

AS/NZS 3500.1 Plumbing and drainage: water services
HB 230–2008 – Rainwater Tank Design and Installation Handbook

GET IT RIGHT

1 What is the main difference between the two rough-ins?

2 What are the advantages and disadvantages of each one?

WORKSHEET 1

To be completed by teachers

Satisfactory ☐

Not satisfactory ☐

Student name: _____

Enrolment year: _____

Class code: _____

Competency name/Number: _____

Task: Review the section 'Design considerations' and answer the following questions.

1   Name four important points to consider when planning and preparing to set out and install a water service from the meter outlet to the fixture connection point.

_____

_____

_____

_____

2   Name four important considerations when designing a service.

_____

_____

_____

_____

3   List four situations in which non-drinking (recycled) water may be used.

_____

_____

_____

_____

4   List five situations in which non-drinking (recycled) water cannot be used.

_____

_____

_____

_____

_____

5 What can be done to prevent water from freezing in a water service?

_____

_____

_____

6 How can a hot water pipe be insulated, and why should it be insulated?

_____

_____

_____

7 Materials for use in water services are often referred to by their abbreviation. Give the full name of the following five polymer products:

a PB _____

b PE-X _____

c PP _____

d PE _____

e PVC-U _____

8 What is the most important consideration when choosing a hot water heater location?

9 What does the term 'polymer pipe' mean?

10 Name an important consideration when installing a water service on a metal roof.

 **WORKSHEET 2**

To be completed by teachers
Satisfactory ☐
Not satisfactory ☐

Student name: _____

Enrolment year: _____

Class code: _____

Competency name/Number: _____

Task: Review the section '3.4 Installing the service' and answer the following questions.

1   What is the common name given to the task of installing a hot and cold water service inside the walls and ceiling of a building?

_____

2   State the recommended minimum cover for a water service located below ground in private property in the following conditions:

   a   When subjected to vehicular traffic: _____ mm

   b   When located under a concrete slab or house: _____ mm

   c   In all other locations: _____ mm

3   When hot and cold water outlets are installed horizontally, on which side is the hot installed?

_____

4   When hot and cold water outlets are installed vertically, is the hot water outlet installed in the higher or lower location?

_____

5   What is the standard setback for a noggin supporting recess bodies and why is this depth very important?

_____

_____

_____

_____

6   What is the maximum size hole to be drilled for a DN 20 pipe through a timber stud?

_____

7   When installing water pipes in metal wall frames, the frame comes with pre-drilled holes. Polymer insulation grommets are also provided. Why are these grommets provided?

_____

_____

_____

8 Sometimes water service pipes are installed in masonry walls. Explain why these pipes have to be lagged.

_____

_____

9 List five factors that must be considered when selecting pipe supports.

_____

_____

_____

_____

_____

10 Where would a plumber find out the minimum allowable spacing for water service pipe supports?

_____

11 What is the standard set-out height of a shower tap set and a shower rose?

_____

12 What are the standard set-out measurements for a bottom inlet water connection to a low-level WC?

_____

13 What is the common name for the section of cold water service from the water meter to the building?

_____

14 Why is it better to use screws instead of nails when fixing noggins, recess tees, lugged elbows and saddles?

_____

_____

15 What is a reason to use stand-off brackets?

_____

_____

WORKSHEET 3

Student name: _____

Enrolment year: _____

Class code: _____

Competency name/Number: _____

To be completed by teachers

Satisfactory ☐

Not satisfactory ☐

Task: Review the sections '3.5 Sizing a water service', '3.6 Testing the service' and '3.8 Cleaning up' and answer the following questions.

1   When installing both hot and cold water pipes for a water service, some pipes are limited in both length and what they provide. Complete the following statements by filling in the missing information.

    a   A DN _____ branch may not exceed _____ m and may supply water to one fixture or outlet as well as:

       ■   a make-up tank supplying a gravity-fed hot water system

       ■   a flushing device.

    b   A DN _____ branch may not exceed 3m and can supply water to a single fixture only as well as a combination bath and shower unit, or a laundry trough and washing machine, or a kitchen sink and dishwasher.

    c   A DN _____ branch may not exceed 1m and may be used to connect hot and cold water to a mixing tap. It must be installed in a reasonably straight run.

    d   A standpipe (hose tap) must not be smaller than DN _____ and must be installed a minimum of _____ mm above finished ground level.

    e   For a storage water heater in excess of 170kPa, the minimum sizes are:

       ■   a DN _____ service from the heater to the first branch

       ■   a DN 15 branch to a _____ or basin

       ■   a DN _____ branch to a sink and laundry

       ■   a DN _____ branch to a bathroom and one other room

       ■   a DN _____ branch picking up all fixtures within a _____.

2   Describe in your own words the potential problem associated with the fact that water is a virtually non-compressible liquid.

_____

_____

_____

_____

3   State the recommended maximum pressure and the maximum velocity for water in a piping system as per current Australian Standards.

_____

4   What is the minimum flow rate for a shower?

_____

5   Name five factors that affect pressure loss in a water service.

_____

_____

_____

_____

_____

6   What are the most common uses for stored rainwater in an urban area?

_____

_____

_____

_____

_____

_____

_____

_____

7   What is the purpose of a first-flush device when storing rainwater?

_____

_____

_____

_____

8   What is the test pressure and test time for water services?

_____

9   What are three conditions that should be observed when testing a water service, after completing the
     work and prior to concealment?

     _____

     _____

     _____

10   How is the hot water pipework temporarily connected for testing?

     _____

     _____

11   How far must a drinking water service be apart from a non-drinking water service above ground?

     _____

12   How far must a drinking water service be apart from a non-drinking water service below ground?

     _____

13   Why is it good practice to keep the hot and cold water installation charged during the construction
     process?

     _____

     _____

14   Why should leftover material be kept and reused?

     _____

     _____

WORKSHEET RECORDING TOOL

Learner name		Phone no.	
Assessor name		Phone no.	
Assessment site			
Assessment date/s		Time/s	
Unit code & title			
Assessment type			

Outcomes

Worksheet no. to be completed by the learner	Method of assessment WQ – Written questions PW – Practical/workplace tasks TP – Third-party reports SC – Scenarios RP – Role plays CS – Case studies RW – Report writing PF – Portfolio	Satisfactory response	
		Yes ✓	No ✗
Worksheet 1	WQ	☐	☐
Worksheet 2	WQ	☐	☐
Worksheet 3	WQ	☐	☐

Assessor feedback to the learner

Feedback method (Tick one ✓):	☐ Verbal ☐ Written (if so, attach) ☐ LMS (electronic)
Indicate reasonable adjustment/assessor intervention/inclusive practice (if there is not enough space, a separate document must be attached and signed by the assessor).	
Outcome (Tick one ✓):	☐ Competent (C) ☐ Not Yet Competent (NYC)
Assessor declaration: I declare that I have conducted a fair, valid, reliable and flexible assessment with this learner, and I have provided appropriate feedback.	
Assessor name:	
Assessor signature:	
Date:	

Learner feedback to the assessor

Feedback method (Tick one ✓):	☐ Verbal	☐ Written (if so, attach)	☐ LMS (electronic)

Learner may choose to provide information to the RTO separately.

		Tick one:	
Learner assessment acknowledgement:		Yes ✓	No ✗
The assessment instructions were clearly explained to me.		☐	☐
The assessment process was fair and equitable.		☐	☐
The outcomes of assessment have been discussed with me.		☐	☐
The overall judgement about my competency performance was fair and accurate.		☐	☐
I was given adequate feedback about my performance after the assessment.		☐	☐

Learner declaration:
I hereby certify that this assessment is my own work, based on my personal study and/or research. I have acknowledged all material and sources used in the presentation of this assessment, whether they are books, articles, reports, internet searches or any other document or personal communication.
I also certify that the assessment has not previously been submitted for assessment in any other subject or at any other time in the same subject and that I have not copied in part or whole or otherwise plagiarised the work of other students and/or other persons.

Learner name:	
Learner signature:	
Date:	

4

INSTALL WATER SERVICE CONTROLS AND DEVICES: VALVES AND ANCILLARIES

Chapter overview

This chapter addresses the operating principles for the most common types of tapware and valves for water control. It also has information on the installation of more complex valves such as thermostatic mixing valves (to prevent scalding) and backflow prevention devices. This chapter also explains the basic adjustment for correct flow operation and maintenance of flushing devices, control valves and appliances.

Learning objectives

Areas addressed in this chapter include:
- planning and preparing for work
- identifying installation requirements
- various categories of system controls and devices
- principal types of valves that fit each category
- installing and adjusting devices
- cleaning up.

4.1 Background

AS/NZS 3500.0 defines a valve as a device for controlling the flow of fluid, with an aperture that can be wholly or partially closed by the movement relative to the seating of a component in the form of a plate or disc, door or gate, piston, plug, ball or flexing diaphragm. AS/NZS 3500.0 defines a tap as a valve with an outlet used as a draw-off or delivery point.

> ### ✓✓ AS/NZS 3500.0 PLUMBING AND DRAINAGE: GLOSSARY OF TERMS

Valves are an extremely important part of any plumbing system. They perform a variety of functions ranging from the simple task of controlling the water flow filling a basin to more complex functions such as automatically controlling the flow and pressure within a water mains system or high-rise buildings. Valves are also installed to ensure the safe operation of a system, such as pressure relief valves for hot water heaters, backflow prevention devices to protect the community's drinking water supply, and thermostatic mixing valves to ensure that infants, children, the elderly and the sick are protected from the potential of scalding injuries from hot water.

Without valves in a system there would be no means of conserving or using water wisely. If water was allowed to flow freely and unrestricted from an outlet, it would soon become obvious the amount of water being wasted. Without valves there would also be no way of reducing over-pressurised water services and mains, which could cause the pipe to burst or leak and lead to further wastage of water. Valves play an important role in ensuring the longevity and reliability of an installation by limiting the pressures within the ranges that are specific to the class and type of each material. They also provide a means of isolating a section of pipework for maintenance, thus reducing the impact of a shutdown on an installation by allowing other sections of the service to remain in operation while the maintenance is being carried out.

With Australia's increasing population and the threat of drought, there is more emphasis being placed on the conservation of resources. This has created a need to create more environmentally friendly regulations such as the Building Sustainability Index (BASIX) certificate and the Water Efficiency Labelling and Standards (WELS) star rating for new and renovated projects and products. These initiatives reduce water use, placing greater importance on the amount of water that is delivered through fixture taps and valves with the use of flow restrictors and non-drinking water services.

> ### GREEN TIP
> Valves installed in water services play a major role in water conservation through pressure and flow control.

The BASIX model

The Building Sustainability Index (BASIX) is a model introduced in NSW to ensure that homes are designed to use less drinking (potable) water and reduce greenhouse gas emissions by setting energy and water reduction targets for homes and units. BASIX is a sustainable planning measure to deliver equitable and effective water and greenhouse gas reductions.

WELS rating

The Water Efficiency Labelling and Standards (WELS) Scheme operates just like the star ratings on electrical appliances. It allows the designer, consumer or plumber to assess at a glance the water efficiency of the tapware being installed: the higher the water consumption, the fewer the number of stars on the label (see Figure 4.1). WELS ratings are explained more fully in Chapter 6.

> ### GREEN TIP
> The introduction of BASIX and WELS schemes has resulted in a large reduction in water and energy usage.

Source: Images provided by DSEWPaC.

FIGURE 4.1 WELS ratings labels

4.2 Planning and preparation

Valves are an integral part of any water service system so it is important that the installation of valves is coordinated with the rest of the works. Planning ahead

HOW TO PLAN OUT THE JOB STEP BY STEP

1 Drawings and specifications must be obtained from the job supervisor or client.
2 Work health and safety (WHS) requirements – including a risk assessment associated with installing and adjusting water service controls and devices, and the workplace environment – must be identified and adhered to throughout the work.
3 Quality assurance requirements should be identified and carried out in accordance with workplace, company and statutory requirements. (Note that valves should carry the WaterMark symbol as proof that they are manufactured to comply with the appropriate standards.)
4 Tasks should be planned and sequenced with others involved in or affected by the work.
5 Tools and equipment required to carry out specific tasks need to be organised, including all PPE necessary.

FROM EXPERIENCE

Being organised and planning ahead demonstrates motivation and initiative, which will earn a good reputation and generate greater profit.

is essential so the required valves and ancillaries are delivered on time, avoiding delays. A site inspection is necessary to see if any special requirements are needed, such as site access and coordination of deliveries. In larger installations, a method of lifting and handling the valves may need to be allowed for.

It is essential that the job be sequenced in a logical manner. This can be achieved by breaking the job down into smaller, more specific tasks.

Obtaining drawings and specifications from the job supervisor

When preparing for work it is essential that the correct information is obtained. In most cases for smaller projects, the plumber will consult with the owner/client to select the appropriate tapware and valves required. It is important that the selection of valves to perform specific tasks complies with the appropriate standards, which is a responsibility of the plumber. The plumber will need to select and install valves (bearing the WaterMark standard) necessary for the safe and reliable operation of the installation. So the knowledge of the latest available products, requirements and regulations is essential. This information can be obtained from the manufacturer of the controls and devices being installed. (The manufacturer's website will usually provide any required specifications.)

Information for larger projects is usually listed in the plans and specifications. Plans and specifications are generally prepared by a hydraulic services engineer/consultant in conjunction with the architect and the client, and should be strictly adhered to. When there is conflicting information between the plans and specifications, *the specifications should be taken as the preferred option*. If in doubt, consult the architect, client and engineer for clarification before the work proceeds.

When reading the plans and specifications or selecting valves for a particular purpose, it is important that the details are carefully considered and properly interpreted. Special note should be taken of the following when reading plans and specifications in relation to installing and adjusting water service controls and devices:

- the quantity of each type of valve or valves
- the type of valve specified and end use
- the size of the valves
- available pressure and amount of fixtures
- whether the design and the intended use of the valves complies with the appropriate regulations
- the orientation of the devices (for example, whether it is appropriate to install the valves horizontally or vertically)
- the direction of flow
- the type of connection required (such as flanged or threaded)
- the size, class and material of pipework
- the location of pipework (for example, exposed, wall, ceiling, buried, and above or below ground)
- access for maintenance and servicing
- any special requirements in relation to a specific installation or purpose.

WHS requirements

As WHS procedure is mainly about hazard and risk management, the work and the tasks associated with the installation of any valves must be properly assessed and the risk levels determined. These risks must then be eliminated or controlled to prevent or reduce the likelihood of accidents and injuries occurring during the performance of the work.

A job safety analysis (JSA) should be completed before any tasks commence so the hazards are identified, the risks are assessed and the control measures are in place.

Quality assurance requirements

More and more companies are adopting quality management systems. The systems contain policies and procedures that assure the quality of the product and/or an installation for the customer. The quality assurance policies are designed to control the processes used in a company to ensure that there is uniformity and reliability in record-keeping, the quality of the materials used and the way the work is carried out. Essentially, it is a standardisation of the processes that will guarantee the customer a quality job that conforms to all the relevant standards and specifications required.

Planning and sequencing tasks in conjunction with authorities

All plumbing installations must have a permit, and local authorities must be advised before work is started. In most instances an application for a permit must be submitted at least two working days before the work is begun. When carrying out work on some water service controls and devices, such as the installation of backflow prevention devices and pumps, notification and approval must be sought from the relevant water utilities before the work is carried out.

Select the appropriate tools and equipment

When installing and adjusting water service controls and devices, some standard plumbing tools and some specialised equipment may be needed. These include, but are not limited to, those tools mentioned in Chapter 7 of *Basic Plumbing Skills*.

Prepare the work area

To ensure a safely executed and quality outcome, it is essential to prepare properly for the work, as this will help to reduce delays and/or prevent accidents from occurring. During the site inspection any special requirements and any existing services that may affect the work should be noted and factored into the preparation of the work area. A services search via Dial Before You Dig (http://www.1100.com.au or telephone 1100) should be undertaken if excavation work is needed.

Valves need to be correctly located and accessible during the design and set-out stage of the works. Valve service ducts, chambers and pits also need to be allowed for.

Identify service design requirements

The installation of water services, controls and devices is very much dependent on a range of factors that are related to the installation as a whole, which may include the size and pressure of the service, the flow rate and the internal temperature. Some water service controls and devices are installed purely to control pressure and temperature, while others control or stem flow and/or prevent adverse flow conditions. Some valves are capable of both controlling pressure and controlling (or regulating) flow and/or preventing adverse flow conditions. There are a number of combined valves that are available for this purpose; however, in most instances each valve has a specific purpose.

FROM EXPERIENCE

It is extremely important that plumbers have an understanding of valve requirements, their function and the factors that influence their installation.

COMPLETE WORKSHEET 1

LEARNING TASK 4.1

1 What are the benefits of having a site visit before starting work?
2 How can a valve be easily identified to know it is approved for use?

4.3 Effects of water pressure on services and devices

Water pressure is a very important issue for water supply utilities in order to conserve water. Water utilities are taking steps to balance out the supply pressure in their reticulation systems and reduce the supply pressure. The balancing of pressure throughout the system will ultimately save water due to a reduction in burst mains piping and leakage. AS/NZS 3500.1 states that the maximum static pressure at any outlet other than a fire service is 500kPa.

AS/NZS 3500.1 PLUMBING AND DRAINAGE: WATER SERVICES

Some areas in a reticulation system have higher pressures due to the amount of head between the service reservoir and the level of the property. This difference occurs because of the topography of the land and the differences between the level of the property and the reservoir. Water pressure will often vary at different locations and is dependent on how far the property is from the service reservoir and its elevation in relation to the service reservoir (see **Figure 4.2**). Put simply, water mains in low-lying areas are usually subject to higher pressures than those at a higher elevation.

Valves often experience some pressure and flow loss across them caused by frictional loss or where the water moving through the valve is required to do some work, such as opening a jumper against a spring in a backflow prevention device. It is important to read the valve specifications to determine the pressure loss across the valve so pressure and flow requirements are met.

Source: © Sydney Water.

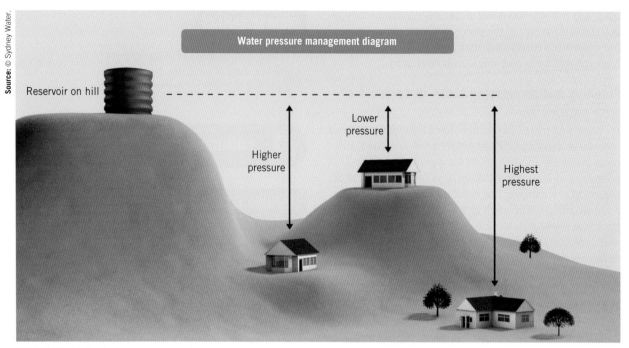

FIGURE 4.2 Mains in low areas are subject to higher pressure than those in high areas

Excessive pressure in a water service may damage valves and tapware. Excessive pressure has been known to contribute to noise and/or leaks in a system over time. It is important that pressure requirements are observed and kept within the specified ranges for the products and classes of materials used. According to AS/NZS 3500.1, the maximum and minimum pressures for hot and cold water services are 50kPa minimum and 500kPa maximum. Also, the available pressure for the water main servicing the property can be obtained from the utility responsible for the supply of water. Pipe sizing must be based on the minimum available head.

Excessive velocity and flow rate may damage connected valves and tapware. This may also cause excessive noise in the system. A water service may deliver differing quantities of water in litres per second depending on pipe size. The amount of water a pipe will deliver in a given amount of time is determined by:

- the pipe size
- the water pressure
- the speed at which the water is passing through it.

AS/NZS 3500.1 states that the maximum velocity (speed) water shall travel in a water service is 3m per second, except for fire services.

AS/NZS 3500.1 PLUMBING AND DRAINAGE: WATER SERVICES

Frictional pressure loss

The flow of fluid through pipes and valves causes a decrease in the fluid's energy level (that is, pressure/head loss), which is caused by the friction between the fluid and inner surface of the pipe. This energy level

decrease shows up as pressure loss in fluid and needs to be considered when selecting valves and devices.

Noise and damage in water service controls and devices may be reduced by the following:

- limiting the velocity of the water flow by increasing the pipe size
- reducing water pressure by installing pressure-reducing, pressure-limiting or pressure ratio valves in pipelines, particularly in areas that have high water pressure.

Most water utilities install pressure control devices in their mains piping systems to balance pressure.

LEARNING TASK 4.2

1 Why do houses in lower-lying areas have higher water pressure than houses in higher areas?
2 What causes friction loss?

4.4 Water hammer and noise transmission

Water hammer is sometimes referred to as line hammer and it is due to the sudden arrest of a column of water moving at speed through the pipes. This occurs because water is non-compressible. The condition is identified by a noise similar to a hammer blow on the pipes and over time can weaken or even burst them. Water hammer and noise transmission are commonly caused by excessive water pressure/velocity as well as poorly constructed or supported pipework. Water hammer also occurs from jumper valves vibrating in an open-valve situation, such as a washing machine tap, dishwasher tap, cistern tap or in-line isolating valves. Fast-closing valves – such as solenoid valves on clothes washing machines and dishwashers, single-lever mixer taps and quarter-turn ceramic disc tapware – also can create water hammer.

Water hammer arrestors

Water hammer arrestors or line hammer arrestors are devices that are installed on water services to eliminate water hammer. They are chambers in which air is sealed, and operate similarly to a shock absorber, either via a flexible diaphragm or via a piston. This reduces the shock of the water stopping quickly, which reduces the noise transmission. Some arrestors have air recharged through a Schrader valve (tyre valve), but the most common types are completely sealed and are not rechargeable. They can be spherical or cylindrical in shape and may be mounted either vertically or horizontally (see Figure 4.3). They must be installed as close as possible to the offending source of line hammer and preferably be without any intermediate bends.

Source: © 2020 Watts.

FIGURE 4.3 Water hammer arrestor

FROM EXPERIENCE

Using problem-solving skills to locate and eliminate water hammer impresses the client.

Float valves

Even though there are a variety of designs of float valves, their basic operation remains essentially the same. The float, which is usually a hollow plastic ball, is attached to a lever arm that operates a plunger fitted with a rubber washer. The float valve is designed to regulate the flow of water into a flushing cistern or tank to a predetermined level. When the required water level has been reached, the float forces the rubber-faced plunger against the seat of the inlet valve to shut off the water supply to the tank. If the water level in the tank drops, the float also drops along with the water level, and the lever pivoting on its fulcrum pin follows the float, lifts the plunger off its seat and so allows water to enter and top up the tank. As the water level gradually rises, the process is reversed: the float with its arm gradually closes the inlet valve, until the required water level has been reached (see Figure 4.4).

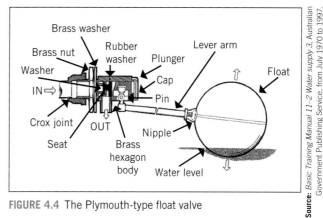

Source: *Basic Training Manual 11–2 Water supply 3*, Australian Government Publishing Service, from July 1970 to 1997.

FIGURE 4.4 The Plymouth-type float valve

Stabilising the ball float valve

Float valves sometimes produce line hammer due to the bouncing of the float on waves created on the water surface. This causes the inlet valve to close and open in quick succession, creating line hammer in the pipes. This condition can be remedied by soldering a piece of copper sheet to the copper float or by cementing a piece of plastic sheeting to a plastic float (see Figure 4.5). The sheet acts as a stabilising fin, which reduces and slows down the float bounce. Proper support of service pipes with suitable clips will also help to eliminate line hammer.

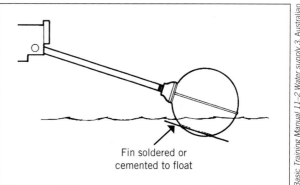

Fin soldered or cemented to float

Source: *Basic Training Manual 11–2 Water supply 3*, Australian Government Publishing Service, from July 1970 to 1997.

FIGURE 4.5 Stabilised ball float valve

LEARNING TASK 4.3

1 Why do some valves in the open position cause water hammer?
2 Where are float valves used?

4.5 Service controls and mixing devices

Service controls and mixing devices should be selected in accordance with the regulations outlined in the relevant standards, utilities' requirements and job specifications.

Isolating valves

The flow from any water main to any water service pipe must be controlled by isolating valves such as stop taps, stop valves and non-return valves. All isolating valves on water services must be opened and closed slowly to prevent shock waves in the piping, which can cause leaks and burst pipes. This is because water is non-compressible. Isolating valve components are made from a wide range of materials including brass, cast iron, steel, stainless steel, various polymers, rubbers and urethanes, and are available in a variety of designs to meet various requirements. The main types are:

- ball valves
- gate valves
- butterfly valves (these valves are used predominantly on large services and allow for full or partial closure of a service; they are not further elaborated on within this section)
- screw-down stop taps.

Isolating valves must be installed at the following water service locations:

- at the water main (at either a tapping or a tee insertion)
- at the water meter or within 1m of the property boundary if no water meter is fitted
- at each flushing device/tank
- at each appliance
- at each backflow prevention device
- at each thermostatic mixing valve
- at each pressure-limiting valve
- at each commercial/industrial appliance or apparatus
- at each pumping apparatus
- at each storage tank inlet
- at each storage tank outlet (where capacity exceeds 50L).

Isolating valves must be installed in multiple building and multistorey construction, as follows:

- at each branch serving individual buildings
- at each branch serving each floor in buildings of two or more storeys
- at each group of fixtures
- at each standpipe.

AS/NZS 3500.1 covers the function of isolating valves. Isolating valves must be installed on fire services, as follows:

- at each water main connection
- at or near the property boundary
- at each hose reel
- at each pumping apparatus.

AS/NZS 3500.1 PLUMBING AND DRAINAGE: WATER SERVICES

Ball valves

Ball valves are part of the rotary movement family of valves, which are often called quarter-turn valves and include ball valves and butterfly valves.

Ball valves are opened and closed by turning a handle attached to a ball inside the valve body. The ball has a hole, or port, through the middle so that when the port is in line with both ends of the valve, fluid flow will occur. When the valve is closed, the hole is perpendicular to or across the ends of the valve, and the flow is blocked. When the valve is open, the handle or lever is usually in line with the port; its position therefore indicates whether the valve is open or closed (see Figure 4.6).

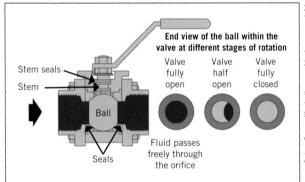

Source: Spirax Sarco. Illustrations and text are copyright, remain the intellectual property of Spirax Sarco and have been used with their kind permission; http://www.spiraxsarco.com/resources/steam-engineering-tutorials.aspx.

FIGURE 4.6 Cross-section of reduced bore ball valve

There are two basic designs of ball valves:

1 the floating ball design, which relies on the valve seats to support the ball
2 the trunnion-mounted ball, which uses a trunnion to support the ball. Trunnion mounting is used on larger valves, as it can reduce the operating effort required to open and close the valve to about two-thirds of that provided by a floating ball.

Ball valves are available as either reduced bore or full bore. Full bore valves have an orifice that is the same size as the diameter of the pipe, whereas in reduced bore valves the orifice diameter is less than that of the pipe. Full bore valves are generally more expensive than reduced bore valves, and they should be used where the pressure drop across the valve is crucial.

Ball valves are durable and usually work to achieve perfect shut-off even after years of non-use. However, they do not offer the fine control that may be necessary in throttling applications, nor do they offer any backflow protection.

The body of ball valves may be made of metal and metal alloys such as brass, plastic or metal with a ceramic centre. The ball is often chrome plated or manufactured from stainless steel to make it more durable.

Gate valves

Gate valves (see Figure 4.7) are widely used valves in industrial applications (although they are not suitable for use as a throttling valve that regulates flow). They are mainly used as stop valves; however, their primary purpose is to fully shut off or fully turn on flow slowly and provide full flow, such as when fitted to a storage tank outlet. Fluid flow moves in a straight line and with minimal resistance when the wedge is fully raised.

FIGURE 4.7 Gate valve

Seating and operation is at right angles to the line of flow where the fluid flow is met head on. Seat types include a metal seat and/or a resilient seat.

A gate valve usually requires more turns to open it fully. Larger-diameter gate valves, or sluice valves as they are often called, open and close in the opposite direction to conventional screw-down valves. The direction for opening and closing is usually marked with an arrow on the valve stem or hand wheel.

Gate valves can be opened and closed via either a rising stem or non-rising stem. For the non-rising stem gate valve, the stem is threaded on the lower end into the gate.

Screw-down stop taps and valves

Stop taps are screw-down pattern taps with a horizontal inlet and outlet connections. They usually incorporate a loose jumper valve, permitting flow in one direction only.

Stop valves are commonly fitted in line and can be used to stop the flow in a pipeline. The types of stop/isolation valves available include ball, gate and butterfly valves.

Valve operation

The most common type of valve used in a water service is a screw-down type; however, other types include ball, gate and butterfly designs. The design of the valve incorporates a disc (washer/jumper valve), which is lifted and lowered onto the body seat by a stem with its axis perpendicular (90°) to the face of the body seat.

When installing or replacing jumper valves (washers), care should be taken to ensure that they are rated at the appropriate temperature. Some plastic jumper valves may fail under hot water conditions and result in injury.

Screw-down type valves are commonly installed as a stop tap. They may be used in service as an isolating valve or a stop tap for a hot water heater, or installed as recess taps, pillar taps, bib taps (see Figure 4.8), hose taps, meter taps, main taps, cistern taps and washing machine taps. All of these taps may be installed as loose jumper valve-type taps. The screw-down tap with a horizontal male threaded inlet socket and a curved spout (or bib), as shown in Figure 4.8, is called a bib tap. The valve shown in Figure 4.9 has a threaded inlet and outlet for connection to pipes, and is called a stop tap. These valves are installed in line and are also available with capillary ends for installation directly onto copper tube up to DN 20.

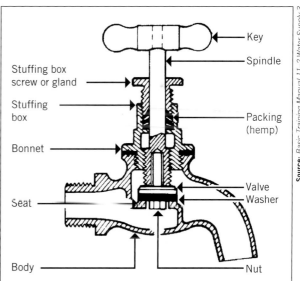

Source: *Basic Training Manual 11–2 Water Supply 3*, Australian Government Publishing Service from July 1970 to 1997.

FIGURE 4.8 Bib tap with valve closed

Care is needed when installing a loose jumper valve as flow can be in only one direction. As a consequence, it is necessary to ensure that the water enters the tap through the lower passage side, so the water will push against the bottom of the jumper valve lifted by its pressure from its seat. Installing the valve the wrong way will cause water pressure to hold the valve on its seat when the tap is opened, resulting in no flow through the valve (see Figure 4.10). Most stop valves have the direction of water flow marked on the body by an arrow to assist in correct installation (see Figure 4.11).

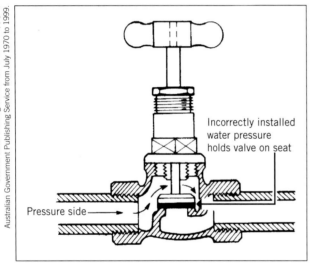

Source: *Basic Training Manual 11–2 Water Supply 3*, Australian Government Publishing Service from July 1970 to 1998.

Can be threaded externally for union connector or with sockets

Correctly installed water pressure lifts valve off seat

Pressure side →

Lower passage

Upper passage

FIGURE 4.9 Stop valve

Source: *Basic Training Manual 11–2 Water Supply 3*, Australian Government Publishing Service from July 1970 to 1999.

Incorrectly installed water pressure holds valve on seat

Pressure side →

FIGURE 4.10 Stop valve

FIGURE 4.11 Stop valve showing flow direction on the valve body

Older stop valves have a stuffing box screw, or gland, which screws into the stuffing box to compress the packing gland around the spindle to prevent water leaking past it (see Figure 4.12). In many bonnets, an O-ring replaces the gland (see Figure 4.13). Most modern taps are made with an O-ring; however, some stuffing box taps are still available, such as main taps.

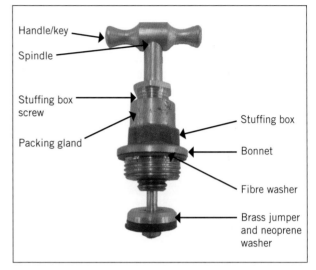

Handle/key

Spindle

Stuffing box screw

Packing gland

Stuffing box

Bonnet

Fibre washer

Brass jumper and neoprene washer

FIGURE 4.12 Stuffing box assembly and jumper valve

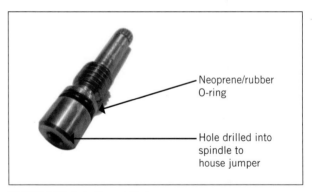

Neoprene/rubber O-ring

Hole drilled into spindle to house jumper

FIGURE 4.13 Spindle and O-ring

The bonnet holds the working parts and screws into the body. Water leakage between the bonnet and the valve body is prevented by inserting a fibre washer between them (see Figure 4.14). The bonnet has an internal screw thread into which the threaded part of the spindle engages. The lower part of the spindle is bored to house the valve stem and act as a guide.

By turning the key/handle in an anticlockwise direction the spindle moves upwards and away from the valve seat when the valve is opened. Water pressure lifts the valve off its seat against the upward-moving spindle, allowing water to pass through the opening and discharge through the upper passage of the body to the tap outlet. The opposite action closes the valve to prevent water flow.

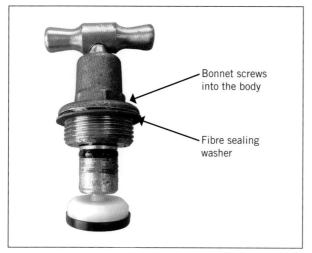

FIGURE 4.14 A typical jumper valve with its stem housed in the bored-out spindle

Labels on figure: Bonnet screws into the body; Fibre sealing washer

Installing screw-down taps

The bodies of screw-down taps can have female (internal) threads for screwing directly to threaded pipe ends or they can have male (external) threads for use with sockets or union connections. They also are available in capillary configuration for installation directly onto copper tube up to DN 20.

It is important to flush the hot and cold water service of any dirt and debris before fitting any tapware or valves to prevent any damage to the components.

FROM EXPERIENCE

A convenient time to flush the hot and cold water service is when the pressure test is being carried out upon completion of the rough-in.

Before fitting the tap, it is necessary to bind the male thread with teflon tape or teflon cord, or to wind hemp in a clockwise direction to ensure watertight sealing of the threads. Other approved liquid-type thread sealing agents such as Slic-tite, Loctite or Loxseal can be used. Note that when using any thread sealing compound on a drinkable water supply, the compound must be approved for use on a drinkable water supply and comply with any regulations, codes and standards set for drinkable water.

HOW TO

HOW TO SCREW TAP TO FITTING

1 Engage the tap body thread with the pipe thread or socket thread and ensure proper engagement of the threads occurs, avoiding cross-threading as this could damage the threads and render the valve useless.
2 Screw in the tap by hand until no further movement is possible.
3 Using a well-fitting spanner, tighten the tap fully.

When screwing a tap into a socket, it is good practice to hold another spanner on the socket to hold against yourself while tightening the tap so other fittings or pipe are not disturbed.

If the outlet does not line up with its chosen location, do not use excessive force to achieve it as this may cause the cast brass to stretch or split. If misalignment occurs, remove the tap, then reapply a small amount of hemp or teflon tape and re-screw the tap to the final line-up position.

Sometimes the joint can be slightly loosened safely to achieve the correct position without the danger of it leaking, especially when hemp or Loctite is used.

Ceramic disc technology

Ceramic disc taps use ceramic discs instead of conventional washers/jumper valves to give a quick action and are typically used in single-lever or half-turn taps. Ceramic washers are more resistant to wear and tear and therefore have a longer life. They are also less likely to leak and are not affected by any lime or solids in the water (see Figures 4.15, 4.16, 4.17 and 4.18). These types of taps are often easier to operate as they require less effort to turn the tap and usually require only a half turn of the handle to fully open or close the tap. When the tap is turned on or off, the lower of the two discs remains stationary while the upper disc turns with the movement of the tap handle. Water will flow only when the upper and lower slots match up in the open

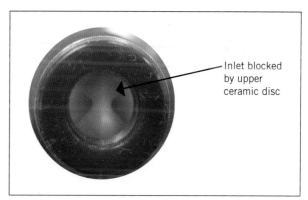

FIGURE 4.15 Ceramic disc inlet closed

Label on figure: Inlet blocked by upper ceramic disc

FIGURE 4.16 Ceramic disc cartridge outlet closed

Label on figure: Outlet blocked by upper ceramic disc

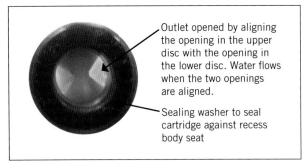

Outlet opened by aligning the opening in the upper disc with the opening in the lower disc. Water flows when the two openings are aligned.

Sealing washer to seal cartridge against recess body seat

FIGURE 4.17 Ceramic disc open

position. The quick-acting shut off can cause water hammer. Consequently most ceramic disc taps now have a half-turn action instead of a quarter-turn action to help eliminate water hammer.

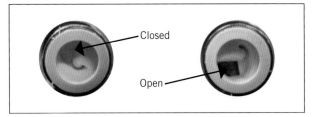

Closed

Open

FIGURE 4.18 Operation of ceramic disc cartridges

Ceramic disc cartridges come in varying shapes; however, their operation is basically the same where the upper disc is moved to allow or stem flow.

Installation of ceramic disc tapware

It is important to prepare the system before fitting ceramic taps.

HOW TO

HOW TO PREPARE BEFORE FITTING CERAMIC DISC TAPWARE

1 Flush the system thoroughly.
2 Turn off the water supply and drain the system.
3 If installing new tapware in an existing installation, remove existing handles and spindles and ensure the seat is free of any foreign matter.

4 Ensure that the tap seat has a uniform, undamaged surface; if not, this may inhibit the seal between the cartridge and the seat. If the seat is in poor condition, it will need to be re-seated or replaced.
5 Clean off any remaining fibrous washer material from the tap and the tap body surface (where the fibrous washer will seal between the bonnet and the body).

HOW TO

HOW TO INSTALL CERAMIC DISC TAPWARE

1 Remove the lock nut and fibre washer from the cartridge (see Figure 4.19).
2 Screw the cartridge by hand into the tap body until it rests on the tap seat (see Figure 4.20). Use one of the following hand tools placed against the hexagonal flats on the cartridge to tighten until it comes to a firm stop:
   - shifting spanner (wrench)
   - 15mm hexagonal tube spanner
   - 15mm open-ended spanner
   - plastic spanner (if supplied with tap).

Note:
   - Pliers or multigrips are not suitable for this task and must never be used.
   - Extreme care must be taken to ensure that the cartridge is not overtightened as this may damage the discs.
3 Fit the fibre washer over the cartridge and against the grooved surface of the tap body face (see Figure 4.21).
4 Screw the lock nut (by hand) onto the cartridge against the fibre washer (see Figure 4.22). Tighten the lock nut with one of the following tools to approximately 20Nm:

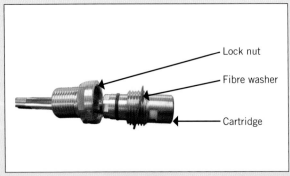

Lock nut

Fibre washer

Cartridge

FIGURE 4.19 Lock nut and cartridge are separated

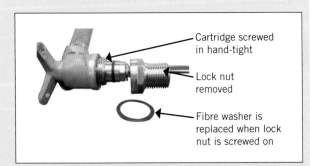

Cartridge screwed in hand-tight

Lock nut removed

Fibre washer is replaced when lock nut is screwed on

FIGURE 4.20 Cartridge screwed in hand tight

>>

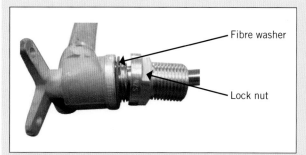

FIGURE 4.21 Fibre washer placed against tap body with lock nut positioned ready for tightening

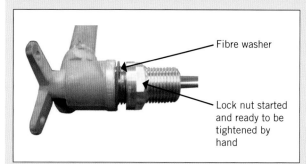

FIGURE 4.22 Screw the lock nut onto the cartridge against the fibre washer

- shifting spanner (crescent wrench)
- 25mm hexagonal tube spanner
- 25mm open-ended spanner.
  Note:
- Pliers or multigrips are *not* suitable for this task and must never be used (see Figure 4.23).

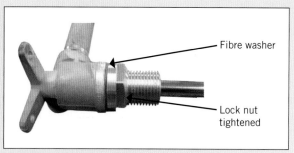

FIGURE 4.23 Lock nut tightened against body

5 Fit the flange and handle to complete the installation (see Figure 4.24). The plastic spindle extensions for spindles are to be fitted deep in the wall.

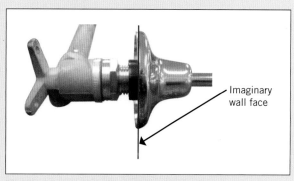

FIGURE 4.24 Flange fitted

 COMPLETE WORKSHEET 2

Other types of control valves

There are many types of valves in the plumbing industry used for different applications. This next section will explain the function and operation of regulating valves, non-return (check) valves, pressure reduction valves, line strainers and solenoid valves.

Globe valves

Globe valves are designed specifically for regulating or throttling water flow in a pipeline. They consist of a movable disc-type element and a stationary ring seat, and are named because of the spherical appearance of their body. An internal baffle separates the two halves of the body. This has an opening that forms a seat onto a movable plug that can be screwed in to close or open and/or adjust the valve. The plug is attached to the stem, which is operated by a screwing action in manual valves (see Figure 4.25).

Globe valves are often installed in flow and return hot water services to balance the flow rates between the flow and return lines in the system. They are also

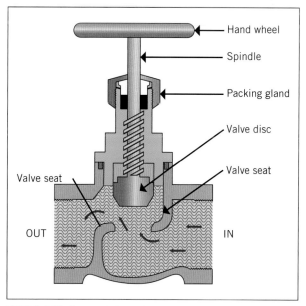

FIGURE 4.25 Cross-section of a globe valve

installed in steam and other specialist fluid applications and can be controlled either manually or automatically. Typically, automated valves use sliding stems, which have a smooth stem rather than a threaded stem and are opened and closed by an actuator assembly. When manually operated, the stem is turned by the use of a hand wheel.

Installation should occur with the inlet below the valve seat in a similar manner to screw-down-type valves. For severe throttling duty, the valve may be installed so that the flow enters over the top of the seat and passes down through it; however, when installed in this manner the packing will be constantly pressurised, causing undue wear. The preferred installation orientation is with the valve in an upright position. Installation upside down is not recommended because it may cause dirt to accumulate in the bonnet.

Duo and trio valves

A duo valve, or combined non-return/isolating valve (see Figure 4.26), offers the dual functions of being an isolating valve and a non-return valve. These valves are used on installations of pressurised storage water heaters and serve as a compact, easy-to-install device.

FIGURE 4.26 Combined stop non-return (duo) valve

There is also a range of basic non-return valves available, as well as a version with an integral strainer. Another versatile valve is the trio valve (see Figure 4.27), which has the combined functions of non-return,

isolator and line strainer. In the trio valve, the line strainer protects both the internal mechanisms of the valve and the devices downstream of the valve from damage due to contaminants in the water supply and is commonly used in hard water areas.

FROM EXPERIENCE

Remember to be aware of the water quality in the area you are working. If the water is high in minerals, usually calcium and magnesium, it is known as hard water. So be sure to install line strainers to protect valves and tapware.

Solenoid valves

A solenoid valve (see Figure 4.28) can be described as an electrically controlled or electromechanical valve for use in the control of the conveyed materials by allowing or preventing an electrical current to pass through a solenoid. A solenoid is basically a coil of wire that creates a magnetic field when an electrical current is passed through it and is connected to a magnet that controls the valve. When the current is active it will open/close the valve, and the opposite will happen when the circuit is broken. Solenoid valves may be used to control automatic flushing systems, infra-red taps and a range of controls in the plumbing industry. They are commonly used in clothes washing machines and dishwashers to control the water cycles. As mentioned earlier, solenoid valves can cause water hammer because they are a fast-closing valve.

Source: Photo courtesy of Hunter Industries Incorporated.

FIGURE 4.28 Solenoid valve

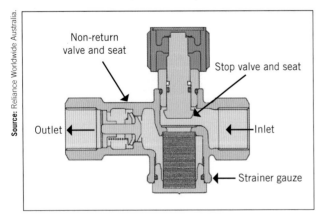

Source: Reliance Worldwide Australia.

Non-return valve and seat

Stop valve and seat

Outlet

Inlet

Strainer gauze

FIGURE 4.27 Trio valve

Ancillary valves

Ancillary valves are valves that are installed upstream of equipment to assist in the operation and performance of a particular fixture. They play an important role by protecting downstream valves and equipment.

Line strainers

Line strainers (see Figure 4.29) are installed to protect downstream valves and fittings from impurities and debris that may be contained in the water supply line. Valves are often expensive and are delicate precision devices. The most common cause of malfunction in most controls and devices is foreign matter such as sand, hemp, teflon, dirt or rust getting stuck under the valve seats. The installation of a line strainer is the most cost-effective way to prevent the foreign matter from entering the valve. In many products line strainers are built into the body of the valve and are available in a number of designs for specific applications. Line strainers are mandatory when installing backflow prevention devices and thermostatic mixing valves.

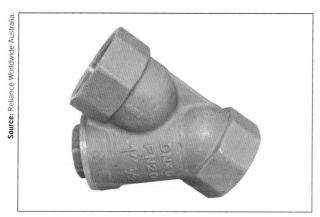

Source: Reliance Worldwide Australia.

FIGURE 4.29 Line strainer

Non-return valves

A non-return valve is a valve to prevent reverse flow from the downstream section of a pipe to the section of pipe upstream of the valve. Non-return valves are also called 'check' valves, and are usually installed on storage water heaters where some backpressure may be experienced due to the water pressure in the heater being higher, because of its temperature, than the supply pressure. Non-return valves are available in several basic types, as shown in Figure 4.30.

Non-return valves are also available for specialised applications, such as a high-temperature version suitable for use in solar water heater systems. Solar and uncontrolled heat source water heaters present an increased hazard to downstream components, as water can become superheated and potentially flash to steam. Only high-temperature non-return valves should be used in these applications.

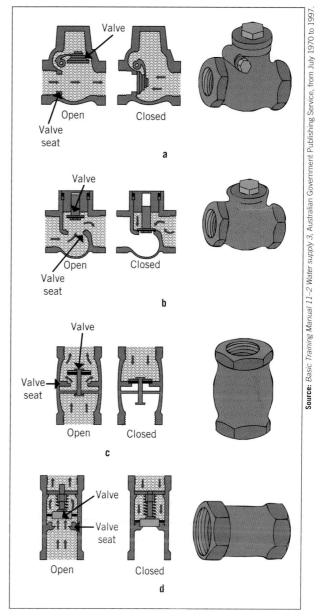

FIGURE 4.30 Types of non-return valve
(a) The swing check – must only be installed in a horizontal position.
(b) The horizontal lift – must only be installed in a horizontal position.
(c) The vertical lift – can only be installed in a vertical position.
(d) The spring loaded (the most common type of non-return valve used today) – may be installed in a horizontal or a vertical position.

Source: Basic Training Manual 11–2 Water supply 3, Australian Government Publishing Service, from July 1970 to 1997.

Pressure-limiting valves

Pressure-limiting valves are designed to reduce a high inlet pressure to a standard specified outlet pressure. When being purchased for installation in a domestic situation they are usually manufactured to limit the outlet pressures to 350kPa, 500kPa and 600kPa settings. They are ideal for use in domestic applications such as water heaters and water meter assemblies.

Pressure-limiting valves control inlet supply line pressure to a fixture at a preset maximum and will remain in the open position if the pressure in the supply line to the valve falls below that preset pressure. These valves are suitable in areas with high water pressure and are installed:

- on the inlet side of the hot water system
- on the inlet side of the dishwasher
- at the boundary of the house on the outlet of the water meter.

They can be installed either horizontally or vertically and are available in either a barrel-design or a T-design (see Figure 4.31). Figure 4.32 shows a cross-section of a barrel-design pressure-limiting valve.

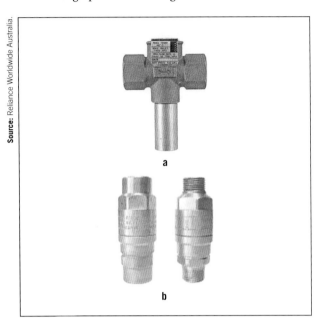

FIGURE 4.31 Pressure-limiting valves
(a) T-design pressure-limiting valve
(b) Barrel-design pressure-limiting valve

Pressure-reducing valves

Pressure-reducing valves deliver a pre-set pressure, which avoids pressure fluctuations throughout the home installation, and are suitable for use in areas of high water pressure. Pressure-reducing valves are designed to be installed at the boundary of the property, in conjunction with the water meter.

Pressure-reducing valves (see Figure 4.33) automatically reduce a higher inlet pressure to a steady lower downstream pressure regardless of any changes in the upstream flow rate or pressure. They are ideal for domestic installations because they create less resistance to water flow.

Pressure-reducing valves come in either a right-angle or a straight-through configuration and are used in residential installations. Using a pressure-reducing valve can also minimise water wastage.

Domestic pressure-reducing valves can deliver high flow rates with minimal head loss. These are available in both adjustable pressure (150–600kPa) and pressure-locked at 500kPa.

Pressure-reducing valves also can be used to reduce pressure upstream of some commercial and industrial devices (see Figure 4.34) such as dosing apparatus, high-pressure cleaners and laboratory equipment. Pressure-reducing valves are favoured in these situations because they deliver a more accurate preset outlet pressure compared to pressure-limiting valves.

Pressure ratio valves

Pressure ratio valves, or mains proportion valves as they are sometimes referred to, are designed to reduce the outlet pressure of the valve by a set ratio to that of the inlet pressure. For example, a valve set at a 2:1 ratio with an inlet pressure of, say, 100kPa would deliver an outlet pressure of 50kPa.

As the outlet pressure varies with the inlet pressure, these valves are not suitable where excessive pressure fluctuations occur, such as in domestic situations or applications where there is a peak demand period. However, they are ideal to break down the pressure from storage tanks supplying water in high-rise buildings.

Installing pressure-limiting, pressure-reducing and pressure ratio valves

When installing pressure-limiting, pressure-reducing and pressure ratio valves, the following requirements should be followed:

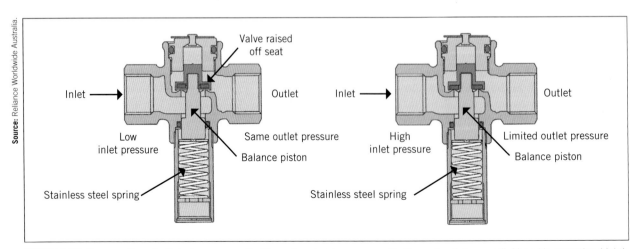

FIGURE 4.32 Cross-section of a T-design pressure-limiting valve, showing pressure-limiting valve: fully open (left) and in operation (right)

FIGURE 4.33 Pressure reduction valve

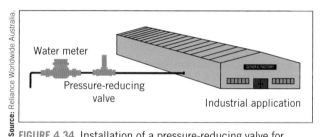

Source: Reliance Worldwide Australia.

FIGURE 4.34 Installation of a pressure-reducing valve for industrial service

- The direction of flow indicated on the valve should always be observed and adhered to.
- Check whether the valve being used is suitable for installation either horizontally or vertically and select a valve to suit the desired orientation.
- Flush all pipelines thoroughly before installation commences.
- Position the valve so as to allow easy viewing and access to the breather hole if required.
- Position the valve to ensure there is easy access to valves for future servicing.
- Screw-down valves should be installed with at least one union to facilitate removal should it be required for maintenance or replacement.
- Protect the valve from heat during welding to prevent damage to the seals.
- The breather hole must be kept free of obstruction. If the line is buried, protect the breather hole from blockage and ensure there is adequate drainage to prevent a cross-connection.
- Valves, pipe systems and fittings should be capable of withstanding the full head pressure they are to be subjected to.
- In some situations where pressures are paramount, such as when balancing between supplies is required,

it is advisable to install pressure gauges in the system between the inlet and outlet of the valve.
- Where two valves per station are installed, such as for a bypass arrangement, the duty between the valves should be rotated to allow the spare valve to function periodically and not become stiff and unreliable.
- Check that the valve has WaterMark accreditation.

Spring-loaded taps

Spring-loaded taps, or self-closing taps, are used mostly for drinking fountains and pillar taps in public places where there is a danger of taps being left open to waste water. Spring-loaded taps are quick-acting valves and their sudden closure may produce water hammer in the pipes. Anti-hammer devices may be installed in the service pipes close to the offending fitting, or they can be built into the fitting itself (for example, dashpots). These devices act as shock-absorbers to the force of water.

Leakage past the spindle requires tightening of the stuffing box gland or replacement of the greased hemp in older types. In more modern types, the hemp and gland have been replaced by a rubber O-ring fitted into a groove that must be lubricated with a silicone-based grease.

COMPLETE WORKSHEET 3

Infra-red sensing tapware

Infra-red sensing tapware is used in many situations for the operation of tapware without having to physically touch the tap. This helps to prevent cross-contamination in sterile areas.

Wall-mounted sensors

Hands-free infra-red tapware uses pulsed infra-red light beams to offer protection from false triggering while being easy for the user to operate (see Figure 4.35). The power supply is usually a 24V AC power supply, which includes surge suppressors. To protect from damage and ensure the tap gives years of reliable service, it is recommended that a strainer is installed before the solenoid valve.

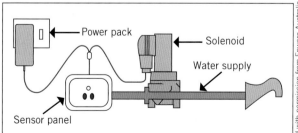

FIGURE 4.35 Typical configuration when installing wall-mounted infra-red-operated tapware

Source: Printed with permission from Enware Australia Pty Limited. For further information email info@enware.com.au or visit http://www.enware.com.au.

Infra-red sensing tapware can help reduce the risk of cross-contamination between operators caused by

physical contact between the operator's hands and a tap handle after another person has touched the same handle. This is of particular importance when a sterile environment must be maintained. Hands-free infra-red tapware is suitable for, and can be beneficial when installed in, the following applications:

- food preparation areas
- factories
- scrub-up areas
- operating theatres
- dental clinics
- laboratories
- sterilising rooms
- abattoirs
- restaurants
- veterinary clinics
- photography labs
- disabled-access bathrooms
- institutions (schools, kindergartens etc.)
- public amenities.

To operate the tap, all the user has to do is wave their hand within 5–10cm of the sensor to turn on the water. A second pass in front of the sensor will turn the water off, or, if the water is allowed to run, the sensor will turn itself off after approximately 45 seconds. An LED indicator on the sensor plate lights up when an object is being sensed, so it is easy for the user to place their hand in the correct position to operate the tap.

When choosing a location to install a wall-mounted sensor, ease of operation, passing traffic and obstructions directly in front or within possible range of the sensor must be considered. Up to 500mm clearance may be necessary from reflective surfaces such as ceramic tiles and stainless steel, directly in front of and parallel to the front face of the sensor. Leads should be installed through a conduit for easy removal. Installation instructions are supplied with every product and the manufacturer should be consulted if any problems with installation arise.

Bench-mounted sensors

The bench-mounted Enmatic tap is also controlled by an invisible infra-red beam (see Figure 4.36). This unit includes simple installation and compact design with the options of battery or mains power. The bench-mounted spout has the sensor at the base of the body, which also encapsulates the electronics and solenoid.

When the sensor detects motion, this triggers the solenoid valve and the water will flow as long as the user's hands remain in the vicinity of the sensor, directly under the spout. When the user removes their hands from the sensor range, the water will turn off after a second. No movement in front of the sensor will stop the flow of water. This facility prevents water being wasted if an object remains in front of the sensor for an extended period. The sensor range is adjustable to suit application requirements.

Source: Printed with permission from Enware Australia Pty Limited. For further information email info@enware.com.au or visit http://www.enware.com.au.

FIGURE 4.36 Bench-mounted sensor

The unit is suitable for installation in most single-hole basins, but is not recommended for stainless steel sinks due to beam reflections. The base diameter is 50mm and the tap includes flexible hose for water connection and fixing plate with surface protection, O-ring seal and hex nut.

When installing the sensor body (see Figure 4.37), make sure the cords are correctly positioned in the cord

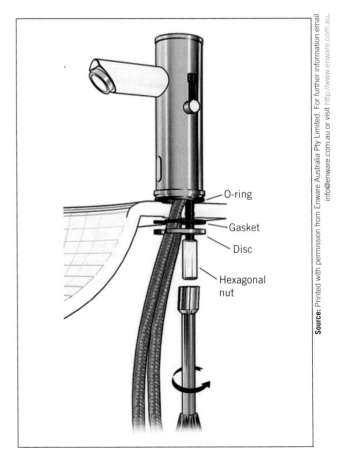

O-ring

Gasket

Disc

Hexagonal nut

Source: Printed with permission from Enware Australia Pty Limited. For further information email info@enware.com.au or visit http://www.enware.com.au.

FIGURE 4.37 Installing the sensor

guides in both the gasket and disc recesses and clear of sharp edges.

Aerators

Essentially, aerators are fitted to tapware outlets to add air to the water stream, resulting in a spray-like flow to reduce the effects of splashing and, more importantly, the flow of the water. Aerators come in a number of sizes and styles and can reduce flow rates by up to 50% while maintaining washing effectiveness.

Aerators have three basic components (see Figure 4.38):

1 *Housing* – the housing connects to the thread of the tap and provides the outer covering for the insert.
2 *Insert* – the insert regulates the tap's water flow. It is the insert that is chosen and fitted to suit the various flow-rate requirements. For example, a 5L per minute aerator should be fitted to hand basins and a 7.6L per minute aerator is best for kitchen and laundry applications.
3 *Rubber washer* – the sealing washer seals the aerator to the tap outlet.

FIGURE 4.38 Three components of aerators

FROM EXPERIENCE

It is important to remove all aerators and flush all debris before refitting when commissioning a water service.

GREEN TIP

The use of aerators significantly reduces water usage.

LEARNING TASK 4.4

1 Why is it important to open and close isolating valves slowly?
2 What is the common name of a 'loose jumper valve'?
3 What is the advantage of ceramic disc taps compared to loose jumper valve taps?
4 Name three types of ancillary valves.
5 Where are solenoid valves commonly used?

4.6 Cisterns and flushing devices

A cistern is a tank in which water is stored at atmospheric pressure. The tank incorporates a means of both controlling the water level and releasing the water when it is required to flush the fixture, with the water entering the cistern controlled by a float valve; it also has an air gap to prevent cross-connection to the water supply. The water leaving the cistern is controlled by a flush valve assembly.

All pans are manufactured to suit a particular flush capacity. The flush capacity of the pan is usually identified with a marking etched into the ceramic finish at the rear of the pan (see Figure 4.39). Today, most cisterns have a 6L full-flush and a 3L half-flush capacity and are only compatible with specifically designed pans.

Capacity etched into the glazed finish on the pan, indicating this pan requires a cistern of 9 litres' flush capacity to enable the removal of soil solids and to make it suitable to take a dual-flush cistern with a 9/4.5L capacity.

9-L

FIGURE 4.39 Pan indicating flushing capacity

There are several kinds of flushing devices. They range from the simple cistern to mains pressure flush valves.

A flushing cistern is a cistern that is capable of discharging a measured quantity of water by manual operation of the flushing mechanism. Generally, a flushing device will supply a predetermined amount of water to a soil fixture for the purpose of cleansing the soil material from the fixture and aiding with its travel along the drainage system. An important point to remember is that all flushing devices should maintain backflow protection between the drinkable water supply and the soil fixture and may be connected to either the drinking or non-drinking water supply.

A flushometer (flusherette) is a flushing device that uses energy from a pressurised water supply system or an elevated storage tank to discharge water. It is also commonly called a flush valve. Put simply, it provides a means of controlling and releasing water to a fixture. Once activated, the flushometer, not unlike a cistern, allows a predetermined amount of water to pass through it into the fixture and then automatically shuts off.

Flushometers are often specified instead of cisterns in high-demand situations. The water supply for flushometer valves can be either supplied by a storage tank, which is usually installed at a higher level to produce the necessary head (a minimum of 3m – approximately 30kPa; and a maximum of 30m – approximately 300kPa), or directly connected to a pressure water supply.

Flushometers can be more expensive to purchase and install than cisterns. However, they have a number of advantages. One is that they can cope with the higher demands of buildings such as office blocks, sports stadiums and theatres. This is because, unlike cisterns, they can be flushed more often as they do not suffer from the time delay required to fill the cistern. Another advantage is that since the water passes through a flushometer under pressure, the fixture is flushed quickly with a scouring action that helps ensure a thorough cleansing, resulting in increased hygiene.

There is one basic type of flushometer, but it may be fitted with either a diaphragm valve or a piston valve.

Flushometers can be activated by either manual force or automatic sensors. Manual flushometers require the user to depress the handle of the unit to activate the valve and complete the flush. Automatic or sensor-operated flushometers require an electronic power source to activate the flushing mechanism and complete the flush.

The sensor is usually set to activate the valve when the user has moved away from the fixture.

An example of a chrome-plated flushometer is shown in **Figure 4.40**.

Mains pressure flush valves are now more common on new installations than on flushometers connected to a storage tank. Mains pressure flush valves are supplied directly from the mains pressure supply and have an integral vacuum breaker valve fitted for backflow prevention. They require a minimum DN 25 pipe to supply the volume of water required for a scouring flush (see **Figure 4.41**).

Vacuum breakers

Some plumbing codes stipulate that every flushometer connected to a drinking water supply for use with a water closet or urinal must be equipped with a vacuum breaker as a backflow preventer and to prevent back siphonage. When flushometers are supplied by a tank system, vacuum breakers may not be required because an air gap is established. A vacuum or negative pressure in the supply line often causes back siphonage when:

■ the water supply fails
■ a main breaks
■ fire hoses draw heavily on the available water supply
■ a line is shut off and drained
■ water pressure is low and a heavy drain occurs at a low point.

Vacuum breakers are designed to allow the passage of water when the fixture is flushed, but if a vacuum situation occurs in the supply line, the vacuum breaker reacts instantaneously to break the siphoning effect.

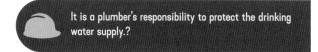

It is a plumber's responsibility to protect the drinking water supply.?

Source: Copyright © Sloan Valve Company. The author's moral rights, including the right of attribution, are asserted.

FIGURE 4.40 Chrome-plated flushometer

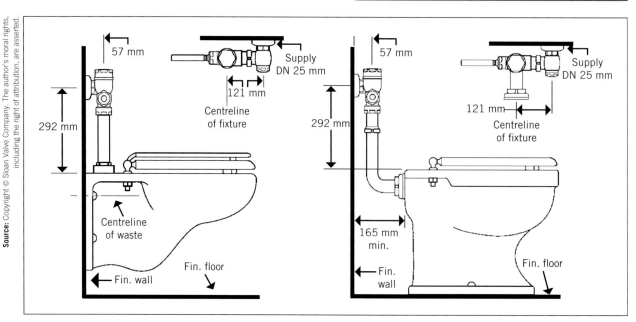

Source: Copyright © Sloan Valve Company. The author's moral rights, including the right of attribution, are asserted.

FIGURE 4.41 Flushometer connected to toilet pans: (Left) mains pressure flushometer top inlet pan, (Right) flushometer rear inlet pan

COMPLETE WORKSHEET 4

Water flow rates and pressure

The required flow rate (litres per minute) is determined by the requirements of the fixture not the flushometer. If the supply pipes are properly sized, water passing through the flushometer allows the fixture to operate efficiently (see Figure 4.41).

Programmed and on-demand sensor flushing

In commercial applications such as office buildings, stadiums, public buildings and educational institutions, flushing to urinals is often electronically controlled either by the use of a timed flushing cycle or on demand after the use of the appliance. These types of flush controls are an essential part of water conservation, particularly in large buildings. Note that due to significant water wastage, automatic or set-cycle cisterns are no longer permitted.

GREEN TIP

Water usage for flushing toilets and urinals has been dramatically reduced to conserve water.

Urinals (individual, grouped or slab) are the most likely fixture to have automatic or programmed flushing systems. It should be noted that urinals have different flushing volumes from toilet pans. AS/NZS 3500.1 states that the quantity of water discharged for sanitary flushing must be not more than 2.5L and not less than 1.5L for each single stall or each 600mm length of continuous urinal wall.

Automatic flushing systems can also be fitted to existing or newly installed cisterns.

AS/NZS 3500.1 PLUMBING AND DRAINAGE: WATER SERVICES

LEARNING TASK 4.5

1 What is another name for a 'flush valve'?
2 Why is a vacuum breaker valve installed in a flush valve?
3 How is cross-connection prevented between a cistern and the water supply?

4.7 Devices to control water temperature

Temperature control devices play an important role in plumbing systems by making them safe and protecting the community from scalding and burns.

Temperature and pressure relief valves

Mains pressure hot water heaters must be fitted with a temperature and pressure relief valve (see Figure 4.42). The purpose of this valve is to relieve the pressure in the hot water heater by venting it to the atmosphere should an unsafe condition arise, such as a thermostat malfunction that may result in the temperature going beyond the specified level or excess pressure occurs. When commissioning a storage hot water heater, the temperature and pressure relief valve must be operated to ensure it is working correctly.

Source: Reliance Worldwide Australia.

FIGURE 4.42 Typical temperature and pressure relief valves

HOW TO

HOW TO INSTALL A SENSOR TO ACTIVATE URINAL FLUSH

1 Position the ceiling sensor not more than 500mm from the urinal wall. Sensor coverage from a 2700mm ceiling is approximately 2400mm × 3600mm.
2 For a single stall urinal, position the sensor above the centre of the stall. Position the sensor slots at right angles to the urinal wall.
3 For a double stall urinal, position the sensor midway between the stalls. Position the sensor slots at right angles to the urinal wall.

4 For a triple stall urinal, position the sensor above the centre of all three stalls. Position the sensor slots parallel to the urinal wall.
5 For more than three stalls, use additional sensors. Remember not to connect the battery or power pack until all plumbing connections are completed. The power must be connected last, as its connection activates the system test mode.

Expansion of water

The corresponding expansion that occurs when water in its liquid form is heated must be allowed for to prevent dangerous situations from occurring. The amount of expansion can be calculated easily using the following formula:

$E = M \times T \times C$

where:

E = expansion in litres
M = mass of water
T = temperature variation in °C
C = coefficient of expansion = 0.000375

(The coefficient 0.000375 is how much 1L of water will expand when raised through a 1°C temperature rise.)

It is for this reason that temperature and pressure relief valves are installed on water heaters. If this volume was not allowed for, an increase in pressure would occur with a corresponding increase in temperature. The expansion that can result approaching vaporisation or boiling point is covered later. However, you must be aware that 1L of water can create approximately 1600L of steam if that water reaches vaporisation or boiling point within a vessel such as a hot water heater.

EXAMPLE 4.1

HOW TO CALCULATE EXPANSION

Calculate the amount of expansion that would occur if the water in a 270L hot water storage unit was heated from 15°C to 68°C.

Formula: $E = M \times T \times C$

where:

E = expansion in litres = ?
M = mass of water = 270
T = temperature variation (68 − 15 = 53)
C = coefficient of expansion = 0.000375

Workings: $E = 270 \times 53 \times 0.000375 = 5.36625$

Thus 5.36625L of expansion would occur. So the total volume of water in a 270L hot water storage unit at 15°C will increase to 275.36625L when heated to 68°C.

Over-pressurisation of the hot water heater can cause a shortened life of the heater. This can cause the tank to bellow – that is, the sides will move in and out each time the pressure drops or rises, such as when a hot water tap is turned on and off. To minimise this effect manufacturers generally rate the maximum temperatures and pressures that are suitable for their product, and these should never be exceeded.

When replacing the TPR valve on a water heater it must be replaced with the exact same model, including maximum temperature and pressure ratings. Under no circumstances should a TPR valve be fitted that has higher maximum pressure or temperature ratings than the original one. They usually have a lifespan of five years.

TPR valves may be used to guard against over-temperature and over-pressure hazards wherever water is stored in unvented containers. This is a requirement of AS/NZS 3500.4.

TPR valves also incorporate a drain vacuum breaker device. This is a further safety precaution and operates to relieve pressure build-up in the cylinder if the drain line becomes blocked or damaged. It will also relieve a partial vacuum in the drain line. In both cases the vacuum breaker will blow out to relieve pressure.

AS/NZS 3500.4 PLUMBING AND DRAINAGE: HEATED WATER SERVICES

Tempering valves

Tempering valves are used to mix hot water with cold water as it leaves the hot water service to supply ablution fixtures (basin, bath and shower) at a temperature of 50°C (see Figure 4.43). AS/NZS 3500.4 requires hot water to be delivered to all domestic fixtures that are used for ablution purposes to be at 50°C to prevent severe scalding. Water at 60°C will cause third-degree burns within five seconds, and at 70°C in one second (see Figure 4.44)

A tempering valve operates where cold water enters the mixing chamber above the piston and hot water enters the mixing chamber below the piston. The thermostatic controlling device (wax element) is positioned in the mixing chamber connected to the piston. The expansion or contraction in the length of the wax element will cause the piston to move either up or down. In the event of an increase in the hot water temperature in the mixing chamber, the element will expand, decreasing the hot water supply. If the water temperature in the mixing chamber decreases, the element will contract, allowing more hot water into the mixing chamber. The sensitivity of the wax element ensures instant movement and therefore a minimum in temperature fluctuation. If there is a failure in cold water supply, the rapid expansion of the wax element will completely shut off the flow of hot water. This is called a thermal shutdown.

Tempering valves are non-serviceable and must be replaced as required by the manufacturer, generally every five years, and cleaning of the valve filters may be necessary to prevent the loss of hot water pressure.

FROM EXPERIENCE

Tempering valves used on solar hot water systems must be of the high performance type to handle higher temperatures. They are easily identified by their orange caps (see Figure 4.45).

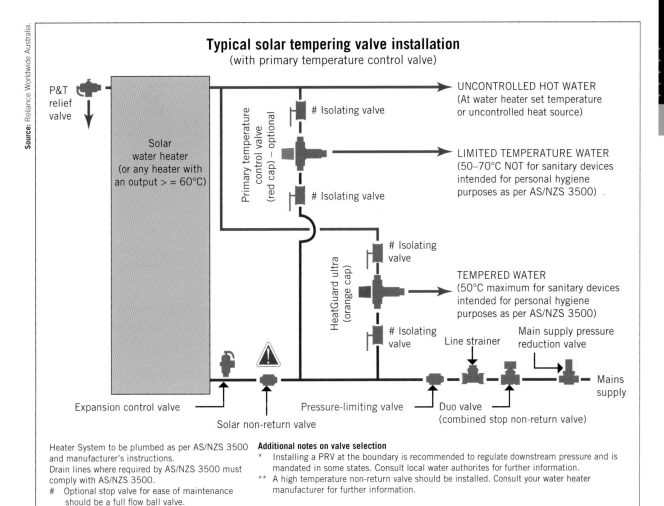

Source: Reliance Worldwide Australia.

Typical solar tempering valve installation
(with primary temperature control valve)

P&T relief valve

Solar water heater (or any heater with an output > = 60°C)

Primary temperature control valve (red cap) – optional

HeatGuard ultra (orange cap)

Isolating valve

UNCONTROLLED HOT WATER
(At water heater set temperature or uncontrolled heat source)

Isolating valve

LIMITED TEMPERATURE WATER
(50–70°C NOT for sanitary devices intended for personal hygiene purposes as per AS/NZS 3500)

Isolating valve

Isolating valve

TEMPERED WATER
(50°C maximum for sanitary devices intended for personal hygiene purposes as per AS/NZS 3500)

Isolating valve

Line strainer

Main supply pressure reduction valve

Mains supply

Expansion control valve

Solar non-return valve

Pressure-limiting valve

Duo valve (combined stop non-return valve)

Heater System to be plumbed as per AS/NZS 3500 and manufacturer's instructions.
Drain lines where required by AS/NZS 3500 must comply with AS/NZS 3500.
Optional stop valve for ease of maintenance should be a full flow ball valve.

Additional notes on valve selection
*   Installing a PRV at the boundary is recommended to regulate downstream pressure and is mandated in some states. Consult local water authorites for further information.
** A high temperature non-return valve should be installed. Consult your water heater manufacturer for further information.

FIGURE 4.43 Typical solar hot water tempering valve set-up

FIGURE 4.44 Blue cap tempering valve

FIGURE 4.45 Orange (high temperature) cap tempering valve

Mixing valves

Mixing valves are used to provide water of a desired temperature at outlets for domestic, industrial or specialised purposes such as hospitals, healthcare facilities, penal institutions and childcare facilities. They are connected to a hot and a cold water supply source, with the desired water temperature at the outlet being obtained through the intermixing of the two water supply sources by the manipulation of one or two valves.

The breeching piece

The simplest way of providing water mixing is by connecting the outlets of hot and cold water with a breeching piece (shower) (see Figure 4.46) to a common outlet. Mixing takes place at the junction of the two branches and the flow to the required temperature is controlled by the degree to which each of the two taps is opened.

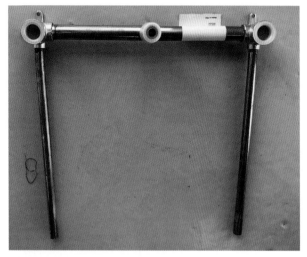

FIGURE 4.46 Typical breeching piece

Single-lever mixing valves

Single-lever mixers, or flickmixers as they are more commonly known (see Figure 4.47), are available for installation on kitchen sinks, laundry troughs, vanity basins and showers (see Figure 4.48). They are available in many styles and qualities. Most single-lever mixing valves use ceramic disc technology to control the water flow. Most of the ceramic disc assemblies are contained in a cartridge that is replaceable.

Source: Printed with permission from Enware Australia Pty Limited. For further information email info@enware.com.au or visit http://www.enware.com.au.

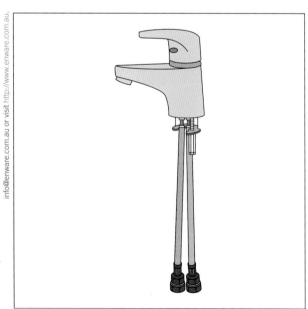

FIGURE 4.47 Single-lever mixing valves

Source: Printed with permission from Enware Australia Pty Limited. For further information email info@enware.com.au or visit http://www.enware.com.au.

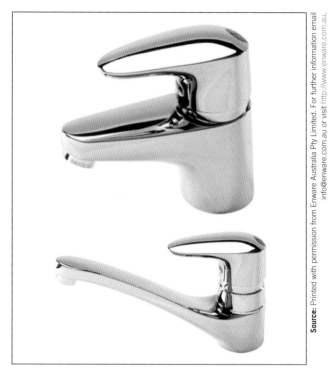

FIGURE 4.48 Basin and sink mixers

The movement of a central lever controls the basic operation of a single-lever mixer. When the lever is lifted, the water is allowed to flow to the outlet. Single-lever mixers should be connected to the supply service in such a manner that when the lever is moved towards the left the water at the outlet should become hotter, and when the lever is moved towards the right the water should become colder.

When installing single-lever mixing valves, they should be connected, as a minimum, so that they are individually controlled by a screw-down-type stop valve. This provides limited backflow prevention and will also allow the mixer to be isolated should the ceramic disc fail. A backflow prevention device such as a check valve should also be fitted, particularly if the mixer has a hose as an outlet, as this can potentially cause a cross-connection.

Thermostatic mixing valve

The word 'thermostatic' means 'being kept at constant temperature'. 'Thermo' means 'temperature', and 'static' means 'remains the same and doesn't move, rise or fall'. For optimum performance it is recommended that the thermostatic mixing valve (TMV) (see Figure 4.49) has a balanced cold and hot water inlet supply pressure to within 10% of each other. TMVs must be installed with check valves, strainers and isolating valves in accordance with AS/NZS 3500.4.

TMVs are installed in healthcare buildings such as hospitals and nursing homes as well as childcare facilities and disabled bathrooms. They must be serviced annually and have their log book updated by

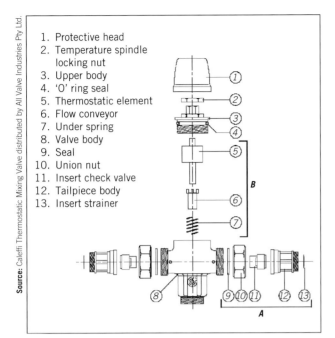

1. Protective head
2. Temperature spindle locking nut
3. Upper body
4. 'O' ring seal
5. Thermostatic element
6. Flow conveyor
7. Under spring
8. Valve body
9. Seal
10. Union nut
11. Insert check valve
12. Tailpiece body
13. Insert strainer

Source: Caleffi Thermostatic Mixing Valve distributed by All Valve Industries Pty Ltd.

FIGURE 4.49 Exploded view of the thermostatic mixing valve

a licensed accredited plumber. **Figure 4.50** shows a TMV installation diagram.

AS/NZS 3500.4 PLUMBING AND DRAINAGE: HEATED WATER SERVICES

A TMV mixes hot and cold water in such a way as to maintain the set temperature of the mixed water at the outlet. To control the temperature a thermostatic element is fully immersed into the mixed water. It then expands or contracts causing movement of the piston, closing either the hot or the cold inlets and so regulating the flow rates entering the valve. If there are variations in temperature or pressure at the inlets, the internal element automatically reacts by attempting to restore the original temperature setting.

In the event of a failure of the hot or cold water supply the piston will shut off, stopping water discharging from the outlet. This is called 'thermal shutdown'. Generally, the manufacturer will specify the maximum and minimum temperatures for the hot water supply.

The operation of a TMV and a tempering valve is very similar. The differences are that a TMV must be serviced annually by an accredited licenced plumber and the temperature settings are more accurate than those of a tempering valve.

'Dead legs' should be kept to a minimum, and the maximum permissible length of pipe from the valve to the furthest fixture must be no more than 10m or 2L in volume. This is to prevent the breeding of *Legionella* bacteria forming at temperatures between 20°C and 45°C (which warm water is delivered at) and also to avoid temperature loss over a longer distance. The minimum length of the branch line is 1m. This is to avoid spiking of hot water in the event of a thermal shutdown.

Backflow prevention devices

Backflow prevention is an extremely important part of any drinking water system design. A cross-connection between the drinkable water supply and a potential source of contamination can pose serious health risks to the individual user and/or the wider community. There are numerous well-documented cases where cross-connections have been responsible for the contamination of drinking water supplies, resulting in the spread of disease. The devices and valves used in the prevention of cross-connection are explained in depth in Chapter 10.

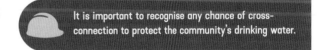

It is important to recognise any chance of cross-connection to protect the community's drinking water.

Source: Caleffi Thermostatic Mixing Valve distributed by All Valve Industries Pty Ltd.

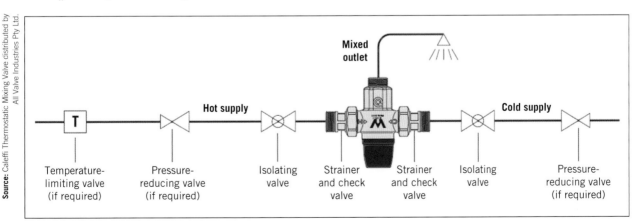

FIGURE 4.50 Thermostatic mixing valve installation diagram

LEARNING TASK 4.6

1 What are the two main functions of a temperature and pressure relief valve?
2 What is the main component of a tempering valve and a thermostatic mixing valve?
3 What three valves must be installed on the hot and cold water inlets to a thermostatic mixing valve?
4 What type of valves prevent cross-connection in a water supply?

4.8 Cleaning up

As with any job, it is important to leave the workplace clean and tidy, and to remove rubbish to the appropriate bins. All recyclable items such as paper, cardboard, plastic and glass need to be disposed of in the recycle bins.

All tools and equipment must be cleaned, checked, maintained and stored in a dry, safe area. Any hire equipment must be cleaned and returned to the supplier.

The documentation involved on completion must to be submitted as per company and local authorities' requirements (for example, 'certificate of compliance').

Finally, the client must be advised of the correct operation, maintenance and responsibility of the valves, flushing devices and pumps installed upon the handover of the job.

 COMPLETE WORKSHEET 5

LEARNING TASK 4.7

1 What types of rubbish can be recycled?
2 What is the name of the certificate issued upon completion?

SUMMARY

- Valves play an important role in the safe and efficient operation of hot and cold water services.
- It is vital that the correct valves are used where required by authorities and specifications.
- Isolation valves control the flow of water in line (stop valve) and at the end of the line (fixture outlet).
- Pressure reduction valves reduce water pressure to protect fixtures and materials, and reduce noise transmission.
- Temperature control valves prevent scalding by providing a blend of hot and cold water at ablution fixtures.
- Temperature pressure relief valves relieve excessive temperature and pressure from hot water systems.

- Infra-red sensing tapware can reduce the risk of cross-contamination by removing the need to touch tapware to operate it.
- Aerators add air at the fixture outlet to reduce water flow and splashing.
- There are several types of flushing devices and valves. Some are connected directly to the water supply and others are independently supplied. They can be set up to flush automatically or manually.
- Backflow prevention valves protect the water supply by preventing cross-connection.

GET IT RIGHT

1   Name the type of valve pictured.

2   Which photo shows the correct installation?

3   What is wrong with the faulty installation?

WORKSHEET 1

Student name: _____

To be completed by teachers

Satisfactory ☐

Not satisfactory ☐

Enrolment year: _____

Class code: _____

Competency name/Number: _____

Task: Review the sections '4.1 Background' and '4.2 Planning and preparation' and answer the following questions.

1   Valves perform a number of important functions. List five different valves and their functions.

_____

_____

_____

_____

_____

2   Explain the term 'BASIX' in relation to water supply.

_____

_____

3   Explain the term 'WELS' in relation to water supply.

_____

_____

4   List four items of information that may be obtained from plans and specifications in relation to installing and adjusting water supply service controls and devices.

_____

_____

_____

_____

_____

5   If a hazard cannot be eliminated, what is the next level of control required?

_____

6   What information can be obtained from specifications?

_____

_____

WORKSHEET 2

To be completed by teachers

Satisfactory ☐

Not satisfactory ☐

Student name: _____

Enrolment year: _____

Class code: _____

Competency name/Number: _____

Task: Review the sections '4.3 Effects of water pressure on services and devices' and '4.5 Service controls and mixing devices' and answer the following questions.

1   What effect can excessive velocity and flow rate have on water controls and devices?

_____

_____

2   Excessive velocity and flow rate can be reduced by which two methods?

_____

_____

3   Name four types of isolating valves.

_____

_____

_____

_____

4   List four situations where isolating valves must be installed on water services.

_____

_____

_____

_____

5   State an advantage that ceramic discs have over the traditional screw-down type of valve.

_____

_____

6 Describe the following valves and where they can be used in a water supply.

    a   Gate valves

      Description

      _____

      _____

      Where can they be used?

      _____

      _____

    b   Ball valves

      Description

      _____

      _____

      _____

      Where can they be used?

      _____

      _____

      _____

7 Which way does the handle turn to allow a valve to rise off its seat?

    _____

8 What does the term 'full bore' mean in relation to ball valves?

    _____

    _____

9 What is another name for a large-diameter gate valve?

    _____

10 Name three ways that stop valves can be fitted in line on an installation.

    _____

    _____

    _____

11 What are ceramic disc taps also known as?

_____

12 What are the maximum and minimum water pressures for a building other than a fire service, as per AS/NZS 3500.1?

_____

_____

13 What is the maximum flow velocity in a water service as per AS/NZS 3500.1?

_____

14 What is the most common cause of water hammer?

_____

_____

15 How can water hammer be eliminated?

_____

_____

WORKSHEET 3

To be completed by teachers

Satisfactory ☐

Not satisfactory ☐

Student name: _____

Enrolment year: _____

Class code: _____

Competency name/Number: _____

Task: Review the sections 'Other types of control valves' and 'Ancillary valves' and answer the following questions.

1   What is the function of a globe valve?

_____

_____

2   Name two situations where a globe valve would be used.

_____

_____

_____

3   What is the function of a line strainer?

_____

_____

4   What two configurations do pressure reduction valves come in?

_____

_____

5   What three functions does a trio valve perform?

_____

_____

_____

6   Where would a solenoid valve be used?

_____

_____

7   In what situation would a pressure ratio valve be used?

_____

8   Name two types of non-return valves.

_____

_____

9   What are the three common preset pressure settings on pressure-limiting valves?

_____

_____

_____

10   What is another name for a non-return valve?

_____

11   Spring-type taps are commonly used for what purpose?

_____

12   What is the difference between a pressure-limiting valve and a pressure-reducing valve?

_____

13   What pressure would a 3:1 ratio valve deliver if the inlet pressure was 12m head?

_____

14   Where is a duo valve commonly fitted?

_____

WORKSHEET 4

To be completed by teachers

Satisfactory ☐

Not satisfactory ☐

Student name: _____

Enrolment year: _____

Class code: _____

Competency name/Number: _____

Task: Review the sections '4.4 Water hammer and noise transmission' and '4.6 Cisterns and flushing devices' and answer the following questions.

1   Name two main components that are part of infra-red sensing tapware.

_____

_____

2   What type of power supply is common with infra-red sensors?

_____

3   List five applications where infra-red sensitive tapware would be beneficial.

_____

_____

_____

_____

_____

4   What is the purpose of a vacuum breaker valve?

_____

5   What is the function of an aerator?

_____

_____

6   Name two types of flushing devices.

_____

_____

7   What is the minimum head required to operate a flush valve?

_____

8 A ceiling sensor should be located within how many millimetres of the urinal wall?

_____

9 What is the minimum size of a water service supplying a mains pressure flush valve?

_____

10 What type of valve is fitted in a cistern to control the water level?

_____

11 What is the minimum amount of water required to flush a single-stall urinal as per AS/NZS 3500.1?

_____

12 Is it possible to fit an automatic flushing system to an existing manual flush system?

_____

WORKSHEET 5

Student name: _____

Enrolment year: _____

Class code: _____

Competency name/Number: _____

To be completed by teachers

Satisfactory ☐

Not satisfactory ☐

Task: Review the section '4.7 Devices to control water temperature' and answer the following questions.

1   Tempering valves are installed to mix hot and cold water together to produce warm water at a certain temperature.

   a   What is this temperature?

   _____

   b   Why is the warm water limited to this temperature?

   _____

   c   How often should tempering valves be replaced?

   _____

2   What fixtures must be supplied with water from a tempering valve?

   _____

   _____

3   Where are temperature pressure relief valves installed?

   _____

   _____

4   How often must a thermostatic mixing valve be serviced?

   _____

   _____

   _____

5   What is the maximum permissible distance from a thermostatic mixing valve to the furthest outlet?

   _____

6   Explain what a thermal shutdown is.

   _____

   _____

7　What is the formula for the expansion of water?

_____

8　Is it normal for a TPR valve to dump a few litres of water per day?

_____

9　What is the approximate lifespan of a TPR valve?

_____

10　Why is backflow prevention important?

_____

WORKSHEET RECORDING TOOL

Learner name		Phone no.	
Assessor name		Phone no.	
Assessment site			
Assessment date/s		Time/s	
Unit code & title			
Assessment type			

Outcomes

Worksheet no. to be completed by the learner	Method of assessment WQ – Written questions PW – Practical/workplace tasks TP – Third-party reports SC – Scenarios RP – Role plays CS – Case studies RW – Report writing PF – Portfolio	Satisfactory response	
		Yes ✓	No ✗
Worksheet 1	WQ	☐	☐
Worksheet 2	WQ	☐	☐
Worksheet 3	WQ	☐	☐
Worksheet 4	WQ	☐	☐
Worksheet 5	WQ	☐	☐

Assessor feedback to the learner

Feedback method (Tick one ✓):	☐ Verbal ☐ Written (if so, attach) ☐ LMS (electronic)

Indicate reasonable adjustment/assessor intervention/inclusive practice (if there is not enough space, a separate document must be attached and signed by the assessor).

Outcome (Tick one ✓):	☐ Competent (C) ☐ Not Yet Competent (NYC)

Assessor declaration: I declare that I have conducted a fair, valid, reliable and flexible assessment with this learner, and I have provided appropriate feedback.

Assessor name:	
Assessor signature:	
Date:	

Learner feedback to the assessor

Feedback method (Tick one ✓):	☐ Verbal	☐ Written (if so, attach)	☐ LMS (electronic)

Learner may choose to provide information to the RTO separately.

	Tick one:	
Learner assessment acknowledgement:	**Yes** ✓	**No** ✗
The assessment instructions were clearly explained to me.	☐	☐
The assessment process was fair and equitable.	☐	☐
The outcomes of assessment have been discussed with me.	☐	☐
The overall judgement about my competency performance was fair and accurate.	☐	☐
I was given adequate feedback about my performance after the assessment.	☐	☐

Learner declaration:
I hereby certify that this assessment is my own work, based on my personal study and/or research. I have acknowledged all material and sources used in the presentation of this assessment, whether they are books, articles, reports, internet searches or any other document or personal communication.
I also certify that the assessment has not previously been submitted for assessment in any other subject or at any other time in the same subject and that I have not copied in part or whole or otherwise plagiarised the work of other students and/or other persons.

Learner name:	
Learner signature:	
Date:	

INSTALL AND COMMISSION WATER HEATING SYSTEMS: HOT WATER

5

Chapter overview

The aim of this chapter is to address the skill and knowledge required to install and commission water heating systems using various approved materials. The chapter covers the installation requirements for low-pressure storage water heaters, mains pressure storage water heaters, continuous flow water heaters and solar water heaters, as well as the installation requirements for multiple water heater installation (manifold installation).

Learning objectives

Areas addressed in this chapter include:
- types of water heaters and their installation requirements
- planning and preparation
- material selection
- sizing requirements
- testing, commissioning and maintaining water heating systems
- cleaning up.

Refer to Chapter 4 for descriptions of the different types of valves used in a hot water installation and Chapter 3 for the installation of pipework to and from the hot water inlet or outlet that forms part of the rough-in. Refer to Chapter 1 for the installation requirements for materials.

5.1 Background

A hot water installation is classed as an installation of one or more water heaters and the required hot and cold piping system to supply hot water to a number of fixtures, appliances and outlets. Water heaters are generally divided into four classes:

1 continuous flow water heaters
2 storage water heaters
3 heat exchanger–coil heaters/calorifiers
4 commercial water heaters (boilers).

A continuous (instantaneous) flow water heater is designed to heat water only at the time it is being used and supplies hot water continuously while the hot tap is turned on. Continuous flow water heaters are generally connected to a mains pressure water supply.

A storage water heater is designed to hold a designated quantity of hot water in an insulated container ready for use as required. These units can be designed to both store and supply hot water at mains pressure (above 350kPa), or to store water at atmospheric pressure and deliver it by gravity.

Heat exchanger–coil heaters/calorifiers and commercial water heaters (boilers) are not part of the learning for this unit and will be covered later in your training.

Water heaters are further classified by storage pressure:

■ **Mains pressure units.** A mains pressure unit is designed to store and deliver water at mains pressure (recommended to be above 350kPa); this provides the hot and cold water at the same outlet pressure.
■ **Reduced-pressure unit.** A pressure-reducing valve (see Chapter 4) or an overhead storage (feed) tank connected to the cold water connection of a water heater reduces the delivery pressure to below that in the utility's water main. This may deliver a different pressure in the cold and hot water supply, making end-use balancing or mixing of hot and cold water more difficult.
■ **Gravity unit (low pressure).** A gravity feed hot water system has a cold water feed (cistern) tank fitted to the storage tank. It is designed to store water at atmospheric pressure and deliver it via gravity to the required hot water outlets. These units are no longer commonly installed due to problems with balancing the supply pressures on single-lever mixing valves, thermostatic mixing valves and tempering valves incorporated into the warm water supply. Also there is a reliance on high usage of electricity as the fuel medium for heating.

GREEN TIP

Using electricity to heat water with traditional 1.8kW, 2.4kW, 3.6kW or 4.8kW elements consumes a high amount of greenhouse-gas-intensive energy.

Water heaters are further classified by their delivery method:

■ **Single-point unit.** These units are designed to supply water to one tap/outlet only; they may be continuous flow or storage design.
■ **Multipoint unit.** These units are designed with sufficient water flow capacity and thermal input to provide a consistent supply of hot water to several outlets at the same time. They may be of a storage or continuous flow type.
■ **Push-through unit.** These units are also known as free outlet under-sink water heaters. They store a small quantity of water at atmospheric pressure; when the hot tap is opened, water is delivered at mains pressure. This will be covered later in this chapter.

FROM EXPERIENCE

It is necessary as a plumber to have the knowledge to select the correct water heater for the right application.

5.2 Water temperature

Hot water must be stored at a minimum temperature of 60°C; this is designed to inhibit the growth of *Legionella* bacteria.

All new hot water installations must be designed to deliver heated water used for personal hygiene (ablution fixtures), such as baths, showers, basins and bidets. As per AS/NZS 3500.4, they must not exceed:

■ 45°C for buildings such as:
  – early childhood centres
  – primary or secondary schools
  – nursing homes or similar facilities looking after young, elderly or sick people or people with a disability (a check of state or territory temperature requirements for the above should be made prior to any installation proceeding)
■ 50°C in all other buildings.

Laundries and kitchen sinks are not a part of these water temperature requirements.

AS/NZS 3500.4 PLUMBING AND DRAINAGE: HEATED WATER SERVICES

LEARNING TASK 5.1

1 What is the difference between a storage water heater and a continuous water heater?
2 What is an ablution fixture?

5.3 Types of hot water heating systems

Understanding the right choice of water heating system for different buildings and situations is vital for plumbers to help clients make an informed decision for the most efficient and economical choice.

Mains pressure storage water heaters

Mains pressure storage water heaters can be heated by gas, electricity, solar or heat pump. The water connection requirements are generally the same. The types of valves required (see Figure 5.1) are explained in Chapter 4.

Mains pressure storage water heaters work on the displacement principle. As hot water is less dense than cold water it sits on top of the cold water. As a hot tap is opened, cold water enters the cylinder from the bottom, forcing the hot water out at the top of the cylinder. The cold water continues to flow into the water heater until the hot tap is turned off and the cylinder is re-pressurised.

These water heaters need to hold enough hot water for the requirements of all the occupants of the building. The water is stored in an insulated container and is ready for immediate use. As heated water is used it is replaced with cold water, which is then heated (unless it is outside the off-peak heating times). The heating continues after the cylinder is re-pressurised, until the whole contents reach the prescribed temperature, which is a minimum of 60°C and is controlled by an adjustable thermostat.

Temperature and pressure relief valve

Mains pressure water storage water heaters require a temperature and pressure relief valve and drain to be installed as part of the installation requirements. The drain line for all water heaters installed within a building is usually installed at the rough-in stage, with the final connection between the water heater and drain line being made at the fit-off stage. Temperature and pressure relief valve drains are sized no smaller than the valve outlet and will be of copper or other approved materials. They must not exceed the maximum length, as shown in Table 5.1.

AS/NZS 3500.4 PLUMBING AND DRAINAGE: HEATED WATER SERVICES

TABLE 5.1 Maximum length of temperature and pressure relief valve drains

Maximum relief drain length (m)	Maximum number of bends greater than 45°
9	3
8	4
7	5
6	6

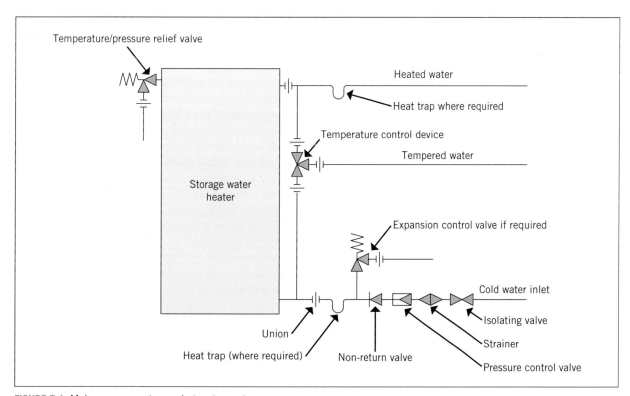

FIGURE 5.1 Mains pressure storage hot water system

Where it is necessary to install a drain line outside these requirements it must discharge via a tundish (see Figure 5.2).

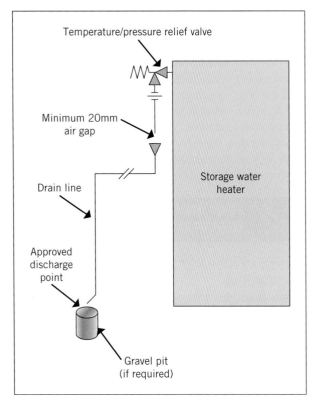

FIGURE 5.2 Temperature and pressure relief drain discharging via a tundish

Temperature and pressure relief valve drains must be installed:

- with no taps, valves or restrictions installed within the drain
- with a continuous fall from the valve to the approved point of free discharge
- so as not to discharge into a safe tray
- to discharge to a point readily visible
- so as not to cause any damage to a building
- so as not to cause injury to persons
- so as not to cause a nuisance by the release of steam or hot water
- with an air gap of twice the diameter of the drain if discharging over a tundish, and be a minimum of 20mm
- so as not to discharge onto any roof covering.

Where the drain terminates outside a building, it must have its point of free discharge between a minimum of 200mm and a maximum of 300mm above an unpaved area, and a minimum of 75mm and a maximum of 300mm above an overflow/disconnector gully or a DN 100 gravel pit in paved areas. This is so visual contact can be made when the temperature and pressure relief valve is discharging. The drain line from a tundish must be a minimum of DN 20 and be one size larger than the largest drain line discharging into the tundish. If the water heater is installed externally, the drain line must discharge away from the water heater to protect any person operating the temperature and pressure relief valve and so as not to cause damage to building footings or foundations.

Extreme hot water discharging via the temperature and pressure relief valve will scald a person very quickly, so it is important that the discharge requirements are met.

In areas subject to water pipe freezing, the drain line must not exceed 300mm in length and must be insulated. It must discharge into a tundish with an air gap of between 75mm and 150mm. The tundish must discharge as described above.

Under no circumstances is the temperature and pressure relief valve to be blocked off or removed. This would lead to the water heater failing and under certain circumstances would create an extremely dangerous and explosive situation.

FROM EXPERIENCE

Always operate the temperature pressure relief valve when commissioning the hot water heater to ensure it is functioning correctly.

Continuous flow water heaters

Continuous flow water heaters may be connected to either electricity or gas. This is illustrated in AS/NZS 3500.4.

When a hot tap is opened, the flow of water activates a valve (for a gas model) or switch (for an electric model), allowing the heating medium to heat the water as it passes through a coil in the water heater and shuts down when the hot tap is closed. Continuous flow water heaters can be a single-point-of-use water heater, as in a bath heater, or designed to serve several outlets.

They are relatively inexpensive to purchase, and the new models available can be fitted with remote electronic controllers to allow the user to adjust the end-use water temperature. Energy is saved as water is not 'overheated' and then cooled to a safer temperature by adding cold water at the mixer or shower set. While continuous flow water heaters are seen to be rather water efficient, they are higher-level energy consumers with an average gas rate of 188MJ/h compared with a storage unit's average rate of 38MJ/h.

Instantaneous hot water heaters have been superseded by continuous flow hot water heaters because the continuous hot water service can provide a better flow rate, particularly when more than one hot tap is on at a time.

COMPLETE WORKSHEET 1

Solar water heaters

There are two main types of solar water heating systems: close coupled (storage tank and collectors on roof together) and split systems (storage tank at ground level and collectors on roof). These systems can be further described in two other categories:

1 Direct solar water heating system – the water is directly heated by the solar collectors and is circulated between the storage tank and the collectors.

2 Indirect solar water heating system – a heat exchange fluid (glycol) is heated by the solar collectors and circulates between the heat exchanger within the storage tank and the collectors. Heat is transferred from the heat exchange fluid to the water via conduction in the heat exchanger. The heat exchange fluid is resistant to freezing at temperatures as low as –28°C. Indirect solar water heating systems are used in areas subject to freezing temperatures.

GREEN TIP

The development of technology using the sun to harness renewable energy is increasingly becoming more efficient and is constantly improving over time.

There are three main components to a solar hot water system: the collectors, the water storage tank and the booster. Australia is in the Southern Hemisphere so the collectors should be installed facing north to point more directly at the sun to collect the most heat. Extreme care must be taken with the placement of a roof storage system, as the roof structure may not be designed for the added weight of the storage cylinder and water. The roof structure may need additional supports prior to placement of the storage cylinder and filling with water. Any penetration of a roof covering must be made waterproof in an approved manner.

Where the building is not occupied, all solar collectors not filled with water (or heat exchange fluid) *must* be covered to prevent overheating of the system until the building is occupied and the solar system is being used.

Only metal pipe, usually copper, is to be used between the solar collectors and the storage cylinder or between the water heater outlet and the temperature control valve.

The water temperature gets extremely high between the collectors and the storage cylinder. Therefore, plastic pipe is not suitable.

Solar energy has a greater input into the proportion of water heated when the system is installed closer to the equator. Thus, Darwin can produce between 90%

and 95% of all water heating requirements, whereas Hobart will produce only 50% to 55%. Sydney will produce between 65% and 75% of all hot water requirements.

A solar non-return (SNR) valve is installed to prevent backflow of heated water from the water heater to the mains line. This serves the dual purpose of protecting the mains supply from contamination and the loss of water that has already been heated. It must be installed between the cold water branch to the tempering valve and the cold water heater inlet (see Figure 5.3).

A further SNR valve is installed on the hot water line between the solar collectors and the hot water storage tank; this will restrict hot water leaving the tank (mainly at night). Without this additional SNR valve, hot water tends to flow back to the top of the solar collectors where the hot sensor probe will sense the warmer water. In turn, the hot sensor may activate the circulating pump to move water around the entire system (at night) when it is not required, ultimately cooling it down.

Solar and uncontrolled-heat heating systems present an increased hazard to downstream components as water can become superheated and potentially flash to steam.

Only high-temperature non-return valves, such as solar non-return valves, should be used on solar hot water installations.

The positioning of the collectors is critical to the efficiency of the system. Collectors need to take advantage of the maximum solar efficiency between the hours of 9 am and 3 pm (Eastern Standard Time) and should therefore be positioned so that excessive shading from trees or buildings does not occur during this time period. Solar collectors have a matt-black finish to obtain the greatest absorption of the sun's radiation. Collectors need to be installed a minimum of 150mm below the bottom of the storage container for close coupled storage systems. For maximum solar efficiency, the collectors need to be installed at a similar angle to the area's latitude angle. So, the further from the sun, the higher the angle. The optimum angle for each Australian city may be found in AS/NZS 3500.4.

AS/NZS 3500.4 PLUMBING AND DRAINAGE: HEATED WATER SERVICES

Solar systems may have either a gas or an electric booster system installed to supplement the heating requirements.

All solar hot water systems, as with any type of water heater, must be installed to meet the manufacturer's installation instructions. Failure to do so may result in the cancellation of the product warranty.

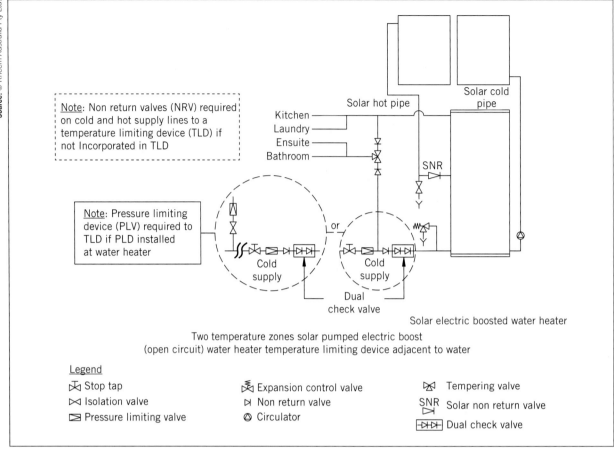

Source: © Rheem Australia Pty Ltd.

Note: Non return valves (NRV) required on cold and hot supply lines to a temperature limiting device (TLD) if not Incorporated in TLD

Note: Pressure limiting device (PLV) required to TLD if PLD installed at water heater

Kitchen
Laundry
Ensuite
Bathroom

Solar hot pipe

Solar cold pipe

SNR

Cold supply

or

Cold supply

Dual check valve

Solar electric boosted water heater

Two temperature zones solar pumped electric boost
(open circuit) water heater temperature limiting device adjacent to water

Legend

Stop tap   Expansion control valve   Tempering valve
Isolation valve   Non return valve   SNR Solar non return valve
Pressure limiting valve   Circulator   Dual check valve

FIGURE 5.3 Solar Lo-line installations showing correct positioning of valve train and cold water take-off to tempering device

Close coupled solar systems

Close coupled solar systems (see **Figures 5.4** and **5.5**) use natural thermosiphon circulation to circulate the water or heat exchange fluid for heating through the system. Further information on close coupled solar systems may be found online at http://www.rheem.com.au/products/residential/solar and https://www.solahart.com.au/landing-pages/solar-hot-water.

Source: © Rheem Australia Pty Ltd.

FIGURE 5.5 Solahart close coupled solar system

Thermosiphon circulation

The heating of the water or heat exchange fluid relies on the three forms of heat transfer: radiation, conduction and convection. This system relies on the fact that hot water rises (convection), which allows water to circulate from the solar collectors to the storage cylinder and back again. This system can be installed either as a close coupled or split system. Whichever type is preferred, they both must have the collectors installed below the storage tank to enable the thermosiphon system to work; a circulating pump is not required.

Source: © Rheem Australia Pty Ltd.

FIGURE 5.4 Rheem close coupled solar system

The system's fluid (either water or heat exchange fluid) in the collectors is heated by radiation from the sun; this in turn heats the fluid in the collectors via conduction. The heated fluid then flows up to the top of the collector and then into the storage cylinder (convection). The fluid in the collector is replaced by cooler (denser) fluid from the base of the storage cylinder; this fluid enters the bottom of the solar collector allowing circulation to start again.

The system control valves (see Figure 5.6) must be installed in a readily accessible location from ground or floor level. It is good practice to install an isolation valve at the inlet to the cylinder on the roof for servicing; however, it must be a full way valve only, such as a ball valve.

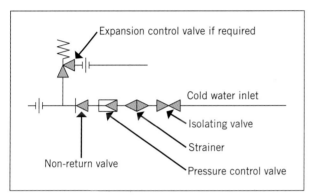

FIGURE 5.6 Valve requirements for a close coupled system to be located at ground or floor level

These systems, like a mains pressure storage water heater, have a temperature and pressure relief valve installed to prevent overheating or pressurising the water heater. It must terminate as described earlier in this chapter and must not discharge onto the roof covering.

Split (forced circulation) solar systems

Split (forced circulation) solar systems use a pump to circulate the system's fluid (water or heat exchange fluid) for heating through the system (see Figure 5.7). The position of the solar collectors in relation to the storage cylinder is not important as there is a pump installed to create a forced circulation of the fluid between the solar collectors and the storage cylinder. This type of system is much more flexible.

The flow of fluid between the collectors and the storage cylinder is controlled by a temperature sensor. This senses if the temperature in the collectors is hotter (about 4°C) than that in the storage tank, turning on a pump to circulate fluid between the two parts.

Solar indirect (gas or electric) system (frost-protected)

Solar indirect systems are a conventional split (mains pressure water heater) indirect (glycol) heat exchange solar system with a continuous flow (mains pressure) water heater attached (see Figure 5.8). The split indirect

Source: © Rheem Australia Pty Ltd.

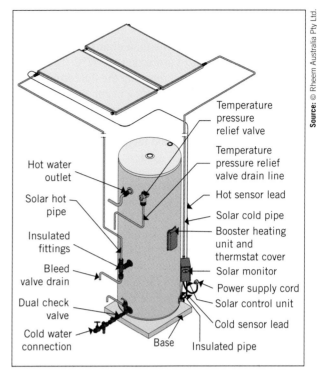

FIGURE 5.7 Rheem forced circulation solar system

Source: © Rheem Australia Pty Ltd.

FIGURE 5.8 Gas-boosted solar indirect system

solar system is installed as you would a conventional indirect solar installation.

When a hot tap is opened, the water from the storage tank passes through the continuous flow gas water heater. If the water is below the required outlet temperature, the continuous flow gas water heater cuts in, heating the water to the required temperature.

This type of system has a very high energy efficiency rating as it has the advantage of solar energy use and only boosts the water temperature as required as it is being used. During periods of solar gain, the heat exchange fluid is pumped through the collectors to capture the sun's radiation. When heating is complete, or during periods of frost, the fluid drains back to the tank providing total frost protection to –17°C. This works like a car radiator, protecting the fluid from freezing in winter and overheating in summer. The solar cold water (flow) and hot water (return) (see Figure 5.9) connect at the top of the solar storage tank. Flow and return lines must be insulated metallic pipes. Polymer (plastic pipes) *must not* be used.

Evacuated tube solar collectors

Instead of using flat solar collectors, evacuated tube solar collectors can be used to heat water (see Figure 5.10). They consist of two glass tubes fused at top and bottom installed in series. The space between the two tubes is evacuated to form a vacuum. A copper pipe (called a heat pipe) running through the centre of the tube meets a common manifold that is then connected to a slow flow circulation pump that pumps water to a storage tank below, thus heating the water during the day.

The evacuation tube systems are more efficient than the flat plate collectors at high water temperatures. Due to the vacuum inside the glass tube, the total efficiency of the glass surface is higher and increased performance is achieved in early morning and late afternoon.

Solar systems installed in problem areas

When installed in a frost-prone area, a solar hot water system must be fitted with a frost protection system or device as recommended by the manufacturer and approved in AS/NZS 2712.

When installing solar systems in cyclone-prone areas, the collectors and mounting system must be approved by the manufacturer and the local authority.

Source: Shutterstock.com/Pavel Vakhrushev.

FIGURE 5.10 Evacuated tube solar collectors

COMPLETE WORKSHEET 2

Low-pressure (gravity feed) water heater systems

Low-pressure (gravity feed) hot water systems (see **Figure 5.11**) are not very common these days because of their low pressure operation. They are designed to take advantage of electrical off-peak rates. The cold water is supplied via a feed tank or cistern (make-up tank), the water in which is controlled by a float valve. The feed tank is usually mounted on the side or top of the storage tank (water heater) to supply a small amount of pressure, making the water flow through the water heater.

The tank-attached and side-feed models are suitable for use in a roof of normal pitches. Low, squat water heaters are made for installing in the roof space area on

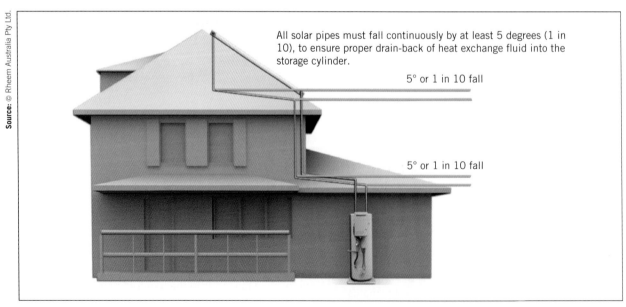

Source: © Rheem Australia Pty Ltd.

All solar pipes must fall continuously by at least 5 degrees (1 in 10), to ensure proper drain-back of heat exchange fluid into the storage cylinder.

5° or 1 in 10 fall

5° or 1 in 10 fall

FIGURE 5.9 Indirect split system flow and return line

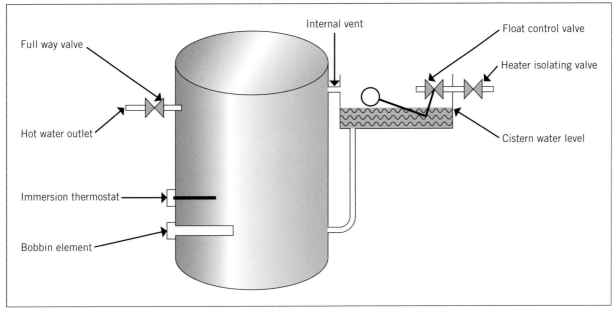

FIGURE 5.11 Low-pressure (gravity feed) hot water system

a support and safe tray. These water heaters retain the water that expands by directing it through a vent pipe and by having the make-up tank large enough to accept the extra water. The cold water feed to these storage tanks requires an isolation valve located on the inlet to the tank and a full way isolation valve on the hot water outlet connection; this is to enable the water heater, its ancillaries and tapware to be serviced. A further control valve must be provided in a position that is readily accessible from the floor or ground level to control the cold water feed to the system, and so enable the property owner to turn off the water supply without needing to access the roof area.

GREEN TIP

Recycling expanded water in low-pressure systems can save up to 3000L of water annually.

The advantages of low-pressure (gravity feed) hot water heaters are that they:
■ are long-lasting
■ have a minimum of operating parts and are trouble-free
■ are easy to service
■ occupy no floor space.
The disadvantages are:
■ the low pressure means more than one outlet cannot be used at a time
■ the installation may require larger diameter piping (sizing)
■ replacement costs are high

■ they always need a safe tray and safe waste if installed in a roof area
■ initial installation costs are high
■ the warm water blend on single-lever mixing taps is difficult because of the high pressure difference between the hot and cold water supply.

The water in a falling level displacement water heater (see Figure 5.12) is heated by an off-peak electrical supply. The heated water may be used as required, but replacement water will not enter the water heater until there is a power supply to operate the solenoid valve.

This type of water heater not only may run out of heated water, but also may run out of water altogether, until it refills when power is available.

The hot water pipework for low-pressure (gravity feed) hot water systems must be sized correctly to give an adequate volume of water to the various hot water outlets. (This is because the supply pressure is created from the storage tank due to the difference in height between the tank inlet and the various outlets.) Generally, the most disadvantaged outlet is the shower outlet. AS/NZS 3500.1 requires a minimum supply pressure to an outlet of 50kPa.

Formula: P = 9.81 – approximately 10kPa per 1m head.

Suggested minimum pipe sizes

For a low-pressure water heater less than 85kPa:
■ a DN 20 (15.0mm internal bore) service from the water heater to the first branch
■ a DN 15 (10.0mm internal bore) branch to a kitchen sink or basin

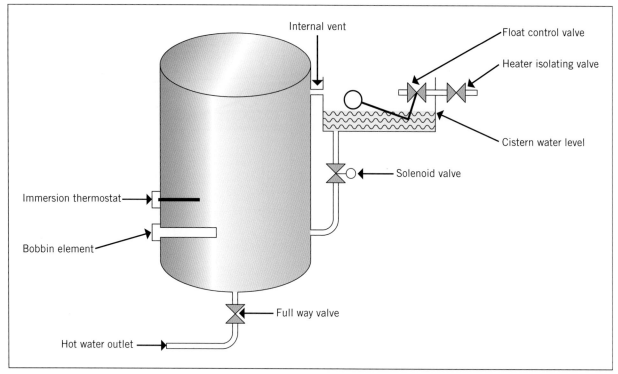

FIGURE 5.12 Falling level displacement water heater

- a DN 15 (10.0mm internal bore) branch to a sink and laundry
- a DN 20 (15.0mm internal bore) branch to a bathroom and one other room
- a DN 18 (12.5mm internal bore) branch picking up all fixtures within a bathroom.
  For a low-pressure water heater 85–170kPa:
- a DN 18 (12.5mm internal bore) service from the water heater to the first branch
- a DN 15 (10.0mm internal bore) branch to a kitchen sink or basin
- a DN 15 (10.0mm internal bore) branch to a sink and laundry
- a DN 18 (12.5mm internal bore) branch to a bathroom and one other room
- a DN 15 (10.0mm internal bore) branch picking up all fixtures within a bathroom.

The above sizes are a suggested minimum only; the service must be sized to meet the minimum flow rate outlet requirements. This may require a larger pipe to be used.

Both gravity feed and falling level displacement systems must have a full way valve installed on the outlet pipe located at the water heater to allow for servicing of the system without requiring the system to be drained. The hot water pipes should be installed with continuous fall from the tank outlet to the fixture outlet to avoid airlocks occurring, as this can block the hot water flow.

FROM EXPERIENCE

The large difference in pressure between the hot and cold water causes major problems in achieving a blend with single-lever mixer taps and tempering valves when installed with gravity feed and falling level displacement systems.

Other types of water heaters

As part of the Australian state and territory governments' sustainability energy push, manufacturers have now produced more sustainable energy water heaters. Some examples of these water heaters are given below.

Push-through 'free outlet under-sink' water heaters

These units are designed to service a single outlet only, usually a sink (see Figure 5.13). Cold water is connected to the hot tap, which acts as the cold water inlet valve for the water heater, as shown in AS/NZS 3500.4. When the hot tap is turned on, cold water enters the water heater, pushing hot water out the free outlet (sink spout). As these water heaters are not under pressure and have a free outlet allowing for hot water expansion, they do not need a temperature and pressure relief valve, nor to be installed in a safe tray.

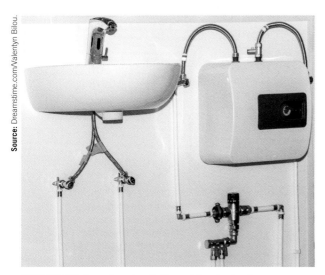

FIGURE 5.13 A modern 'push through' water heater

Heat pump air-sourced water heater (close coupled system)

A heat pump air-sourced unit (see Figures 5.14 and 5.15) is another form of solar system.

FIGURE 5.14 Mains pressure heat pump water heater

In a simple form it is a conventional water storage tank with a heat pump attached. The heat pump converts the outside air temperature to heating energy in a similar manner to the way that an air conditioner heats or cools a house or car.

These types of systems are an ideal solution for the replacement water heater market, allowing property owners to take advantage of converting external air temperature into heat energy, without the expense and inconvenience of having solar collectors and pipework installed, to reduce their energy consumption. A typical heat pump system uses up to two-thirds less energy than a traditional electric water heater. The water connection to this type of system is the same as for mains pressure water heaters.

For best results these systems should ideally be installed in an outdoor location, and will work better if connected to a continuous tariff.

GREEN TIP

A heat pump hot water system is claimed to reduce the carbon footprint of a domestic property by up to 2.7 tonnes of carbon dioxide (CO_2) per year.

A heat pump air-sourced system must have a 300mm clearance around the air louvres of the module (see Figure 5.16).

Water heater construction

Water heaters available today may be constructed from the following materials:

1 *Copper.* While copper has been traditionally used for the construction of low-pressure water heaters, it is not suitable for pressure systems above 85kPa. Its use is therefore restricted to gravity feed systems. Copper provides a natural corrosion resistance and it is not unusual for these types of water heater to last for 20 or more years.

2 *Stainless steel.* Stainless steel is being used more often these days by manufacturers; it provides for a lighter water heater and gives a similar life to a quality manufactured, glass-lined steel water heater. However, it is more expensive.

3 *Mild steel vitreous enamel.* Mild steel vitreous enamel water heaters are the most common type of water heater; they are manufactured as a lower-cost alternative to stainless steel. The vitreous enamel coating used on water heaters is similar to that applied to cooking equipment and used on barbecues.

Vitreous enamel water heaters require a sacrificial anode to be installed to help protect and prolong the life of the water heater. Anodes are manufactured of either magnesium or aluminium with a mild steel wire core; some water heaters may have two anodes installed. When selecting the anode, particular attention should be given to its compatibility with the local water supply. Water heater manufacturers can advise on the correct anode for their water heater in an area. Anode replacement forms part of the servicing requirements for vitreous enamel water heaters to ensure the longevity of the water heater's life.

Manufacturers may offer two grades of vitreous enamel coating. For example, Rheem Australia offers two types of vitreous enamel water heaters: Rheemglas water heaters, which have a seven-year cylinder

Source: Dreamstime.com/Valentyn Bilou.

Source: © Rheem Australia Pty Ltd.

Source: © Rheem Australia Pty Ltd.

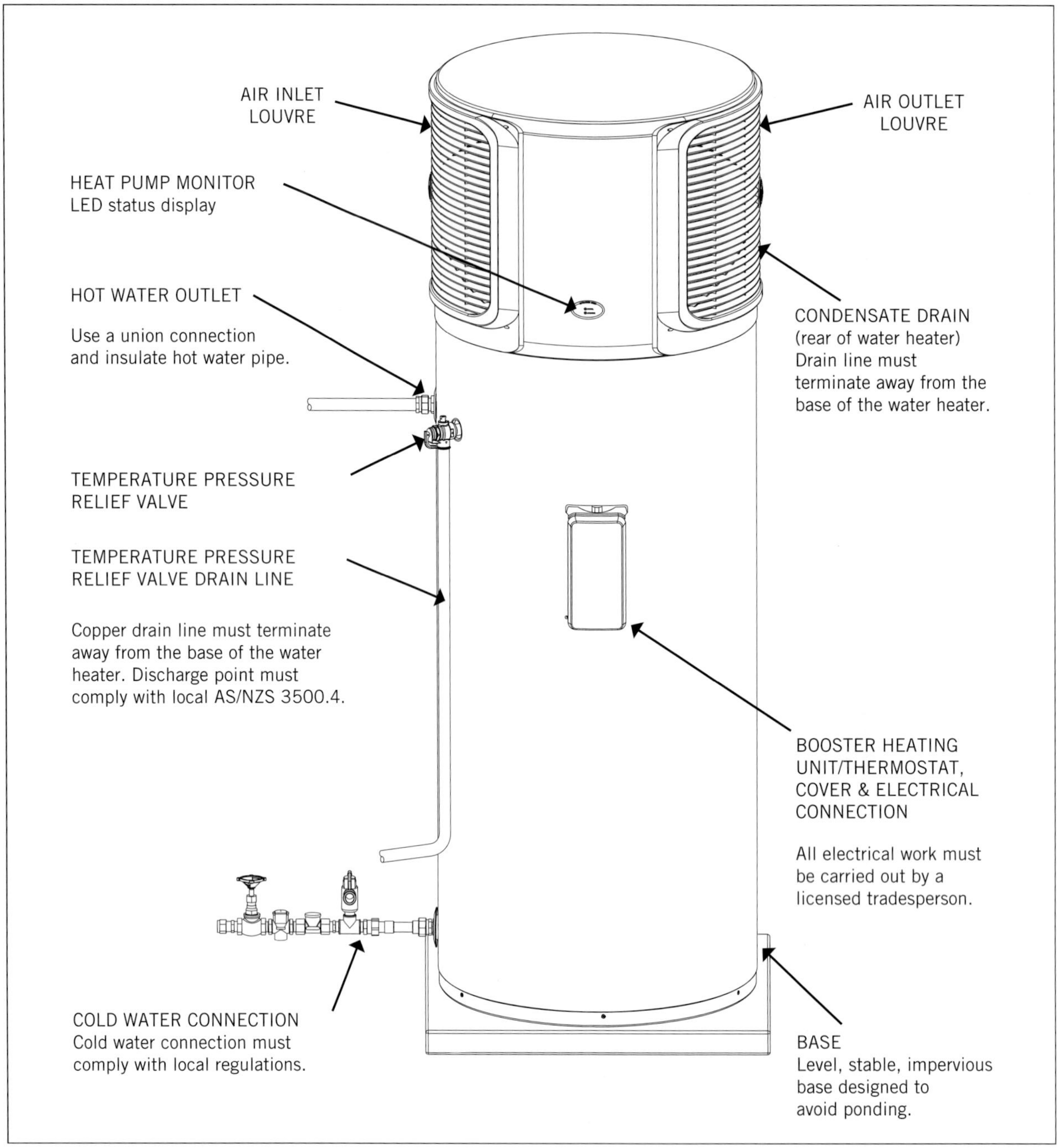

AIR INLET
LOUVRE

AIR OUTLET
LOUVRE

HEAT PUMP MONITOR
LED status display

HOT WATER OUTLET

Use a union connection
and insulate hot water pipe.

CONDENSATE DRAIN
(rear of water heater)
Drain line must
terminate away from the
base of the water heater.

TEMPERATURE PRESSURE
RELIEF VALVE

TEMPERATURE PRESSURE
RELIEF VALVE DRAIN LINE

Copper drain line must terminate
away from the base of the water
heater. Discharge point must
comply with local AS/NZS 3500.4.

BOOSTER HEATING
UNIT/THERMOSTAT,
COVER & ELECTRICAL
CONNECTION

All electrical work must
be carried out by a
licensed tradesperson.

COLD WATER CONNECTION
Cold water connection must
comply with local regulations.

BASE
Level, stable, impervious
base designed to
avoid ponding.

FIGURE 5.15 Typical installation for heat pump air-sourced electric boosted water heater (outdoor locations)

warranty, and Optima and Stellar water heaters, which
have a 10-year cylinder warranty.

Hot water energy sources

A number of different energy sources are available with
each type of water heater:

■ Gas:
  – natural gas
  – liquefied petroleum gas.
  The characteristics of gas and design requirements
  will be included in *Gas Services*, which covers the
  gas stream of your training.

■ Electricity:
  – off peak (rates one and two)
  – quick recovery (domestic tariff 240V)
  – continuous (instantaneous) flow (domestic tariff
    usually 415V).
  The use of electricity as a water heating energy
source is being phased out in favour of more
efficient supplies.

■ Solar energy – there are three types of solar heating
systems:
  – closed couple (thermosiphon circulation) systems
  – split or forced circulation systems
  – heat pump air-sourced water heater (close
    coupled systems).

300mm   300mm

Source: © Rheem Australia Pty Ltd.

FIGURE 5.16 Minimum installation clearances

Solid fuels such as wood and coal may also be used. This type of installation is mainly used in country regions of Australia; however, it is not covered in this chapter.

LEARNING TASK 5.2

1 How can an airlock be avoided on a gravity-fed hot water system?
2 What type of solar hot water system does not require a circulating pump?
3 What does a heat pump water heater rely on to heat water?
4 What is an 'off-peak' rate?

📋   **COMPLETE WORKSHEET 3**

5.4 Installing the water heating system

When installing hot water heaters, it is important to check the plans and specifications to ensure that the installation complies with what has been requested. If there are any issues, they need to be negotiated at this time.

Sequencing the task

The installation of most water heaters is carried out either at the fit-off stage or just prior to the property being occupied. As with any plumbing task, it is important to sequence the task into logical steps for successful completion. This sequencing is an ideal opportunity to carry out a risk assessment and prepare a safe work method statement (SWMS) or job safety analysis (JSA) for the task. Recording and reviewing these procedures can help you to refine and enhance your company's quality assurance procedures. The sequencing will also allow you to identify safety equipment, including personal protective equipment (PPE), and tools relevant to the task:

■ *PPE.* This may include, but is not limited to, any of the in-depth list of PPE found in Chapter 6 of *Basic Plumbing Skills*.
■ *Tools and equipment.* The type of plumbing tools that may be used for the installation and commissioning of a water heater can be found in Chapter 7 of *Basic Plumbing Skills*.

Electrical safety

When carrying out any work on a metallic water service connected to the water utility main, consideration needs to be given to electrical safety precautions and earthing to protect against potential electrical shock. Refer to the references at the end of this chapter for further information.

Always test for stray current and use bonding straps when disconnecting a metallic service to avoid electrocution.

Planning and preparation

It is important to consult with the builder/owner as to what type of water heater is going to be used; this will need to be determined prior to the rough-in stage and take into account AS/NZS 3500 and any state or territory requirements.

It is good practice to have a maximum draw-off of 2L of cold water from a hot water tap to conserve water (this equals about 25m of DN 15, or 11m of DN 20 for copper pipe), which requires particular attention to be given to:

- the location of the water heater
- the installation of a flow and return hot water system
- the type and location of temperature control devices
- the installation of more than one water heater.

AS/NZS 3500.4 PLUMBING AND DRAINAGE: HEATED WATER SERVICES

As in the other chapters in this book, it is important to carry out pre-planning work such as:

- obtaining the required permits from the water utility and any other body having authority over the work
- checking the approved plans for any specific installation requirements
- checking AS/NZS 3500 and any state or territory codes for installation requirements
- selecting the material type to be used
- carrying out a site inspection
- ensuring quality assurance and workplace health and safety
- arranging any special installation requirement for the water heater installation.

FROM EXPERIENCE

Always check the approved plans and BASIX specification for a nominated water heater on new developments.

Sizing the water heater

When selecting the correct type of water heater, the following points need to be considered:

- government regulations
- the heating energy source (such as gas or solar) available

- the consumer's requirements and needs
- running costs
- purchase and installation costs
- ease of installation
- water heater recovery and performance (the WELS rating)
- temperature requirements
- pressure and flow rate requirements
- space available for installation.

Points needing consideration when selecting the correct size of water heater are:

- the type of water heater required (gas, solar or electric)
- the number of bathrooms
- the number of bedrooms
- the number of people using the system and any possible family expansion
- the age of any teenagers, or soon-to-be teenagers
- whether the dishwasher is connected to the water heater
- whether the clothes washing machine uses hot water
- whether there is a spa bath
- the type of energy source (electrical and gas) rates
- the location of the water heater
- the type of warranty (five or 10 years).

Table 5.2 is a general summary of water heater sizing; it is recommended that manufacturers' guides are consulted for more specific details of sizes. Rheem Australia has a web-based sizing system that may help you when sizing water heaters for domestic use (see the references at the end of this chapter). Other suppliers also readily supply information about their heating units.

Location and placement of the water heater

Water heaters should be located as close as possible to the most frequently used outlet; this is generally the kitchen sink. Remember that it is good practice for a hot water service to have a maximum of 2L of cold water

TABLE 5.2 General summary of water heater sizing

Sizing guide							
Electric continuous tariff		Electric off-peak tariff		Gas		Solar	
Heater delivery capacity (litres)	Number of users	Number of users	Number of bedrooms	Storage (litres)	Continuous flow	Solar storage (litres)	Heat pump storage (litres)
25							
50	1						
80	1–2						
125	2–3						
160	2–4						
250	3–5	1–3	1–2	90	18	270	
315	4–6	2–4	3	130	24	325	310
400	5–9	4–6	4–5	160	26	410	325

Source: © Rheem Australia Pty Ltd.

draw-off from a hot tap (this equals about 25m of DN 15 copper pipe). The water heater must be installed so that:

- its rating plate and instructions are readily visible
- there is unobstructed access to its burner, element, thermostat, controls and any equipment requiring maintenance
- its valves and easing gear are accessible
- it has a minimum of 150mm clearance for the removal and replacement of the temperature and pressure control valve
- it can be removed and replaced without major structural alteration to the building or major alteration to the pipework.

An additional point to note is that if a water heater is installed in a known earthquake area, it must be restrained against movement.

AS/NZS 3500.4 PLUMBING AND DRAINAGE: HEATED WATER SERVICES

Gas water heaters

When installing any gas water heater, its location and method of installation *must* comply with all relevant standards and building codes, specifically AS/NZS 5601.1, and is dependent on the unit's gas capacity in MJ/h. (The diagrams in Figure 5.17 are a guide only.)

Concealed water heaters

All water heaters installed in a location such as a roof space or a cupboard must be placed on a safe tray (see Figure 5.18). For an example of this, refer to AS/NZS 3500.4. When installing a water heater on a safe tray,

provision must be made for draining and prevention of overflow. If the safe tray is installed under a sink, a minimum size of DN 25 safe waste drain must be installed, with a minimum of DN 50 (DN 40 in New Zealand only) for all other installations. If a safe waste drain is fabricated from sheet metal it must be lapped in the direction of the flow and all lapped joints must be watertight. Any seam in the sheet metal must be installed on the top.

A safe waste drain must:

- be installed with a constant fall to its point of discharge
- be supported adjacent to the point of connection to the safe tray and at a maximum distance of 1m on a grade and 2.4m when installed vertically
- be readily visible if discharging within a building and not cause any damage to the property or injury to people
- discharge to an external point of a building and within the property boundary that is readily visible and clear of any openings into a building such as windows or doors.

When installing a mains pressure water heater in a safe tray, provision for protection against damage from leaking water can be made by installing an approved shut-off device (such as a terminator valve) instead of a safe waste drain. The terminator valve (see Figure 5.19) is installed between the cold water inlet isolation valve and the water heater inlet, upstream of any expansion control valve.

When a leak is detected (by water entering the safe tray), the valve mechanically shuts off the water entering the water heater.

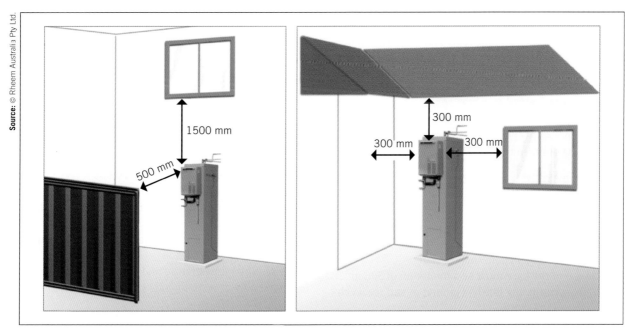

Source: © Rheem Australia Pty Ltd.

1500 mm

500 mm

300 mm

300 mm

300 mm

FIGURE 5.17 Installation location for continuous flow gas water heaters

FIGURE 5.18 Electric water heater installed in a safe tray and hanging brackets

FIGURE 5.19 Terminator valve installed on the cold water inlet of a pressure water heater

An unconcealed water heater located inside any building on or above an impervious floor, and draining to a floor waste gully or an external door, does not require a safe tray.

Water heaters located in a roof space

Any water heater installed in a roof space must be installed on a safe tray and supported by a hardwood platform, as outlined in AS/NZS 3500.4.

Externally installed water heaters

When a water heater is installed in an external location it must be supported by:

■ 75mm thick bonded bricks or concrete (cast in situ) support

■ 50mm precast concrete support slab.

The top of the base must be a minimum of 50mm above the surrounding finished ground or surface level.

AS/NZS 3500.4 PLUMBING AND DRAINAGE: HEATED WATER SERVICES

Materials selection

Unless otherwise stated in the plans or specifications, the type of materials that may be used is covered by AS/NZS 3500 and includes:

■ copper

■ polybutylene (PB)

■ cross-linked polyethylene (PE-X)

■ polypropylene (PP)

■ polyethylene (PE)

■ unplasticised polyvinyl chloride (PVC-U).

When selecting material you need to consider:

■ what the pipework is going to be used for

■ the water quality and its temperature

■ the compatibility of the material and products

■ frost protection

■ the water pressure within the water utility's supply system

■ any special material installation requirements; for example:

– some polymer (plastic) pipes and fittings may not be installed in direct sunlight

– polymer pipes may not be used between the hot water inlet isolation valve and the water heater

– polymer pipes may not be used as part of a temperature and pressure relief valve

– polymer pipes and fittings may not be used within 1m of the outlet of a water heater

– polymer pipes and fittings may be installed on the outlet side of a temperature control device.

Any material or product used as part of the heated water installation must comply with the Plumbing Code of Australia and AS 5200.000–2006. A database of authorised products may be found in the references section at the end of this chapter.

When selecting a material other than copper, it is important to refer to the Australian equivalent pipe size guide in AS/NZS3500.1.

Requirements for water connections

Every water heater must have:

■ a union or other similar coupling connection on its inlet and outlet

■ the required valves as specified by the manufacturer and described below

■ the required temperature and pressure relief valves supplied by or recommended by the manufacturer.

Installation of valves

All valves used to control the water supply to and from a water heater must be installed in accordance with both the water heater manufacturer's requirements and the following requirements:

- The water heater isolation valve must be installed in a location that is readily accessible from the ground or floor level.
- The valves on the cold water inlet, when required by the water heater manufacturer, must be installed in the following sequence:
  1 isolating valve
  2 line strainer
  3 pressure control valve
  4 non-return valve
  5 cold inlet expansion control valve where required by the local water utility or water heater manufacturer.

  This is illustrated in AS/NZS 3500.4.

AS/NZS 3500.4 PLUMBING AND DRAINAGE: HEATED WATER SERVICES

Sizing connections to the water heater

The minimum sizes of hot water branches for a storage water heater in excess of 170kPa are:

- a DN 18 (12.5mm internal bore) service from the water heater to the first branch
- a DN 15 (10.0mm internal bore) branch to a kitchen sink or basin
- a DN 15 (10.0mm internal bore) branch to a sink and laundry
- a DN 15 (10.0mm internal bore) branch to a bathroom and one other room
- a DN 15 (10.0mm internal bore) branch picking up all fixtures within a bathroom.

  For other types of heated water installation refer to AS/NZS 3500.4.

Flow rates

Each tap, valve and outlet is required to meet a minimum flow (discharge) rate (see Table 5.3). A more in-depth list may be found in AS/NZS 3500.4. Note that no pipe sizes are indicated in Table 5.3. Pipe diameter may vary to meet the minimum flow rates and consumer needs.

TABLE 5.3 Minimum outlet flow rate

Outlet	Minimum flow rate
Basin (standard outlet) Shower Sink (aerated outlet)	6L/min
Bath	18L/min
Laundry tub	7L/min

Insulation of hot water pipes

Thermal insulation of hot water pipes must meet the minimum thermal insulation R-value for the area. This information may be found in AS/NZS 3500.4. Insulation is placed on hot water pipes for a number of reasons, including to:

- prevent heat loss
- protect pipes 'chased' into solid walls (also to help prevent adverse effects from expansion when sealed in these walls)
- help prevent burns from exposed pipework.

Areas subject to freezing

If a water heater is to be installed in an area subject to regular low temperatures (below 0°C), the service must be protected by preventing the water from freezing, as outlined in Chapter 3. Installing a water heater in an external location in these conditions is not recommended, unless it has built-in protection against freezing.

LEARNING TASK 5.3

1 What would be the correct size gas continuous water heater for a three-bedroom house?
2 How much clearance is required around a temperature and pressure relief valve so it can be easily replaced?
3 What is the minimum size of a safe waste drain?
4 Why is insulation important on a hot water pipe?

COMPLETE WORKSHEET 4

Large residential and commercial/industrial hot water supply services

If a large volume of hot water is required, water heaters can be manifolded together. It is not uncommon to see two or more water heaters manifolded together, which will dramatically raise the delivery capability of the system. Manifold systems are generally installed as flow and return systems. This piping configuration is known as an equaflow system.

Manifolding storage water heaters

The 'equal-flow' installation method means that the demand on each water heater in the bank is the same.

The manifold must be sized to meet the installation requirements.

All types of systems, such as mains pressure storage, solar and continuous flow, may be manifolded together to supply larger volumes of heated water, meeting the needs of the property owner/occupier.

HOW TO INSTALL MULTIPLE HOT WATER HEATERS

- The heated water manifold must be designed to leave the bank from the opposite direction to the cold water entering the bank, balancing the flow from each unit (see Figure 5.20).
- The cold water manifold must be designed to enter the bank from the opposite direction to the heated water leaving the bank, balancing the flow to each unit (see Figure 5.20).
- The water heaters must have the same storage capacity and energy input.

- The inlet and outlet connections must be the same size and, if possible, the same length.
- Full way (ball or gate) valves must be fitted to the inlet and outlet of each water heater.
- The hot water return line must be connected to the cold water manifold (see Figure 5.21), or to a return connection boss if one has been supplied. It must enter the banks from the opposite direction to the hot water leaving the bank.

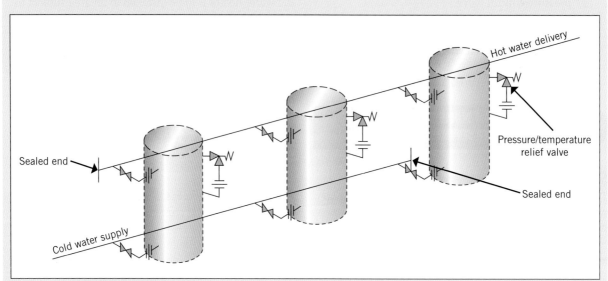

FIGURE 5.20 Manifold (equaflow) system

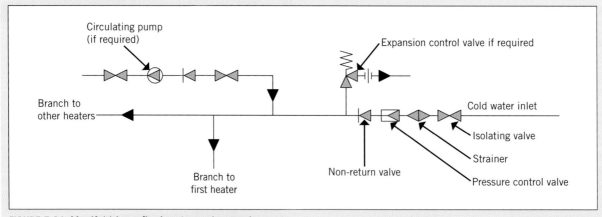

FIGURE 5.21 Manifold (equaflow) system valve requirements

Manifolding continuous flow gas water heaters

Continuous flow water heaters are manifolded in a similar manner to storage water heaters, with the exception that each unit has a staging valve (also known as a pressure responsive control valve) installed on its cold water inlet

(see Figures 5.22 and 5.23). These valves are designed to respond to the flow of hot water through the system by igniting each unit depending on the flow requirements. A check must be made with the water heater manufacturer for any specific design requirements.

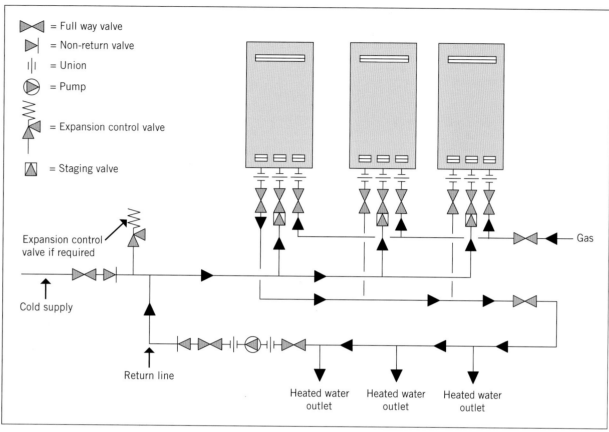

FIGURE 5.22 Gas continuous flow mains pressure manifold (equaflow) system – diagram

FIGURE 5.23 Gas continuous flow mains pressure manifold (equaflow) system

Manifolding of water heaters has several advantages, including:

- allowing individual water heaters on the manifold to be isolated and maintained or replaced while the rest of the system continues to function and provide hot water
- allowing part of the system to be shut down during off-peak periods (such as in caravan parks).

Installing a domestic building ring main for storage water heaters

When installing a building ring main, an additional small mains pressure water heater may need to be installed to maintain the temperature within the piping system, as shown in Figure 5.24. *Do not* return the hot line back through the main storage system for solar, heat pump, electric off-peak or twin (non-simultaneous) water heaters: a separate booster water heater needs to be installed to maintain any heat loss. A 25L or 50L 2.4kW element on a flex and plug storage water heater is more than sufficient.

Most people prefer to use a timer to control the circulating pump, as it suits their lifestyle and needs.

LEARNING TASK 5.4

1. What is the purpose of a flow and return system (ring main)?
2. Explain a manifolded (equaflow) system.

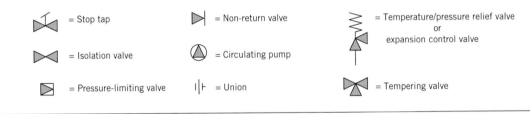

Source: © Rheem Australia Pty Ltd.

Kitchen hot water outlet

Laundry hot water outlet

Circulator
thermostatically
or
time controlled
or
24-hour operation

Electric
water
heater
(continuous)
60°C
(25L or 50L)

Cold water
supply

Bathroom tempered
water outlet

Cold water
supply

Ensuite tempered
water outlet

Y

Cold water
supply

Note: A pressure-limiting valve (PLV) is required to be
installed on the cold water supply line to the tempering
valve if a PLV is installed on the cold water supply line to
the water heater.

= Stop tap   = Non-return valve   = Temperature/pressure relief valve
or
expansion control valve

= Isolation valve   = Circulating pump

= Pressure-limiting valve   = Union   = Tempering valve

FIGURE 5.24 Domestic building ring mains and circulating loops system for solar, heat pump, electric off-peak or twin (non-simultaneous) element water heaters

5.5 Hot water maintenance and testing

Understanding the function and operation of basic components of water heaters will help diagnose and problem solve issues related to maintaining and repairing hot water heaters.

Element selection for electricity

The selection of the element size is critical to the performance of an electric water heater. Elements may be either immersion or ceramic bobbin.

Immersion elements

An immersion element (see **Figure 5.25**) is installed in direct contact with the water within the water heater. Should the element need replacing, the water heater needs to be drained.

If the water heater is connected to an off-peak power supply, then a 4.8kW element must be installed. Elements are also available in other sizes, such as 2.4kW and 3.6kW. Some electrical authorities may allow a 3.6kW element to be used; check with the electrical authority for their requirements.

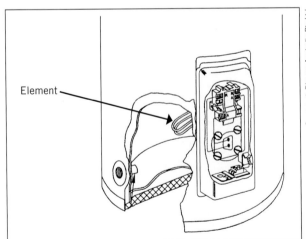

Source: © Rheem Australia Pty Ltd.

Element

FIGURE 5.25 Immersion element

Ceramic bobbin elements

A ceramic bobbin element (see **Figure 5.26**) has an outer watertight sheath; the (bobbin) element is installed inside the sheath. These elements are not as efficient as an immersion element and may be replaced without emptying the water heater of water. They are mainly used in low-pressure (gravity feed) water heaters.

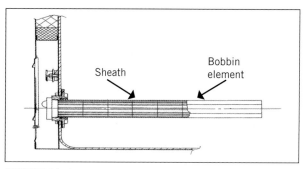

FIGURE 5.26 Ceramic bobbin element

Source: ANTA Learner Guide BCPET3003A Install and commission water heating services. This file is licensed under the Creative Commons Attribution-Share Alike 3.0 Unported licence.

Electric elements are installed at the bottom of the water heater. When the temperature of the water falls below that set on the thermostat, and if power is available (that is, not subject to off-peak limitation), the water is heated. As the water is heated it becomes less dense and rises, with the colder water falling to the bottom to be heated. This is called convection.

Twin booster (non-simultaneous) elements

Manufacturers can supply water heaters with twin elements (see Figure 5.27). The main element is located at the bottom of the tank, with the second one located in the upper section of the water heater. The top element is connected to a continuous power supply. Its function is to maintain the temperature in the top of the storage water heater at or above 60°C. The top element is designed to heat the top 50L to 90L only. The bottom element is normally connected to an off-peak tariff. (The elements are designed to heat individually and at no stage will both elements heat simultaneously.)

Source: © Rheem Australia Pty Ltd.

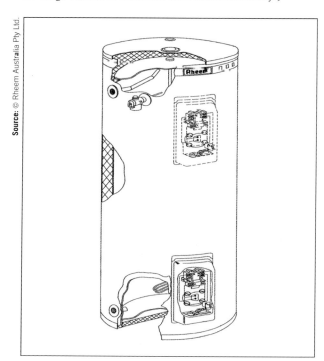

FIGURE 5.27 Twin booster elements

Thermostats

As discussed at the beginning of this chapter, heated water must be stored above 60°C to prevent the growth of *Legionella* bacteria. So all storage water heaters have a thermostat installed to control the water temperature. The thermostat may be of the contact type (see Figure 5.28) or the immersion type (see Figure 5.29). A contact thermostat is attached to the outside wall of the water heater and relies on heat transfer through the wall of the tank. Immersion thermostats are used on gas water heaters; the thermostat extends into the heater, giving a quicker response time.

Source: © Rheem Australia Pty Ltd.

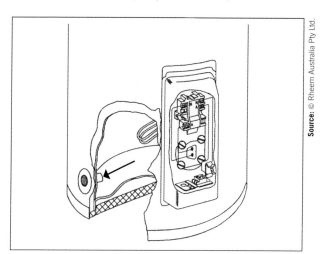

FIGURE 5.28 Contact thermostat

Source: © Rheem Australia Pty Ltd.

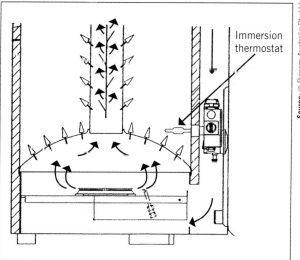

Immersion thermostat

FIGURE 5.29 Immersion thermostat

Sacrificial anodes

Sacrificial anodes are easily corroded materials in the form of a long rod installed in hot water tanks that are sacrificed to corrosion instead of the steel cylinder (see Figure 5.30). Therefore, the steel tank will last much longer.

The anode rod is made from aluminium/zinc or magnesium and slowly dissolves away from galvanic corrosion happening between the water and steel tank. The rod depletes from this corrosion and the steel tank is left alone until there is no more sacrificial metal left.

FIGURE 5.30 Sacrificial anode

Sacrificial anodes generally last around five years in a mains pressure hot water storage unit. So it is good practice to replace them in that time and extend the life of the hot water heater.

Testing the system

LEARNING TASK 5.5

1 What is the function of a thermostat?
2 What is the purpose of a sacrificial anode?

The system should be flushed prior to connecting any water heater to remove any foreign matter. The water service should have been tested at the completion of the rough-in stage and left charged with water during construction. Once the water heater is connected, any interconnection between the hot and cold water services must be removed to prevent any damage occurring to the system valves by the backpressure from the cold water on the hot water system. The water heater installation must be visually checked for leaks. *Do not pressure test the water storage cylinder.*

Commissioning the system

Once the water heater has been installed the system needs to be commissioned. The following is a guide only and the manufacturer's guidelines must be followed.

HOW TO

HOW TO COMMISSION A HOT WATER SYSTEM

1 Fill the system with water and purge it of air prior to the heating medium being turned on.
2 Check the following points for correct operation:
  a temperature and pressure relief valve and expansion control valve (if fitted)
  b stored water temperature
  c water delivery temperature
  d flow and return temperature
  e water level of a gravity system
  f that the correct valves are installed, are opened and operate correctly
  g outlets' minimum flow rates
  h pump operation
  i pump vibration
  j pressures, where a pressure-limiting valve is fitted
  k noise and water hammer.

When installing multiple water heaters in a bank, each water heater must be checked for correct operation independently of the others.

The manufacturer's instructions must be left for the owner/occupier.

 COMPLETE WORKSHEET 5

5.6 Cleaning up

The clean-up process is very important so all rubbish should be disposed of thoughtfully and placed in the correct bins provided. Any leftover material should be stored so it can be used on another job.

Be sure to clean and store all tools and equipment safely so they are ready to use on the next job. Remember to return any equipment on hire in the same condition it was hired out.

The old saying 'the job is not finished until the paperwork is done' is always relevant. Submit all relevant documentation and hand over any product warrantees to the client on completion.

The client must be shown the operational procedure and any maintenance instructions.

LEARNING TASK 5.6

1 Why is it important to remove all the air from a hot water service when commissioning it?
2 When commissioning a hot water system, why is the temperature and pressure relief valve operated?
3 Where is the pressure-limiting valve fitted?
4 Why is the hot water cylinder not pressure tested?

SUMMARY

- The two most common water heaters are mains pressure storage heaters and continuous water heaters.
- Hot water must be stored at a minimum temperature of 60°C to prevent *Legionella* bacteria growth.
- The maximum temperature to an ablution fixture is 50°C.
- It is important to understand the requirements for temperature pressure relief valves.
- There are two types of solar water heating systems: a split system and a close coupled system.
- A heat pump is very energy efficient by transferring the surrounding air into heat.
- A cold water expansion valve discharges expanded water from the bottom of the tank instead of the top, therefore saving hot water.

- Plastic pipe cannot be used within 1m of the hot water outlet and between the isolation valve and cold water inlet.
- Hot water heaters must be located close to fixtures to prevent water wastage.
- A heat trap must be installed on the hot water outlet if the hot water heater doesn't have an integral one to prevent heat loss.
- Pipe insulation prevents heat loss and saves energy.
- A flow and return system prevents water wastage as well as providing hot water instantly.
- Be sure to flush pipework and remove all air when commissioning hot water heaters.
- The client must be informed of the hot water heater operation instructions and maintenance requirements upon completion.

REFERENCES

Energy Safe Victoria: **http://www.esv.vic.gov.au**
Material production standards: **http://www.watermark.standards.org.au**

Rheem Australia: **http://www.rheem.com.au**
Standards Australia ('Earthing of earthing installation using the water reticulation system'): **http://www.standards.org.au**

GET IT RIGHT

1 Which photo shows the correct installation?

2 What is the problem?

3 Why is that a problem?

WORKSHEET 1

Student name: _____

To be completed by teachers

Satisfactory ☐

Not satisfactory ☐

Enrolment year: _____

Class code: _____

Competency name/Number: _____

Task: Review the sections '5.1 Background', '5.2 Water temperature' and '5.3 Types of hot water heating systems' and answer the following questions.

1 Water heaters are divided into four classes. List them.

_____

_____

_____

_____

2 Water heaters are further classified by their delivery method. List and describe the three types of delivery method.

_____

_____

_____

3 Water can be heated by a number of energy sources. List four of these energy sources.

_____

_____

_____

_____

4 Explain the differences between a continuous flow water heater and a storage water heater.

_____

_____

_____

5 Explain the basic operation of a mains pressure water heater.

_____

_____

_____

6   Explain the basic operation of a continuous flow water heater.

_____

_____

_____

7   State three conditions relating to how temperature and pressure relief valve drains must be installed.

_____

_____

_____

8   If a water heater is installed externally, how far must the temperature and pressure relief valve drain line discharge away from the water heater?

_____

9   What is the minimum temperature water must be stored in a hot water tank, and why?

_____

_____

10   What is the maximum temperature of hot water supplying a bath, shower or basin?

_____

WORKSHEET 2

Student name: _____

Enrolment year: _____

Class code: _____

Competency name/Number: _____

To be completed by teachers

Satisfactory ☐

Not satisfactory ☐

Task: Review the section 'Solar water heaters' and answer the following questions.

1   There are two main types of solar water heating systems. What are they?

_____

_____

2   List and describe two other categories of solar water heaters.

_____

_____

3   Name the three main components in a solar hot water system.

_____

_____

_____

4   Is an evacuated tube system more or less efficient than a flat solar panel system?

_____

5   Which direction must the solar collectors face when installed in Australia, and why?

_____

_____

6   Why can't plastic pipe be used between the solar collectors and the storage cylinder?

_____

WORKSHEET 3

Student name: _____

Enrolment year: _____

Class code: _____

Competency name/Number: _____

To be completed by teachers

Satisfactory ☐

Not satisfactory ☐

Task: Review the sections 'Low-pressure (gravity feed) water heater systems' and 'Other types of water heaters' and answer the following questions.

1   What are three advantages of a low-pressure (gravity feed) type of water heater?

_____

_____

_____

2   What are two disadvantages of a low-pressure (gravity feed) type of water heater?

_____

_____

3   What is a 'free outlet' on a push-through water heater?

_____

4   Explain the basic operation of a heat pump air-sourced water heater.

_____

_____

_____

5   How much clearance is required around the heat pump?

_____

6   Why is it preferable to use heat pump storage water heaters rather than the typical electric heated storage water heater?

_____

_____

WORKSHEET 4

Student name: _____

Enrolment year: _____

Class code: _____

Competency name/Number: _____

To be completed by teachers

Satisfactory ☐

Not satisfactory ☐

Task: Review the section '5.4 Installing the water heating system' and answer the following questions.

1   List four points that should be considered when installing a water heater.

_____

_____

_____

_____

2   List two ways in which a water heater must be supported when it is installed in an external location.

_____

_____

3   State the minimum required measurements for the external location for continuous flow gas water heaters. Write the figures on the diagrams.

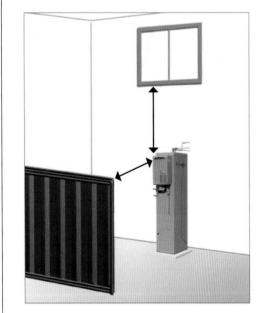
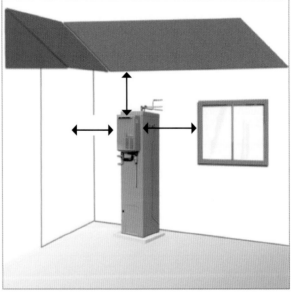

4   Explain the purpose of a terminator valve on the cold water inlet of a pressure water heater.

_____

_____

_____

_____

5   What is the recommended maximum amount of cold water draw-off from a hot water tap?

_____

6   In order to achieve this maximum draw-off, the plumber needs to pay attention to the location of not only the water heater but also some other items. What are they?

_____

_____

_____

7   List four items to note when selecting a material for a hot water service (including any special material installation requirements).

_____

_____

_____

_____

8   Careful consideration should be given when selecting the correct type of water heater. List six of the 10 points to be considered when selecting the correct type of hot water heater.

_____

_____

_____

_____

_____

_____

9   What must always be checked first before disconnecting any hot water heater?

_____

10   How close can plastic pipe connect to the hot water outlet connection?

_____

 **WORKSHEET 5**

To be completed by teachers

Satisfactory ☐

Not satisfactory ☐

Student name: _____

Enrolment year: _____

Class code: _____

Competency name/Number: _____

Task: Review the sections '5.5 Hot water maintenance and testing', 'Large residential and commercial/industrial hot water supply services' and 'Commissioning the system' and answer the following questions.

1 List five points that should be checked when commissioning a water heater.

_____

_____

_____

_____

_____

2 Should the storage tank be pressure tested when commissioning a hot water system? Explain your answer.

_____

_____

_____

3 What is the main principle of the manifold (equaflow) system?

_____

_____

_____

4 What is the function of a thermostat in a water heater?

_____

_____

WORKSHEET RECORDING TOOL

Learner name		Phone no.	
Assessor name		Phone no.	
Assessment site			
Assessment date/s		Time/s	
Unit code & title			
Assessment type			

Outcomes

Worksheet no. to be completed by the learner	Method of assessment WQ – Written questions PW – Practical/workplace tasks TP – Third-party reports SC – Scenarios RP – Role plays CS – Case studies RW – Report writing PF – Portfolio	Satisfactory response	
		Yes ✓	No ✗
Worksheet 1	WQ	☐	☐
Worksheet 2	WQ	☐	☐
Worksheet 3	WQ	☐	☐
Worksheet 4	WQ	☐	☐
Worksheet 5	WQ	☐	☐

Assessor feedback to the learner

Feedback method (Tick one ✓):	☐ Verbal ☐ Written (if so, attach) ☐ LMS (electronic)

Indicate reasonable adjustment/assessor intervention/inclusive practice (if there is not enough space, a separate document must be attached and signed by the assessor).

Outcome (Tick one ✓):	☐ Competent (C) ☐ Not Yet Competent (NYC)

Assessor declaration: I declare that I have conducted a fair, valid, reliable and flexible assessment with this learner, and I have provided appropriate feedback.

Assessor name:	
Assessor signature:	
Date:	

Learner feedback to the assessor

Feedback method (Tick one ✓):	☐ Verbal	☐ Written (if so, attach)	☐ LMS (electronic)

Learner may choose to provide information to the RTO separately.

	Tick one:	
Learner assessment acknowledgement:	**Yes** ✓	**No** ✗
The assessment instructions were clearly explained to me.	☐	☐
The assessment process was fair and equitable.	☐	☐
The outcomes of assessment have been discussed with me.	☐	☐
The overall judgement about my competency performance was fair and accurate.	☐	☐
I was given adequate feedback about my performance after the assessment.	☐	☐

Learner declaration:
I hereby certify that this assessment is my own work, based on my personal study and/or research. I have acknowledged all material and sources used in the presentation of this assessment, whether they are books, articles, reports, internet searches or any other document or personal communication.
I also certify that the assessment has not previously been submitted for assessment in any other subject or at any other time in the same subject and that I have not copied in part or whole or otherwise plagiarised the work of other students and/or other persons.

Learner name:	
Learner signature:	
Date:	

FIT OFF AND COMMISSION HEATED AND COLD WATER SERVICES: FIT OUT

6

Chapter overview

The aim of this chapter is to address the skill and knowledge required to fit off, connect, test and commission hot, tempered and cold drinking and non-drinking water services supplying fixtures and appliances.

Learning objectives

Areas addressed in this chapter include:
- planning and preparation
- identifying installation requirements and product selection
- making connections and testing the service
- commissioning the system
- cleaning up.

Note: Installation of fixtures and appliances is not covered in this book. It is addressed in *Sanitary and Drainage*.

6.1 Background

Although the rough-in phase of the work is critical (that is, the installation of pipes and fittings inside the walls, under the floors and in the ceiling spaces), it does not remain visible to the user. However, the final fit-off and commissioning work is the end product that remains in constant view of the user. This part of the work therefore affects the aesthetics of the overall system and faulty workmanship is readily evident.

If this phase of the work is not carried out in an efficient and neat manner, it can cause a great deal of angst between the client and the plumber, and in some cases a client may refuse to pay until this part of the service is rectified.

FROM EXPERIENCE

The majority of plumbers take extreme care in ensuring that the final fit-off is finished correctly and is aesthetically pleasing. This ensures a good reputation.

6.2 Planning and preparation

Interpreting plans and specifications to determine the final outcome of the plumbing installation must be communicated and understood at the very beginning of the job to avoid costly mistakes.

Product selection

It is important to consult with the builder/owner as to what type of tapware, appliances and fixtures are going to be used. As part of the building application, the type of hot water heater energy source, the taps, appliances and outlets are taken into consideration. The information on the type of tapware, appliances and fixtures may be found in the specifications, and must be installed in accordance with the installation requirements of the manufacturer. This should be determined prior to the rough-in stage and should take into account AS/NZS 3500, any state or territory requirements and WaterMark and Water Efficiency Labelling and Standards (WELS) requirements. Tapware, appliances and fixtures must display a WELS water-efficiency label and/or a WaterMark logo.

GREEN TIP

The WELS scheme is designed to help Australian households recognise and use water- and energy-saving appliances and tapware.

WELS (water efficiency) and WaterMark (quality control)

The Water Efficiency Labelling and Standards (WELS) Scheme is a joint Commonwealth, state and territory regulatory scheme that requires a range of water-using products to be labelled for water efficiency. Under the scheme, product suppliers are required to provide water-efficiency information and star ratings to consumers for the following products:

- clothes washers
- dishwashers
- registered flow controllers (optional)
- shower heads/roses
- taps
- toilets
- urinals (but not waterless urinals).

Just because a product has a WELS label does not mean it may be installed. Products should also display a WaterMark logo (see Figure 6.1). Do not install any plumbing product that does not display this logo without checking with the water utility first.

FIGURE 6.1 WaterMark logo

The differences between the WELS label and the WaterMark logo are that:
- the WELS label relates only to the water efficiency of the product
- the WaterMark logo confirms that the product complies with the requirements of the Plumbing Code of Australia and is fit for the purpose of installation under that code.

What does WELS mean for plumbers?

Plumbers are often in a position to recommend, supply or install products for clients. The WELS scheme can make a plumber's recommendation easier, as it provides labels that identify the water efficiency of a range of products and gives each product a star rating – the more stars, the more water efficient the product.

Plumbers are also bound by law to ensure that they are supplying WELS-rated products. They need to ensure that any WELS products they supply:

1   are WELS-registered and labelled
2   have the labels clearly visible to the customer when offered for supply.

Supplying products

You *are* considered responsible for supplying a product if you:

- include product supply as part of tendering, or part of offering work in building and construction projects, or in building refurbishment and renovation, and/or
- supply products as part of installation or repair work, or for any other purpose.

You *are not* considered to be selling products if you are involved only in installing WELS products.

Tapware

The term 'tapware' covers a wide range of products that may be used by plumbers in the fit-off stage and are used to deliver water at an outlet. They are usually the conventional screw-down type, the single lever type (flickmixer) or the half-turn type (ceramic disc type). Prior to installing any tapware it is important to check with the manufacturer that the tapware is suitable for installation within the system you are installing, as not all tapware is suitable for gravity feed or low-pressure systems.

Some of the tapware that you can come across in the fit-off stage may be supplied by the manufacturer in kits containing handles and the required outlet (fixed or swivel spout), such as for the taps shown in Figures 6.2 to 6.12. Care should be taken not to damage the finished surfaces during installation. You may view other examples of these types of products by visiting the websites in the references at the end of the chapter.

FIGURE 6.3 Bath/shower single-lever mixer tap with divertor valve

FIGURE 6.4 Basin single-lever mixer tap

FIGURE 6.5 Wall sink set

FIGURE 6.2 Shower set

FIGURE 6.6 Wall bath set

FIGURE 6.7 Hob basin set

LEARNING TASK 6.1

1 What is an important consideration at the fit-off stage?
2 Why are equal hot and cold water pressures necessary for certain tapware?

FIGURE 6.8 Kitchen single-lever mixer tap

FIGURE 6.9 Pillar tap

FIGURE 6.10 Washing machine taps

Source: © 2020 Copyright GWA Group Limited.

FIGURE 6.11 Hob spa set

FIGURE 6.12 Bidet mixer

6.3 Identifying installation requirements

Before installing any tap, valve or appliance it is important to check with the manufacturer or refer to the product's technical information. There may be pressure limitations or other specific installation requirements that require a pressure control valve either to be fitted to the whole water system or at least to be installed at the appliance or before the valve. It should be further remembered that AS/NZS 3500.1 requires a maximum service pressure of 500kPa in a water service.

AS/NZS 3500.1 PLUMBING AND DRAINAGE: WATER SERVICES

The types of appliances requiring flow control that plumbers may be required to connect, or supply a water connection to, include (but are not limited to):

- basins
- baths
- bidets
- clothes washing machines
- dishwashing machines
- hot water heaters (covered in Chapter 4)
- kitchen sinks
- laundry troughs
- outside hose taps
- showers
- toilets.

The tapware selected must be suitable for the purpose it will be used for. For example, you may need to install a strainer or water softener if the quality of water is likely to damage the valve mechanism or seat. It should be noted that the quality of water between states and territories, and even within them, varies considerably.

Tapware may be fitted with a 5L/minute flow-regulated aerator insert. This low flow rate may not be suitable for connection to continuous flow (instantaneous) water heaters, some tempering valves, some solar water heaters and some thermostatic mixing valves. Check with the manufacturer of these products before installing any tapware with a flow regulator.

As the final connection to appliances and tapware may be exposed to view by the user, it is important that this work is carried out in a professional manner. You may be required to make the final connection using chrome-plated copper or copper tube. It is very common today to use stainless steel flexible connections (see **Figure 6.13**), commonly called 'easy hookers' or 'flexi connectors'. If these connections are used, they need to meet the requirements as outlined in AS/NZS 3499 and may be subject to temperature limitations. Not all stainless steel flexible connectors are suitable for use on water services above 60°C. Adequate care should be taken not to twist these types of connectors when tightening.

Source: Shutterstock.com/Budimir Jevtic.

FIGURE 6.13 Stainless steel connector used to connect a toilet

If copper tube is used to make the final connection, care must be taken not to kink or flatten the tube when bending. This can be achieved by the proper use of tube benders or internal/external bending springs for annealed tube (see **Figures 6.14** and **6.15**).

When installing any tapware, valve, appliance or fixture, the manufacturer's installation instructions, such as those below, must be followed. Failure to do so may result in cancellation of the product warranty.

- Hot and cold water inlet pressures should be equal.
- Inlet pressure range: 150kPa to 1000kPa. AS/NZS 3500 has a maximum inlet requirement of 500kPa.
- Maximum hot water temperature: 80°C.

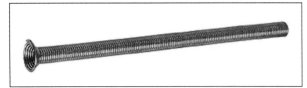

FIGURE 6.14 External bending springs

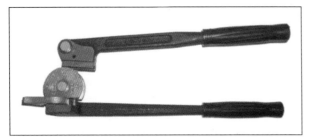

FIGURE 6.15 Lever-type tube benders

Before installing any tapware, all pipework must be flushed thoroughly to remove all foreign matter. Failure to do so may result in failure of the tapware, appliances and fixtures, leading to cancellation of the product warranty.

 Remember that any ablution fixture must be supplied with a maximum temperature of 50°C.

LEARNING TASK 6.2

1   Why is it important to check manufacturers' specifications before installing tapware?
2   What must be done to the pipework before installing tapware?
3   What is the maximum temperature allowed at an ablution fixture?

 COMPLETE WORKSHEET 1

6.4 Making the final connection

There are many different styles of tapware made all over the world that have various installation methods.

In this section, we will look at the methods and techniques used to install different types of tapware.

Installing a tap set

If the taps are not familiar, read the installation instruction before commencing work. This could avoid costly mistakes and save valuable time.

Source: Caroma Industries.

HOW TO

HOW TO INSTALL A WALL TAP SET

1   Check that the face of the recess tap body (1) is 0–13mm from the wall/tile face, as shown in **Figure 6.16**. This is an important detail set up at rough-in stage.
2   If replacing existing tapware, check that the seats in the recess tap bodies (1) are in good condition and free of foreign materials. Re-seat if required.
3   Fully open the tap spindle (3) in the tap assembly (2), ensuring that the fibre washer and jumper valve are in position, as shown.
4   Screw the tap head assembly (2) into the recess tap body (1) by hand-tightening using a suitable tube spanner (do not use multigrips or similar types of tools).

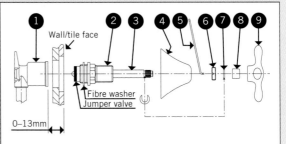

FIGURE 6.16 Example of a Dorf wall tap assembly

5   Screw on the wall flange (4) and tighten it fully by hand against the wall/tile face, taking care not to damage the finished wall.
6   Fit the handle assembly as required by the manufacturer, using the following as a guide:
    a   Place the retaining nut (6) over the spindle (3) with the slot facing the flange (4).
    b   Place the retaining clip (7) into the groove at the bottom of the spline in the spindle (3).
    c   Ensure that the handle insert (8) is in position in the handle (9). Fit the handle (9) to the spindle (3) and position to suit. Push the handle (9) firmly down until it bottoms internally.
    d   Screw the retaining nut (6) into the handle (9) and tighten using the spanner (5) provided.
7   Polish the decorative finish with a clean cloth to remove any marks.

Installing a bath, sink or laundry outlet, a shower rose, a washing machine tap or a cistern tap

Examples of laundry, bath, sink outlets and washing machine taps can be seen in **Figures 6.17**, **6.18** and **6.19**.

Source: Caroma Industries.

FIGURE 6.17 Laundry outlet

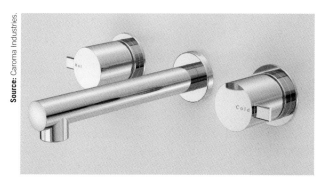

Source: Caroma Industries.

FIGURE 6.18 Bath or sink wall outlet

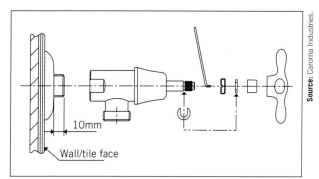

Source: Caroma Industries.

10mm

Wall/tile face

FIGURE 6.19 Example of a Dorf washing machine tap

FROM EXPERIENCE

When tightening spouts that have no square or hexagonal section, a strap wrench (see Figure 6.22) will do the job. Do not use multigrips or footprints because the teeth will damage the finished surface.

HOW TO

HOW TO INSTALL A BATH, SINK OR LAUNDRY OUTLET, A SHOWER ROSE, A WASHING MACHINE TAP OR A CISTERN TAP

1 Check that the fitting with an external thread has the correct length protruding through the cover plate (10mm). Cut it to length (see Figure 6.20) if required, ensuring that the end face is square. Use a purpose-built thread cutting tool to ensure this and chamfer the end (see Figure 6.21).

FIGURE 6.20 Cutting the brass external thread to the required length

2 Rough the external thread with a hacksaw blade or multigrips, giving the thread tape sufficient grip.

3 Wind the thread-sealing (teflon) tape or jointing compound onto the external thread in a clockwise direction. Ensure that the thread tape cannot become dislodged and block the flow or regulating device, if fitted. Fit the required cover plate over the external thread (if required).

4 Screw the spout, arm, tapware or rose inlet socket onto the external thread in a clockwise direction (ensuring not to cross-thread it) and tighten using a shifting spanner or strap wrench (see Figure 6.22). *Do not overtighten.*

5 When adjusting the inclination of a shower rose, it is recommended that the hot and cold taps are turned off. Unscrew the adjustment wing nut one turn, position the shower rose as required and then retighten the wing nut.

6 Polish the decorative finish with a clean cloth to remove any marks.

Note: Low-flow plastic shower heads are available and care should be taken not to overapply the teflon or overtighten to avoid splitting the female socket.

FIGURE 6.21 Purpose-built thread cutting tool

FIGURE 6.22 Strap wrench

Installing a basin/vanity set

Basin/vanity sets may require either a triple-hole set (see **Figure 6.23**) or a single-hole set (see **Figure 6.24**) to be installed.

Source: Caroma.

FIGURE 6.23 Triple-hole basin/vanity set

FIGURE 6.24 Single-hole basin/vanity set

HOW TO

HOW TO INSTALL A BASIN/VANITY SET

1 Measure the centre distance between the tap body holes in the deck and cut the breeching piece (17) to the required length, as shown in **Figure 6.25**. (Manufacturers may include a cutting guide template.)

2 Install the basin outlet (21) with the seal (20). Rotate the outlet (21) so that the manufacturer's marking is facing towards the front. Fit the washer (19) and backnut (18) and tighten.

3 Fit the compression nuts (16) and olives (15) to each end of the breeching piece (17), connect to the tap bodies (14) and tighten the compression nuts (16) slightly.

4 Fit one locknut (13) and one washer (12) to each tap body (14). Install the assembly into the basin, ensuring that the O-rings on the breeching piece (17) are not damaged as they enter the tail of the basin outlet (20). Adjust the locknuts (13) so that the tap bodies (14) protrude 5mm above the deck and tighten.

5 Tighten the compression nuts (16) on the breeching piece (17).

6 To install the ceramic disc cartridges (10), first identify the cartridge spindle rotation by seal (11) colour. Hot tap: use normal rotation cartridge (red seal). Cold tap: use contra rotation cartridge (blue seal).

7 Remove the locknut (8), the cap (9) and the fibre washer from the cartridge (10). Screw the cartridge (10) into the tap body (14). Using a suitable spanner, tighten the cartridge (10) until it comes to a firm stop. Do not tighten the cartridge using the spindle. Fit the fibre washer and the locknut (8), and tighten using a suitable spanner.

8 Fit the handle assembly as required by the manufacturer, using the following as a guide:

a Turn both spindles to the fully closed position.

Source: Caroma Industries.

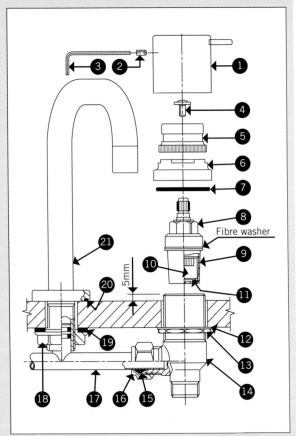

FIGURE 6.25 Example of a Dorf triple-hole basin/vanity set

b Fit the handle insert (5) to the spindle and secure using the screw (4).

c Position each handle (1) on the handle insert (5), then push it firmly down until it bottoms internally.

d Tighten the grub screw (2) with the Allen key (3) provided.

HOW TO CONNECT WATER TO A BASIN SET

1 Check that the external thread has the correct length protruding through the cover plate (10mm). Cut to length if required, ensuring that the end face is square.
2 Fit the required cover plate over the external thread.
3 If connection is to be made using stainless steel flexible connectors, screw one end onto the external thread of the water supply and the other onto the external thread of the basin set, and tighten using a shifting spanner, but do not overtighten. Ensure any stainless steel connector used for hot water is rated at a minimum of 90°C.

4 If connection is to be made using copper tube, bend the tube to the required shape (ensuring not to kink it).
5 Place a compression nut over each end of the tube. Either an olive of copper, brass, nylon or rubber or a croxing tool may be used to make the joint watertight. If a rubber olive is used, a croxing tool must be used on the end of the tube to prevent the pipe from blowing off. A compound or grommet must be used with a croxing tool joint to make it watertight.

HOW TO INSTALL A SINGLE-LEVER BASIN MIXER TAP

1 From the fixing pack fit the O-ring (9) into the groove in the underside of the mixer base (8) as shown in Figure 6.26. Screw the stud (10) into the underside of the mixer (8).
2 Apply a suitable lubricant to the O-rings (11). Pass the flexible tail (15) (with red indication) up through the deck hole and install it into the hole marked 'H' in the underside of the mixer body (8). Pass the remaining flexible tail (with blue indication) up through the basin hole and install it into hole marked 'C'. Tighten both tails (15) firmly by hand.
3 Place the mixer over the basin hole, ensuring that the hot water tail (15) is to the left and the stud (10) is to the back. Fit the gasket (12) and the fixing plate (13) over the stud (10) and screw on the fixing nut (14). Position the mixer as required before tightening the fixing nut (14).
4 Connect the flexible tails (15) to the hot and cold isolating stop taps, ensuring that the flexible tails are not kinked, twisted or in tension as they are tightened.
5 Polish the decorative surface with a clean cloth to remove any marks.

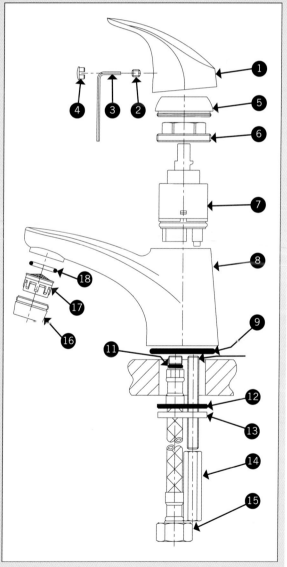

Source: Caroma Industries.

FIGURE 6.26 Example of a Dorf single-hole (flickmixer) basin/vanity set

HOW TO REPLACE THE AERATOR INSERT

1 Carefully remove the aerator housing (16) from the mixer body (8) as shown in Figure 6.26, taking care not to damage the decorative finish. Wearing a dry rubber glove will assist when removing and tightening the housing.

2 Remove the O-ring (18) and aerator insert (17) from the aerator housing. Check that the aerator housing is clean. Deposits of lime can be removed by washing in a vinegar solution.

3 Fit the new aerator insert (17) into the aerator housing (16) followed by the O-ring (18), then screw the assembly into the mixer body (8) and tighten securely (to prevent removal by hand).

HOW TO REPLACE THE CARTRIDGE

1 Turn off the hot and cold water supplies at the isolating stop taps.

2 Remove the indicator plug (4) as shown in Figure 6.26. Using the 2.5mm Allen key (3), loosen the screw (2) and remove the handle (1). Remove the cap (5), taking care not to damage the decorative finish. Unscrew the nut (6) and lift out the old cartridge (7).

3 Ensure that the inside face of the mixer body (8) is clean. Check that the O-rings are in position in the base of the new cartridge (7). Fit the new cartridge (7) into the mixer body (8), taking care that the two lugs on the base of the cartridge (7) fit into the mating holes in the mixer body (8).

4 Screw on the nut (6). The nut (6) must be tightened to a torque of 10Nm.

5 Replace the cap (5), tightening by hand. Fit the handle (1), taking care that it is pushed fully down. Tighten the screw (2) and replace the indicator plug (4), positioning it with red to the left.

6 Turn on the isolating stop taps and check the operation.

Note: This procedure also applies to a wall-mounted bath or shower mixer (see Figure 6.27).

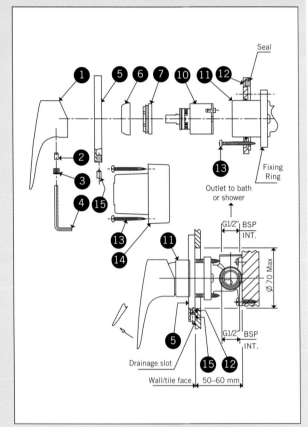

Source: Caroma Industries.

FIGURE 6.27 Example of a Dorf bath or shower flickmixer

Installing ceramic disc taps

The following information is based on installing Caroma Lusso lever handle ceramic disc taps and should be used as a guide only (see Figures 6.28 and 6.29). The tap manufacturer's installation instructions must be followed; failure to do so may result in damage to the tapware and cancellation of the product warranty.

FIGURE 6.28 Bath tap set (ceramic disc)

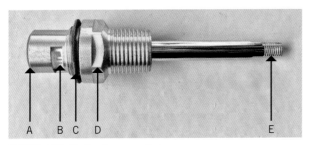

FIGURE 6.29 Parts of a ceramic tap
A = Disc retaining washer
B = Ceramic discs
C = O-ring, which stops any water seepage up to the head of the tap
D = Valve retaining nut
E = Spindle on which the handle sits

FROM EXPERIENCE

For the optimum performance of ceramic disc cartridges, hot and cold water pressure should be equal, with a recommended maximum working pressure of 500kPa and a maximum temperature requirement of 80°C.

HOW TO

HOW TO INSTALL CERAMIC DISC WALL ASSEMBLY TAPS

1 Flush the system thoroughly to remove any foreign matter.

2 Check that the face of the recess tap body (1) is 0–13mm from the wall or tile face, shown in Figure 6.30.

3 Check that the seats in the recess tap bodies (1) are in good condition and free of foreign objects. Re-seat if required.

4 Fully open the tap spindle (3) in the tap head assembly (2). Ensure that the fibre washer and sealing washer are in position.

5 Remove the lock nut and fibre washer and install the ceramic spindle (as explained in Chapter 4).

6 Screw on the wall flange (4) and tighten fully by hand against the wall/tile face, taking care not to damage the decorative finish.

7 Fit the handle assembly as required by the manufacturer, using the following as a guide:

  a Place the retaining nut (6) over the cartridge spindle (3), with the slot facing the flange (4).

  b Place the retaining clip (7) into the groove at the bottom of the spline in the cartridge spindle (3).

  c Ensure that the handle insert (8) is in position in the handle (9). Fit the handle (9) to the spline of the cartridge spindle (3), position the handle levers to the required position and push the handle (9) down firmly until it bottoms internally.

  d Screw the retaining nut (6) into the handle and tighten it with the spanner (5) provided.

8 Polish the decorative surface with a clean cloth to remove any marks.

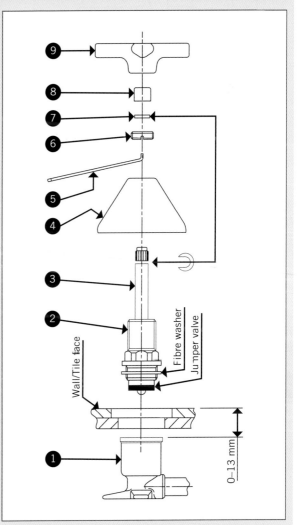

Source: Caroma Industries.

FIGURE 6.30 Example of a Caroma ceramic cartridge wall assembly kit

Basin set assembly (ceramic disc taps)

Installation is the same as previously described for a three-hole basin/vanity set, paying attention to the variances mentioned in the installation of ceramic disc taps (see Figure 6.31).

Source: Caroma Industries.

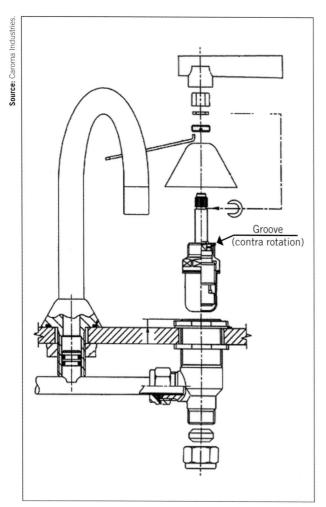

Groove
(contra rotation)

FIGURE 6.31 Example of a Caroma ceramic cartridge basin set assembly kit

FROM EXPERIENCE

It is very important that the spindles are not overtightened or the ceramic discs will be damaged.

HOW TO

HOW TO INSTALL A PILLAR TAP SET

1   Fit the seal (5) into the groove in the base of the pillar tap body (4) as shown in Figure 6.32. Install the pillar tap body (4) together with the seal (5) into the basin in the correct position, ensuring that the lugs on the base of the pillar tap fit into the square hole in the deck.

Source: Caroma Industries.

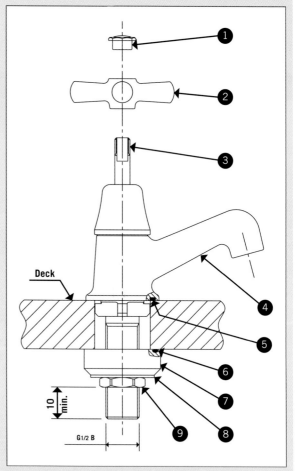

Deck

10 min.

G1/2 B

FIGURE 6.32 Example of a Dorf pillar tap set

2   Fit the O-ring (6), spacer (7) and washer (8) over the threaded tail of the pillar tap body (4). Screw on the backnut (9) and tighten securely, ensuring that the O-ring (6) remains in position in the groove of the spacer (7) during assembly.

3   Connect the water supply to the thread on the tail of each pillar tap body (4). If thread tape is used on the inlet connections, it is important that care is taken to ensure that the tape cannot become dislodged and block the flow regulating device, causing a reduction in water flow.

>>

4   Fit the handle assembly as required by the manufacturer, using the following as a guide:
    ■ Fit the handle (2) directly onto the spindle (3).
    ■ Screw the button (1) onto the spindle (3) and tighten firmly against the handle (2), taking care not to damage the decorative finish.
5   Polish the decorative surface with a clean cloth to remove any marks.

Note: Installation is the same as previously described for a ceramic disc pillar tap set, paying attention to the variances mentioned in the installation of ceramic disc taps (see **Figure 6.33**).

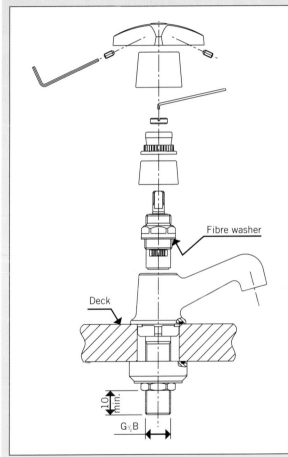

Source: Caroma Industries.

FIGURE 6.33 Example of a Caroma pillar tap (ceramic disc)

Installing a pop-up bidet assembly

When installing any bidet tapware, the manufacturer's installation instructions must be followed. Not doing so may result in failure of the product and cancellation of the product warranty. The following is an installation guide only.

HOW TO

HOW TO INSTALL A POP-UP BIDET ASSEMBLY

1   Fit the bidet tapware to the bidet bowl in a similar manner to installing a single-hole basin set and as recommended by the tapware manufacturer.
2   Fit the plug body (2) together with the seal (1) into the pan/basin as shown in **Figure 6.34** (a). Apply a suitable sealant to the seal (1) if required. Fit the lower seal (3) onto the plug body (2), screw the waste body (4) onto the plug body (2) and position it so that when it is tightened securely, the pop-up rod assembly (10) (when installed) is in a position suitable for connection to the lifting rod (12); see **Figure 6.34** (a). For installations where Dimension 'L' does not allow the waste body (4) to be tightened securely against the lower seal (3), fit the spacer (14) and additional lower seal (15) in the order shown in **Figure 6.34** (b).

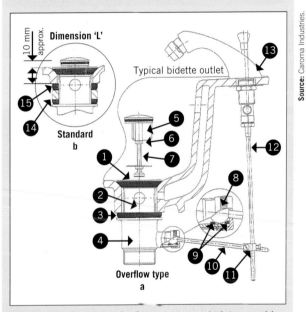

Source: Caroma Industries.

FIGURE 6.34 Example of a Caroma pop-up bidel assembly

3   Install the pop-up rod assembly (10) into the waste body (4) and tighten the nut (8), making sure that the seals (9) are in position, as shown in **Figure 6.34** (a). Slide the swivel joint assembly (11) onto the pop-up rod assembly (10) as shown, install the lifting rod (12) through the hole in the tap outlet (13) and then through the remaining hole in the swivel joint (11). With the lifting rod (12) and the pop-up rod (10) in the fully down position, lock the

>> two screws in the swivel joint (11) to their respective rods.

4 Place the plug assembly (5) into the waste body (2) and check that the plug (5) is approximately 10mm above the waste body when in the open position, as shown in Figure 6.34 (b). To adjust, loosen the locknut (6) and screw (7) in or out to achieve this dimension, then tighten the locknut (6).

5 Check that the plug (5) seats fully in the waste body (2) by lifting the operating rod (12).

6 Polish the decorative surface with a clean cloth to remove any marks. If the bidet bowl needs cleaning, use warm soapy water and dry with a soft cloth. Do not use an abrasive cleaner.

COMPLETE WORKSHEET 2

Installing a hose tap (standpipe) for a drinking water service

Hose tap risers must be fixed and supported by a hardwood or treated pine post (see Figure 6.35) or a wall;

FIGURE 6.35 Hose tap with hose vacuum breaker and riser fixed to a treated pine post

they must not be supported by or fixed to a fence. To prevent any damage to the riser it must be properly supported prior to screwing in the hose tap. Hose taps are required to be a minimum of 450mm above finished ground level. It is good practice to fit a hose connection vacuum breaker valve to all external hose tap outlet threads.

> Plumbers have a direct responsibility to protect the community's drinking water supply from contamination.

AS/NZS 3500.1 PLUMBING AND DRAINAGE: WATER SERVICES

FROM EXPERIENCE

If installing a hose tap into the internal thread coming out of a wall, care must be taken to ensure that no pressure is applied to the pipe or fitting; an additional shifting spanner should be used to hold against the turning action when installing the hose tap.

Installing recycled water and rainwater tapware

The same installation procedures are used to install tapware for recycled or rainwater installation. Extreme care must be taken to ensure that they are connected to the correct supply, and notification signs must be installed in accordance with the local water utility's requirements.

Recycled water hose taps

Recycled water hose taps differ from drinking water taps (see Figure 6.36) and must:

- have a removable handle
- not be installed inside a building
- be a purple (powder-coated) colour
- have a left-hand hose connection thread
- have a permanently fixed warning sign stating 'not for drinking'.

Some water utilities may require the recycled water hose tap to be located behind the drinking water tap with a minimum of 300mm clearance. Check your local installation requirements if this applies in your area.

AS/NZS 3500.1 PLUMBING AND DRAINAGE: WATER SERVICES

FIGURE 6.36 Hose taps: the left-hand tap is for recycled water and the right-hand tap is for drinking water

LEARNING TASK 6.3

1 Which tool is used to tighten a chrome spout that doesn't have a square or hexagonal section to fit a spanner?
2 What careful consideration must be given when screwing in a ceramic disc spindle?
3 Does a pillar tap mix hot and cold water?
4 Why is equal hot and cold water pressure recommended on single-lever mixer taps?

6.5 Commissioning the system

At the completion of the fit-off stage the system must be commissioned. This process requires you to check the operations of all valves, taps, appliances and fixtures, ensuring they are working correctly and that all air is purged from the system. Check local authority websites for extra information on the commissioning processes.

Cold water service

The installation should already have been pressure tested, as required in AS/NZS 3500.1, to a minimum of 1500kPa for a minimum of 30 minutes. Commissioning the cold water service is as follows:

1 All aerators should be removed and the system flushed to remove any foreign matter and air. The aerators may be replaced after the commissioning is complete.
2 A visual check for leaks should be carried out on any final connections, taps, appliances and fixtures.
3 The system's pressure must be checked to ensure that it is not above the requirements of AS/NZS 3500.1 (500kPa).
4 Appliance and hot water unit pressure must be checked to ensure that it is not above the manufacturer's requirements.

5 Check whether all tapware, fixtures, hot water units and appliances have been installed to the manufacturer's installation requirements. Not doing so may result in failure of the product and cancellation of the product warranty.
6 The outlet flow rate should be checked to ensure that it is not less than required by AS/NZS 3500.1 (see Table 6.1). This test should be carried out with at least half the outlets operating at the same time.

TABLE 6.1 Minimum outlet flow rates

Fixture	Minimum flow rate in L/min
Basin (standard outlet) Shower Sink (aerated outlet)	6
Bath	18
Laundry tub	7

✓✓ **AS/NZS 3500.1 PLUMBING AND DRAINAGE: WATER SERVICES**

Appliances and hot water heaters

All appliances and hot water heaters need to be commissioned as per the manufacturer's requirements to ensure that they are operating as required. The client should be shown the correct operation of fixtures and appliances upon hand-over. The client must be advised of any maintenance requirements.

Recycled water service

The commissioning of recycled water tapware is carried out in the same manner as for drinking water.

The following test procedure must be carried out prior to arranging a final inspection with the water utility within a dual reticulation area. The system must be flushed in accordance with the requirements of AS/NZS 3500.1 and tested for a cross-connection with the drinking water supply as follows:

1 Turn off the non-drinking (recycled) water supply to the property at the meter bypass shut-off valve (or the meter isolation valve if the meter is fitted). The drinking water supply remains on.
2 Turn on all outlets *(both hot and cold)* one by one. Outlets that run continuously are connected to the drinking water supply.
3 Toilets should not refill, as they should not be connected to the drinking water supply.
4 Turn on all outside taps. The external drinking water tap should run continuously. Taps that run dry are connected to the non-drinking water service and should be clearly identified with appropriate warning signs.
5 Recharge the recycled water service, ensuring all air is purged.

6   Once these steps have been completed satisfactorily, contact the local water utility to arrange a recycled water final inspection.

Remember that the plumbing installation must meet the installation requirements of AS/NZS 3500 and any state or territory regulations.

AS/NZS 3500.1 PLUMBING AND DRAINAGE: WATER SERVICES

LEARNING TASK 6.4

1 State the testing requirements for a water service.
2 Name two things that are removed when flushing hot and cold water services.
3 What is the flow rate of a shower?
4 What is the recycled water service tested for on the final inspection?

6.6 Cleaning up

Cleaning as you go is important at any stage of the job, but it is most noticeable and effective upon completion of the job. It is at this time that the plumber leaves a final impression with the client, so be sure the fixtures, appliances and the area are clean and tidy, with the floors swept and clean.

Any leftover material can be stored and reused another time, saving cost and waste.

All rubbish must be disposed of and any cardboard packaging should be placed in the recycling bin, along with anything else that is recyclable.

All tools and equipment must be cleaned and put away, ready to used next time. Replace or sharpen any blunt blades so they are ready for use. Any tools and equipment on hire should be cleaned and returned as soon as possible to avoid extra costs.

All documentation and warranties must be submitted to the client and regulatory authorities as required upon completion.

Quality assurance policies should be adhered to, which will ensure the work has been carried out in line with WHS requirements and finished in a professional manner. Working within these guidelines demonstrates a high level of professionalism and helps to maintain a good reputation for the company.

LEARNING TASK 6.5

1 Why is the final clean-up the most important?
2 What information should be given to the client upon completion?

 COMPLETE WORKSHEET 3

SUMMARY

- Prepare and plan well by obtaining all the information required before starting the job (for example, type of tapware and fixtures).
- It is the plumber's responsibility to ensure the WaterMark logo is displayed on all plumbing fixtures and tapware before installation.
- Be sure to understand manufacturer's installation instructions and requirements when installing tapware (for example, minimum and maximum pressure). Then the warranty remains valid.
- Single-lever mixer tapware requires equal hot and cold water pressure to maintain a warm water blend.
- Hose taps must be well supported and a minimum of 450mm high from ground level.

- A recycled tap must have a removable handle, be outside, be purple in colour, have a left-hand hose connection thread, and have a permanent sign stating 'not for drinking'.
- Flush all air and debris from pipework before tapware installation.
- Be sure to show the client how to operate and maintain tapware and fixtures.
- The recycled water service must be tested for cross-connection on final inspection.
- Complete all the relevant documentation and hand it over to the client and regulatory authorities.

REFERENCES

Caroma: **http://www.caroma.com.au**
Dorf: **http://www.dorf.com.au**
Sydney Water – Working with recycled water: **https:// www.sydneywater.com.au/web/groups/publicwebcontent/ documents/document/zgrf/mdq3/~edisp/dd_047457.pdf**

Victoria Building Authority – Hose Connection Vacuum Breakers: **https://www.vba.vic.gov.au/__data/assets/pdf_file/0009/ 98496/5.04-Cold-Water-Plumbing-Hose-Connection- Vacuum-Breakers.pdf**

GET IT RIGHT

1 Which photo shows correct installation?

2 Why is it correct?

WORKSHEET 1

Student name: _____

Enrolment year: _____

Class code: _____

Competency name/Number: _____

To be completed by teachers

Satisfactory ☐

Not satisfactory ☐

Task: Review the section '6.2 Planning and preparation' and answer the following questions.

1 What do the initials WELS stand for?

_____

2 Describe what the WELS scheme is designed for.

_____

3 List four products to which the WELS rating applies.

_____

_____

_____

_____

4 Describe the main difference between the WaterMark logo and the WELS label.

_____

5 List two situations where some types of tapware may not be suitable.

_____

_____

6 Describe what will happen if a stop valve (loose valve, jumper valve) is installed in reverse.

_____

7 What pressure (kPa) is available at a water outlet with a head of 24m?

_____

8 What is the legal definition of 'supplying WELS goods'?

_____

_____

_____

9   What is the maximum inlet pressure requirement as per AS/NZS 3500?

_____

10   Why is it important to flush the hot and cold water services before installing tapware?

_____

_____

11   Why is it important to install tapware as per manufacturer's installation instructions?

_____

_____

12   Why is it important to have equal hot and cold water pressure for a mixer tap?

_____

_____

 **WORKSHEET 2**

Student name: _____

Enrolment year: _____

Class code: _____

Competency name/Number: _____

Task: Review the section '6.4 Making the final connection' and answer the following questions.

1   List three different methods of connecting the water supply to a vanity basin.

_____

_____

_____

2   List two things that must be adhered to so that the product warranty is not voided or cancelled by the manufacturer.

_____

_____

3   In which direction should teflon tape be wound onto a thread?

_____

4   To what temperature should stainless steel connectors be rated?

_____

5   What type of joint should be used when connecting a basin to the water supply using a Kinco nut and olive to prevent it from blowing off?

_____

6   Where would a pillar tap be used, and for what purpose?

_____

_____

7   A 5L/minute flow rated aerator may not be suitable for use in conjunction with which types of hot water heaters?

_____

8   What can happen if the spindle for a ceramic disc tap is overtightened?

_____

_____

9 What problem is associated with fast-closing taps such as quarter-turn ceramic disc and single-lever mixer types?

_____

_____

10 What position must the spindle be in when installing jumper valve taps?

_____

_____

WORKSHEET 3

Student name: _____

Enrolment year: _____

Class code: _____

Competency name/Number: _____

Task: Review the section '6.4 Making the final connection' and answer the following questions.

1   List five conditions for the installation of hose taps on a recycled water service.

_____

_____

_____

_____

_____

2   List the mandatory inspection stages required in recycled water areas.

_____

_____

_____

3   What should be done to the water service prior to the fitting-out stage of the job?

_____

4   What is the required flow rate for a bath?

_____

5   Which two tools are best suited to tightening chromed fittings against a wall?

_____

6   What must hose taps not be fixed onto?

_____

7   What is the minimum pressure and time required to test a cold water service?

_____

8   What fitting can be fitted to a hose tap to prevent back siphoning?

_____

9   What is the minimum height at which a non-recycled water hose tap should be installed?

_____

_____

10 What is the minimum height at which a hose tap should be installed?

_____

_____

11 Why are the aerators removed when commissioning tapware?

_____

_____

12 What documentation is handed to the client upon completion?

_____

_____

13 Why is quality assurance important?

_____

WORKSHEET RECORDING TOOL

Learner name		Phone no.	
Assessor name		Phone no.	
Assessment site			
Assessment date/s		Time/s	
Unit code & title			
Assessment type			

Outcomes

Worksheet no. to be completed by the learner	Method of assessment WQ – Written questions PW – Practical/workplace tasks TP – Third-party reports SC – Scenarios RP – Role plays CS – Case studies RW – Report writing PF – Portfolio	Satisfactory response	
		Yes ✓	No ✗
Worksheet 1	WQ	☐	☐
Worksheet 2	WQ	☐	☐
Worksheet 3	WQ	☐	☐

Assessor feedback to the learner

Feedback method (Tick one ✓):	☐ Verbal	☐ Written (if so, attach)	☐ LMS (electronic)

Indicate reasonable adjustment/assessor intervention/inclusive practice (if there is not enough space, a separate document must be attached and signed by the assessor).

Outcome (Tick one ✓):	☐ Competent (C)	☐ Not Yet Competent (NYC)

Assessor declaration: I declare that I have conducted a fair, valid, reliable and flexible assessment with this learner, and I have provided appropriate feedback.

Assessor name:	
Assessor signature:	
Date:	

Learner feedback to the assessor

Feedback method (Tick one ✓):	☐ Verbal	☐ Written (if so, attach)	☐ LMS (electronic)

Learner may choose to provide information to the RTO separately.

	Tick one:	
Learner assessment acknowledgement:	**Yes** ✓	**No** X
The assessment instructions were clearly explained to me.	☐	☐
The assessment process was fair and equitable.	☐	☐
The outcomes of assessment have been discussed with me.	☐	☐
The overall judgement about my competency performance was fair and accurate.	☐	☐
I was given adequate feedback about my performance after the assessment.	☐	☐

Learner declaration:
I hereby certify that this assessment is my own work, based on my personal study and/or research. I have acknowledged all material and sources used in the presentation of this assessment, whether they are books, articles, reports, internet searches or any other document or personal communication.
I also certify that the assessment has not previously been submitted for assessment in any other subject or at any other time in the same subject and that I have not copied in part or whole or otherwise plagiarised the work of other students and/or other persons.

Learner name:	
Learner signature:	
Date:	

7 FABRICATE AND INSTALL FIRE HYDRANT AND HOSE REEL SYSTEMS

Chapter overview

The installation of fire hydrants and hose reels plays a major role preventing the spread of fire and protecting people and property throughout the community. The information in this chapter will help you to gain the skills and knowledge to competently fabricate and install fire hydrants and hose reel systems using different materials and jointing systems.

Learning objectives

Areas addressed in this chapter include:
- planning and preparing work activities
- identifying and determining installation requirements
- fabricating, installing and testing the system
- cleaning up.

7.1 Background

Firefighting existed before the hydrant, and the idea of getting the wet stuff onto the red stuff is very old. The inventor of the first device that we'd recognise today as a fire hydrant can't be told, because the hydrant was developed over a period of many years by many people.

We do know that the first pillar or post street fire hydrant was developed in America by Frederick Graff in 1801. Prior to that, 'cisterns' or underground tanks were used to supply water for firefighting, and these are still used today.

The first practical fire hose was invented in Holland by Jan van der Heyden in 1673. The hose was made from 50-foot lengths of leather or sail cloth sewn together in a single seam.

Fire hydrants, fire hose reels, fire sprinklers and fire extinguishers are essential for effective fire protection in order to save lives and protect properties.

Fire hydrants are installed within properties for use by the fire brigade. Fire hydrant systems should only be used for firefighting purposes.

7.2 Planning and preparing work activities

Planning and preparing the workplace provides for a smooth-flowing, efficient and cost-effective job. This involves interpreting plans and identifying hazards so a risk assessment can be implemented. Planning and sequencing tasks with the required tools and equipment will help to ensure quality assurance requirements are met.

Design drawings and job specifications

Hydraulic plans and design drawings for fire hydrant and hose reel system installations are designed, drawn and endorsed by hydraulic consultants. They specify the location of hydrants and hose reels as well as the pipe and fitting material to be used. The pipe sizes are also determined from the available water main supply pressure and flow rate.

The Australian Standards state the minimum requirements for the fabrication and installation of fire hydrants and fire hose reels:

- AS 2419.1 Fire hydrant installations
- AS 2441 Installation of fire hose reels
- AS 1221 Fire hose reels (design, construction and performance)
- AS 2118 Automatic fire sprinkler systems
- AS 4118 Fire sprinkler components
- AS/NZS 3500.1 Water supply
- AS/NZS 3500.0 Glossary
- National Construction Code (NCC).

Work health and safety (WHS) and environmental requirements

Legislation requires that work health and safety requirements (WHS) be observed and adhered to. A thorough risk assessment combined with a safe work method statement (SWMS) is necessary to:

- identify the hazards
- assess the risk
- apply control measures.

Working at heights and working in confined spaces are major risks involved in the fabrication and installation of fire hydrants and hose reel systems. Therefore, the correct procedures must be carried out. The correct techniques for safe trench excavation must be carried out to avoid any chance of trench collapse.

Environmental requirements involve taking the appropriate measures to reduce excessive noise and dust when drilling, cutting and sawing different materials. Also care must be taken when excavating so that the spoil and fill is stockpiled safely so as not to cause silt and sedimentation damage to the surrounding area.

The risk of injury can be reduced by the appropriate use of personal protective equipment (PPE). The types of PPE include the following:

- *Safety boots* have a steel capped toe to protect your feet from falling objects, and a hard-wearing sole that sharp objects cannot pierce.
- *Safety glasses or goggles* will protect your eyes from injury when drilling, cutting or sawing with hand or power tools.
- *Ear plugs or ear muffs* will protect your hearing when using power tools and working in noisy work sites.
- *Dust masks or respirators* will protect your respiratory system from breathing in dust and fine particles when drilling, cutting or sawing and working in a dusty environment.
- *Gloves* will protect your hands from cuts, abrasions and burns when working with different materials, tools and equipment.
- *Overalls* will cover your skin and protect you from exposure and different conditions.
- *Highly visible clothing* allows workers on-site to easily identify where each other are and helps to avoid accidents.

Manual lifting and handling techniques must always be followed according to WHS requirements. A correct size-up of the load to be lifted is important to determine whether a mechanical lifting device, a crane or a two-person lift is required. Try to avoid excessive manual handling and double handling of materials wherever possible.

When working at heights or in confined spaces, the correct procedures must be in place. Workers must be appropriately trained and accredited to perform these duties. Heavy fines apply to those who do not follow the correct procedures.

Quality assurance requirements

Information from specifications and plans can determine the type of pipe and fittings to be used as well as the brackets, clips and fixing types for the fabrication and installation of fire hydrants and hose reel systems.

Most companies have quality assurance procedures in place to promote a good standard of workmanship and efficient operations. This also helps to prevent mistakes from happening.

The requirements of this policy would include:

- controls and procedures in the workplace (WHS)
- use and maintenance of power tools and equipment
- interpreting plans and specifications
- quality of materials selected
- handling procedures
- recording information
- clean-up of job site.

Tasks are planned and sequenced

Installing fire hydrant and hose reel systems can involve trench excavation, working at heights and working in confined spaces, so it is vital the correct procedures are taken. This involves planning and sequencing the order of tasks. Preliminary tasks include the following:

- Ensure all fees and permits have been dealt with (for example, a road opening permit, footpath opening permit and water main drillings).
- Carry out a 'Dial Before You Dig' enquiry.
- Be certain a thorough risk assessment has been done with a completed safe work method statement (SWMS).

 When excavating trenches on-site, be sure to communicate this with the site supervisor or builder well in advance to avoid any clash of trades or deliveries on-site. Trench support also may be necessary depending on the soil type and depth. It is essential that a risk assessment takes place.

It is essential when planning tasks to inform the site supervisor of the nature and timing of the particular task to avoid clashes with other trades and any access issues. This will help everyone on-site work together and prevent unnecessary arguments. Reputation and respect is earnt for those who work this way, resulting in future work prospects.

Tools and equipment

There are specialised tools and equipment necessary for different jointing systems used in the fabrication and installation of fire hydrant and hose reel systems.

Crimp and press-fit tools used on copper and plastic pipe and fittings need calibrating regularly to ensure the tool is pressing or crimping to the correct tolerances, so the joint is sound. Depending on the tool, recalibration should take place after 50 000 cycles. Some tools have a warning light that alerts when recalibration is due.

Electro-fusion welding machines used for jointing polyethylene pipe also need to be regularly calibrated to ensure they are heating to the required tolerances.

All electrical tools and leads must be tagged for safety regularly to the local authorities' rules. The general requirement for building and construction is to inspect and tag electrical equipment every three months because of the harsh environment the electrical equipment is exposed to.

Work area is prepared

Having the job site well prepared helps the work progress more safely and smoothly. A site visit before starting work will help determine access for deliveries, plant and equipment. A storage area can be established for materials, tools and equipment. Ensure site inductions have been carried out. Having the job set and marked out at this time will help identify any problems associated with the installation.

FROM EXPERIENCE

Setting out the job at the site inspection identifies problems early so they can be dealt with more easily. For example, other services could be in way of the fire hydrant service. By seeing this at the planning stage, a better solution can be negotiated.

LEARNING TASK 7.1

1 What is the Australian Standard code number for the installation of fire hose reel systems?
2 Name one task to be carried out prior to starting work on a fire hydrant service installation.
3 How often does electrical equipment need to be tagged and tested in the building and construction industry?

 COMPLETE WORKSHEET 1

7.3 Identifying installation requirements

This section is aimed at understanding different types of fire service systems and installations, and the requirements necessary to meet the authorities' standards.

Types of systems and requirements

There are several different systems that incorporate fire hydrants and fire hose reels. This system is shown in the appropriate plans. As specified in AS/NZS 2419, fire hydrant systems must be a wet pipe system. This means that the service is fully charged with water at all times at the required pressure, or is allowed to be boosted by an approved pumping system.

Hydrant system

The hydrant system is a separately dedicated service connected to the water main. This service supplies water only to fire hydrants via a booster assembly and is generally a minimum of 100mm in diameter (see Figure 7.1). The water supply to the fire hose reels connects to the domestic water supply metered system.

Combined fire hydrant and fire hose reel system

On certain sites the fire hose reel service is combined with the fire hydrant service (see Figures 7.2 and 7.3). This is appropriate where the fire hose reels are located near the fire hydrants to keep pipework to a minimum. The service type is designed by the hydraulic consultant and sized accordingly to supply the minimum pressure and flow rates. A combined fire hydrant and fire hose reel system is designed to deliver water at the minimum required pressure and flow rate for four hours.

Combined sprinkler and hydrant system

This system combines the fire sprinkler service with the fire hydrant service and is designed to protect a wide range of building types. Flow in the system is controlled by valves, switches, actuating devices, alarms and pumps. The fire sprinkler system is designed to deliver water at a minimum flow rate and pressure for 30 minutes. By then the fire brigade should have connected to the system and increased the pressure through the booster assembly.

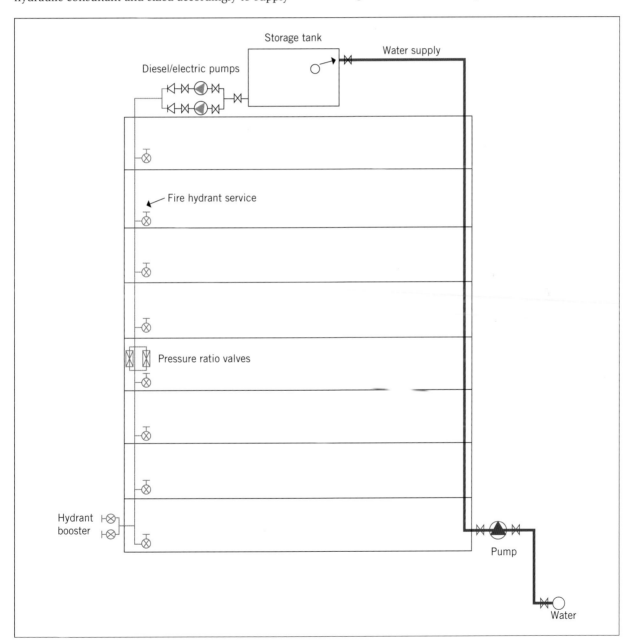

FIGURE 7.1 High-rise hydrant installation

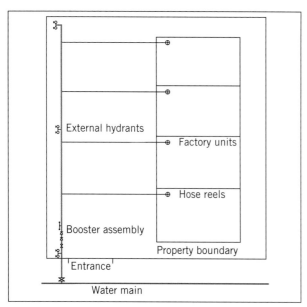

FIGURE 7.2 Low-rise fire hydrant and fire hose reel installation

FIGURE 7.3 Combined fire hydrant and fire hose reel service

Fire hydrant system components

A fire hydrant is a valve with a fire hose connection suited to connect to the hose coupling used by the firefighting agency responsible for the area.

A feed hydrant supplies water via a hose to the inlet of the pump on a fire truck. A hose carries water at an increased pressure from the outlet of the fire truck pump and connects to the *booster inlet*, which charges the fire hydrant service on the property.

An attack hydrant is the hydrant to which a firefighter connects a hose to extinguish the fire (see Figure 7.4).

FIGURE 7.4 External attack fire hydrant

A fire hydrant system can be supplied to either a potable or recycled water main with a tee and valve. Where the service enters the property, a fire brigade booster and suction connection is set up to supply the fire brigade with water and increase the system pressure with the fire brigade pump. AS 2419.1 states that certain building classes require a ring supply main so water is fed from both directions (see Figure 7.5). This helps prevent the water supply from being starved at any hydrant operating.

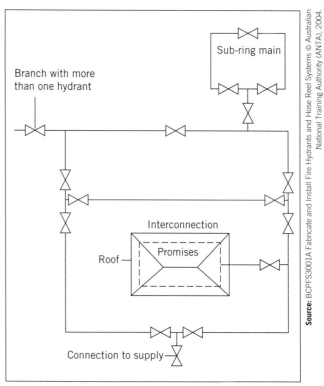

FIGURE 7.5 Fire hydrant ring main

Source: BCPFS3001A Fabricate and Install Fire Hydrants and Hose Reel Systems © Australian National Training Authority (ANTA), 2004.

An isolating valve is installed inside the property boundary downstream of the water authority valve. It is locked or strapped in the open position so it cannot be tampered with, and is normally a component of the booster assembly. Its purpose is to isolate the fire hydrant system from the water supply. Additional isolation valves are located so that hydrant system zones can be isolated in 25% sections and so half of the available hydrants can protect each fire compartment.

Pressure gauges in fire hydrant systems as per AS 2419.1 must be installed with a gauge cock (isolating valve) to permit removal for testing and replacing. The dial face must have a minimum diameter of 100mm. The gauge must read 25% higher than the hydrostatic test pressure (see Figure 7.6).

FIGURE 7.6 Pressure gauge

Fire hydrant pressure gauges shall be installed:
- on the inlet and outlet of any pump
- next to the booster inlet connection
- on the delivery side of any jockey pump
- at each pressure switch
- at any test point.

Internal or external *cabinets* may be required to house fire hydrants and booster connections or for hose reels. The layout and dimensions of the cabinets must comply with AS 2429.1 (see Fire hydrant installations. Part 1: System design, installation and commissioning, Standards Australia Limited, p. 97: https://sso.standards.org.au/idp/startSSO.ping?PartnerSpId = hub&usertype = comuser&entityID = hub&returnID Param = idp). Cabinets protect equipment from the weather, unauthorised interference and vandalism.

The cabinet must have a lever-style lock compatible with local fire brigade requirements, and be clearly labelled FIRE HYDRANT or FIRE HYDRANT – HOSE REEL or HYDRANT – BOOSTER as appropriate. The words must in capital letters and a minimum of 50mm in height (see Figure 7.7).

FIGURE 7.7 Hydrant booster cabinet

Metal cabinets remote from the building shall be mounted on legs to provide not less than 50mm space between the bottom edge of the cabinet and the finished surface level (see Figure 7.8).

FIGURE 7.8 Tank water supply with booster inlet

All other cabinets, enclosures or recess shall have a sloping floor from the rear of the cabinet to the front so any water on the floor will run off.

Classes of buildings

There are different fire hydrant systems for different classes of buildings. Table 7.1 lists how the National Construction Code (NCC) classes buildings.

AS 2429.1 shows the number of hydrants required to flow simultaneously depending on the building classification, number of floors and floor area. (see Fire hydrant installations. Part 1: System design, installation and commissioning, Standards Australia Limited, p. 18: https://sso.standards.org.au/idp/startSSO.ping? PartnerSpId = hub&usertype = comuser&entityID = hub& returnIDParam = idp).

Fire hydrant coverage shall be determined by distances that are measured along the most direct laid-on-ground or floor route to a car park, open yard or marina, and within the protected building. The main measurements are as follows:
- External hydrants are hydrants located outside the building and should be positioned to cover the whole site.
- A firefighters' hose is 30m in length. External coverage is based on using 60m of hose (two

TABLE 7.1 Classes of buildings

Classes of building		
Class 1	Class 1a	A single dwelling being a detached house, or one or more attached dwellings, each being a building, separated by a fire-resisting wall, including a row house, terrace house, town house or villa unit.
	Class 1b	A boarding house, guest house, hostel or the like with a total area of all floors not exceeding 300m^2, and where not more than 12 reside, and is not located above or below another dwelling or another Class of building other than a private garage.
Class 2	A building containing 2 or more sole-occupancy units each being a separate dwelling.	
Class 3	A residential building, other than a Class 1 or 2 building, which is a common place of long term or transient living for a number of unrelated persons. Example: boarding-house, hostel, backpackers accommodation or residential part of a hotel, motel, school or detention centre.	
Class 4	A dwelling in a building that is Class 5, 6, 7, 8 or 9 if it is the only dwelling in the building.	
Class 5	An office building used for professional or commercial purposes, excluding buildings of Class 6, 7, 8 or 9.	
Class 6	A shop or other building for the sale of goods by retail or the supply of services direct to the public. Example: café, restaurant, kiosk, hairdressers, showroom or service station.	
Class 7	Class 7a	A building which is a car park.
	Class 7b	A building which is for storage or display of goods or produce for sale by wholesale.
Class 8	A laboratory, or a building in which a handicraft or process for the production, assembling, altering, repairing, packing, finishing or cleaning of goods or produce is carried on for trade, sale or gain.	
Class 9	A building of a public nature.	
	Class 9a	A health care building, including those parts of the building set aside as a laboratory.
	Class 9b	An assembly building, including a trade workshop, laboratory or the like, in a primary or secondary school, but excluding any other parts of the building that are of another class.
	Class 9c	An aged care building.
Class 10	A non-habitable building or structure.	
	Class 10a	A private garage, carport, shed or the like.
	Class 10b	A structure being a fence, mast, antenna, retaining or free standing wall, swimming pool or the like.
	Class 10c	A private bushfire shelter.

Source: National Construction Code (2005), Volume One, pp. 35–6.

lengths) and a 10m hose stream (70m in total) when connected to a dual hydrant. When using a single hydrant, coverage is based on a 30m hose and 10m hose stream (40m in total).

- Where internal hydrants are used to protect parts of the building, all points must be covered by a 10m hose stream from a 30m hose (40m in total), connected across the floor to the hydrant.
- In multistorey buildings, one storey down and one storey up can be covered, provided:
  - there is at least one hydrant supplying each level
  - no more than 30m of hose is laid in a stairway
  - a minimum of 1m of hose projects into the area being covered by the hose stream.

AS 2419.1 explains how it is possible to have a hydrant on a stair landing that is different from the finished floor level it is supplying (see Fire hydrant installations. Part 1: System design, installation and commissioning, Standards Australia Limited, p. 166: https://sso.standards.org.au/idp/startSSO.ping? PartnerSpId = hub&usertype = comuser&entityID = hub&returnIDParam = idp).

- For roof areas that have occupant access, the area of access must be covered by a 10m stream from a 30m hose.

- Open yards must be covered by a 60m hose and a 10m water stream.
- Hydrants should be in accessible locations at least 10m from any building and with hardstanding areas for pump appliance vehicles.
- Feed hydrants should be a maximum 20m from the hardstand area.
- The minimum size pipe supplying a single fire hydrant is DN 80 and a dual hydrant is DN 100.
- The minimum size hydrant valve inlet is DN 65.
- An approved fire brigade quick-connect adaptor must be fitted to the hydrant valve outlet.
- The fire hydrant is installed at a height between 750mm and 1200mm.
- The angle of the fire hydrant valve is 35° from horizontal so the hose does not kink as it drops from the fire hydrant.
- There must be 100mm of clearance around the valve handle for easy access, and 300mm clearance around the valve hose connection (see **Figure 7.9**).

AS 2419.1 FIRE HYDRANT INSTALLATIONS

FIGURE 7.9 Fire hydrant clearances

Fire brigade booster connection

A booster connection point is an assembly of valves on a fire hydrant or fire sprinkler service to allow the fire brigade to connect and pump water into the system at a boosted pressure (see Figure 7.10). This helps the fire brigade to work more effectively in controlling a fire, saving lives and protecting property (see Figure 7.11).

FIGURE 7.10 Booster assembly

A fire booster assembly must be installed where:

- the fire hydrant service has internal fire hydrants installed
- two or more external (including street) hydrants are needed to provide coverage
- the fire hydrant service has external hydrants located more than 20m from the fire brigade pumping appliance hardstand
- a pump set is installed
- a storage tank is installed
- more than one external fire hydrant serves a building with a fire compartment floor area greater than 2000m^2
- the fire hydrant service has a ring main installed.

AS 2419.1 FIRE HYDRANT INSTALLATIONS

Booster assemblies are available in different patterns depending on the number of valves and the space required (see Figures 7.12, 7.13 and 7.14).

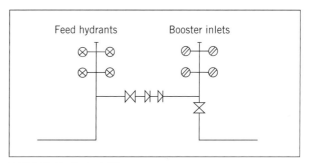

FIGURE 7.12 Typical 'H' pattern booster assembly

A block plan is a permanent plan located at the booster assembly that is weather resistant and cannot fade. It is essentially a site plan with a schematic layout of the fire service showing the location of all the fire hydrants, isolation valves, booster connections, pumps, storage tanks, available pressure and flow rate, along with the highest, most disadvantaged outlet and the location of the main electrical switch room and substation, if applicable (see Figure 7.15 for an example).

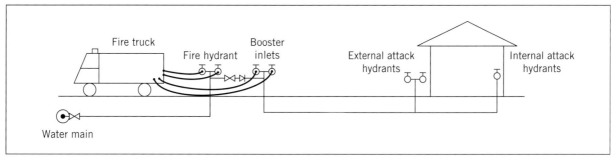

FIGURE 7.11 Example of a booster assembly connected to a fire brigade appliance

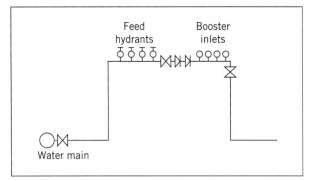

FIGURE 7.13 Typical 'in-line' booster assembly

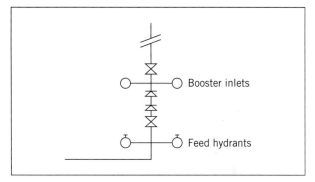

FIGURE 7.14 Typical 'I' pattern booster assembly

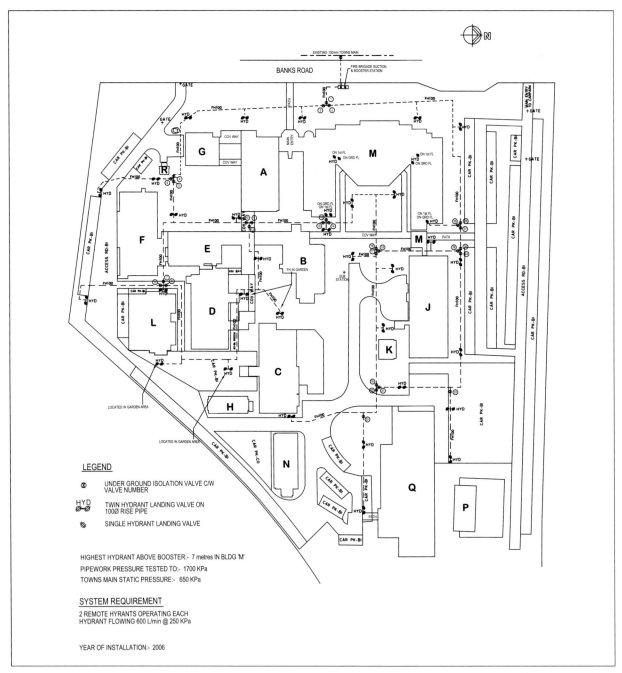

FIGURE 7.15 Block plan for Miller TAFE college

The block plan enables firefighters on arrival at the scene to quickly assess the site layout and the location of all the firefighting, water supply and electrical equipment.

A hardstand area is an area for the fire brigade pumping appliance to park on and safely carry out firefighting operations. It must be a reasonably level all-weather surface that is strong enough to take the weight of a fire truck with total height clearance. It must be located no more than 20m from an external feed fire hydrant.

An open yard is a designated area greater than 500m² used for storing or processing combustible material.

Fire hose reels

A fire hose reel is used for an initial attack on a Class A fire such as paper, wood, textiles, most plastics and rubber. Unlike a fire hydrant and booster assembly, a hose reel is meant to be used by any member of the public responding quickly to fight the fire at an early stage.

Hoses are made of a non-kinking rubber that is 36m in length and is 19mm in diameter. They are permanently connected to a water supply and have an isolation valve, a hose guide, a hose and an adjustable shut-off nozzle.

The National Construction Code of Australia (NCC) nominates when and where fire hose reels are required. This depends on the class of building and its uses, the number of floors and the total floor area.

Fire hose reels (see Figure 7.16) must be mounted at a height between 1400mm and 2400mm from finished

FIGURE 7.16 A typical fire hose reel

floor to the centre of the hose reel. The recommended height is 1500mm.

The stop valve shall be mounted between 900mm and 1100mm from the finished floor level.

The stop valve cannot be mounted more than 2000mm away from the hose reel and must have a clearance of at least 100mm around the valve handle so it is easy to access for operation (see Figure 7.17).

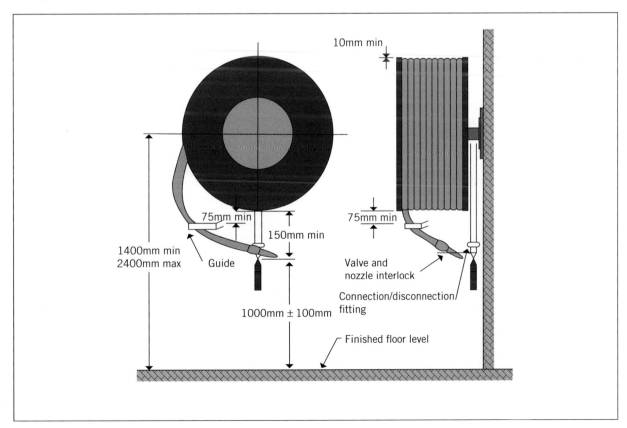

FIGURE 7.17 Fire hose reel measurements

The minimum size pipe supplying a fire hose reel is DN 25, with a minimum pressure of 220kPa and a minimum flow rate of 0.33L per second.

The hose reel assembly must comply with AS 1221 for design, construction and performance. For installation, the hose reel assembly must comply with AS 2441.

Piping and material types

The piping material used for fire hydrant and fire hose reel services can vary depending on the application, and will depend if the piping is below or above ground. Piping materials used are:

- copper pipe
- polyethylene
- unplasticised polyvinyl chloride (PVC-U)
- galvanised steel
- stainless steel.

Other factors to consider when selecting a piping system are:

- compatibility of materials, fittings and fixings
- the corrosive nature of the ground and environment
- whether the area is frost prone
- access for service and repair.

External pipework

As per AS 2419.1, wherever possible external pipework shall be located below ground. Any external pipework above ground that's subject to freezing shall be suitably protected.

External above-ground pipework that is attached to an external wall or roof must be protected from fire damage by the supporting structure having either a fire-resistant level (FRL) of 60 minutes or an automatic fire sprinkler system as per AS/NZS2118.1.

Internal pipework

Any internal above-ground pipework shall be protected by the supporting building structure and pipe supports having a FRL of 60 minutes or an automatic fire

sprinkler system as per AS/NZS2118.1. The pipework also can be protected in a fire-isolated stairwell or a fire-resistant shaft.

Below-ground piping

The most common piping materials used below ground for fire hydrant and hose reel systems are:

- copper
- polyethylene
- PVC-U
- cement-lined cast iron.

Where the ground is subject to ground movement, including mine subsidence, provision must be made to support the pipework.

Galvanised steel pipe and fittings can only be used for hydrant risers or short connection pieces no more than 1.5m long.

Any section of the pipe below the ground shall be double-wrapped in petroleum tape.

Galvanised steel pipe cannot be used to connect the inlet service from the water agency main to the property backflow prevention device.

Ordering and delivery of materials and equipment

It pays to take the time to work out an accurate order and to have the materials on-site in a timely manner. Be sure to refer to the plans and specifications to get the correct material required. It is important that the specified material complies with the relevant Australian Standards. This should not be a problem, as the plans and specifications must be approved before starting any work.

Depending on company procedures, the order may be made direct or through a purchasing officer.

HOW TO

HOW TO SET OUT AND INSTALL BELOW-GROUND PIPEWORK

1. Check your Dial Before You Dig search is complete.
2. Make sure all road/footpath opening permits are paid.
3. Pay and book your water drilling permit.
4. Have necessary barricades and signage in place.
5. Locate the water main in the street in line with the building connection point.
6. Mark out a straight line using a string line and marking paint.
7. Excavate the trench to the required depth and width.
8. Install trench support if required.
9. Install pipework and bedding as per plans and specifications.
10. Pour concrete thrust blocks as required.
11. Test the installation as per AS 2419.1 prior to backfilling.
12. Backfill as soon as possible.

A number of suppliers may be involved to supply the variety of materials, valves and fittings required.

Deliveries must be checked against the order on arrival and missing items documented, along with any items on back-order. This should be part of the company policy assurance process.

Having this process in place shows professionalism and will also help to prevent delays and arguments with suppliers.

LEARNING TASK 7.2

1 Describe a feed hydrant.
2 Describe an attack hydrant.
3 What is the purpose of a booster valve assembly?
4 Name two pipe materials used below ground for fire hydrant services.

📋 **COMPLETE WORKSHEET 2**

7.4 Fabricating, installing and testing the system

Fabricating, installing and testing fire services must be done with sharp attention to plans and specifications. If not, the results can be disastrous. Working with large pipework equates to large volumes of water and major leaks can cause major damage.

Setting out job from plans, specifications and job instructions

Fire systems must be designed by a qualified person, usually a hydraulic consultant. This is to ensure the system is suitable for firefighting purposes for any specific site. The fire system design must comply with the National Construction Code (NCC) and the relevant Australian Standards as a minimum requirement.

Water supply agencies have control over connections to their supply, with backflow prevention being a major consideration (see Figure 7.18). In some instances, on-site storage tanks must be installed if the

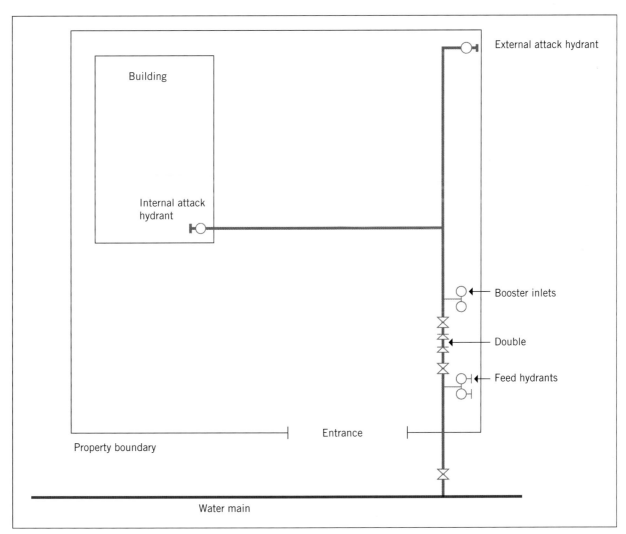

FIGURE 7.18 Fire hydrant service connection to water main with external and internal attack hydrants

main water supply is not suitable for firefighting purposes. This could be due to lack of pressure and low flow rate. The water authority has control over the materials and products used in fire hydrant and fire hose reel installations.

Therefore, it is important that the plans, specifications and job instructions are set out as designed for the fire service to comply and perform to the specific site installation.

When setting out the job it is important to check plans for dimensions and levels. For underground pipework, check excavation depth and plan trench support, if required. Accurately mark out deck penetrations for vertical pipework to avoid costly core hole drilling and pipe offsets. Set out and mark horizontal pipework effectively to minimise pipe offsets and avoid clashing with other services.

Pipe supports and fixings

Fire system pipes, valves and ancillary equipment must be adequately supported. The pipe supports and fasteners should be capable of supporting twice the weight of the pipework filled with water, with an additional load of 115kg at each point of support. There is a wide range of brands available to suit attaching pipework to structural steel, concrete, masonry (brickwork) and timber.

Other important considerations are as follows:

- Is the structure strong enough to support a pipe full of water?
- Does the bracket and fixing type meet the requirements of AS 2419?
- Is the required bracket spacing adequate as per AS 2419?
- Is extra protection such as galvanising, plastic sheathing or bituminous paint needed because the fire system is exposed to a corrosive/aggressive environment?

AS 2419.1 FIRE HYDRANT INSTALLATIONS

Fixings and fasteners

Generally, the fixings and fasteners used on fire services are made from steel (see Figure 7.19). They are bolted or screwed to the supporting structure and have a lifespan of the building. Wooden or plastic fixing inserts are not suitable for fire services.

Nails or explosive powered fasteners are not approved for fixing brackets to fire services. Preferred fixings are:

- expansion bolts
- screw fitting inserts either drilled or cast into concrete
- bolted fixings
- bolted clamps or clips.

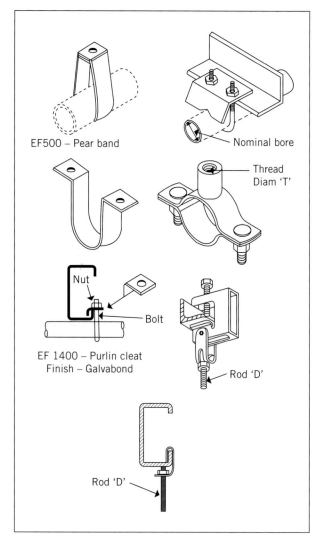

FIGURE 7.19 Examples of brackets and clamps

Thrust blocks and anchors

Below-ground water/fire services using an elastomeric joining ring (rubber ring joint), as used on PVC-U pipe, must have concrete thrust blocks installed to prevent any movement (see Figure 7.20). Thrust blocks must bear against a firm side of the trench excavation. They are installed at:

- bends and tees
- end caps
- valves
- reducers
- grades in excess of 1:5.

AS/NZS 3500.1 WATER SERVICES

Installation of piping and materials

Plans and specifications will determine what piping material is to be used. This also depends on whether the

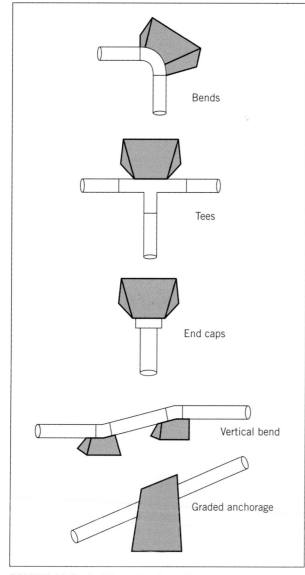

FIGURE 7.20 Typical thrust block positions

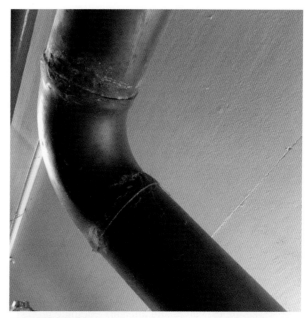

FIGURE 7.21 Silver-soldered joint

Depending on the work site, a 'hot works permit' may need to be completed and signed off before any silver soldering is carried out.

The most common method used to join copper tube today is press-fit or crimp fittings (see Figure 7.22). A special tool crimps the fitting onto the pipe. Although the fittings are quite expensive, much time is saved on the installation and no heat needs to be applied to the pipe or fittings, so certain hazards can be eliminated.

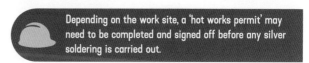

Source: Kembla.

FIGURE 7.22 Press-fit joint

FROM EXPERIENCE

Where copper and steel may come into contact, insulation must be installed to prevent electrolysis (corrosive reaction) from occurring.

PVC-U pipe
Thick-walled PVC-U pipe is commonly used below ground for fire hydrant services for pipe diameters of 100mm or larger. It is also commonly used for water

pipe is installed above or below ground and the size of the fire service.

There is a range of methods used to join pipes, fittings and valves. This depends on the type of piping material used, the pipe diameter and the location.

Copper tube
The installation of copper tube can be above or below ground. It must be either type A or B grade for the appropriate wall thickness. A common method of jointing copper tube and fittings is by press-fit or silver soldering. Manipulated joints such as an expanded joint or branch formed joints are silver soldered (see Figure 7.21). Copper and brass fittings can be either press-fit or silver soldered. The silver content in the rod must be a minimum of 1.8% as per AS/NZS 3500.1. Valve assemblies larger than 50mm are usually joined with flanges.

Soft solder cannot be used on fire services.

mains (see Figure 7.23). The material is cost effective in purchase price and installation time. An elastomeric rubber ring jointing method is used for this material (see Chapter 1 for the jointing method). This is a similar method used in joining cement-lined cast iron and cement-lined ductile iron pipes. As mentioned earlier, concrete thrust blocks must be installed to prevent any movement from the rubber ring joints and must comply with AS/NZS 3500.1.

FIGURE 7.23 Elastomeric ring joint

AS/NZS 3500.1 WATER SERVICES

Polyethylene pipe

Polyethylene pipe is used underground for fire hydrant and fire hose reel services. This material is joined most commonly using the electro-fusion jointing method (see Figure 7.24). It can also be joined using the heated tool

Source: Shutterstock.com/serato.

FIGURE 7.24 Electro-fusion jointing method

socket method or the butt welding method. This is explained in more detail in Chapter 1. AS 2419 explains that any polymer (plastic) pipe must have detectable (metal wire) laid in the trench so it can be located.

FROM EXPERIENCE

A common problem with electro-fusion jointing is the joint leaks because pipe is not shaved and cleaned with an alcohol wipe prior to electro-fusing the joint.

Steel pipe – above ground

Usually galvanised steel pipe is used for fire hydrant and fire hose reel systems downstream of the main isolating valve. The galvanised steel pipework is commonly joined using the roll grooved method for larger services and screwed for smaller diameter services. Medium black steel grade pipe is used for fire sprinkler services. Black steel pipe is usually prefabricated off-site with welded joints and screwed together on-site. Flanged or roll grooved jointing is used on larger services. The roll grooved jointing technique (see Figure 7.25) is set out in Chapter 1.

FIGURE 7.25 Roll grooved joint

Steel pipe – below ground

For fire hydrant services, only 1.5m of galvanised steel pipe is allowed below ground. It must be double-wrapped in a petroleum tape or sleeved with a polyethylene coating. It can only be used downstream of the main isolating valve. Below-ground steel pipe is either flanged (see Figure 7.26) or rubber ring jointed.

Screwed joints are made on smaller-diameter steel pipe usually up to 50mm in diameter.

FIGURE 7.26 Flanged joint

HOW TO

HOW TO MAKE A SCREWED JOINT

1 Thread the end of the pipe with a threading machine or a manual stock and die.
2 Wrap the thread with hemp and soap. Alternatively, thread sealant or teflon tape can be used.
3 Screw the pipe firmly into the fitting, being careful not to cross the thread.
4 Tighten with a Stillson wrench.

Water supply

Water supply for fire services can come from several sources:

- town main water
- recycled water – Class A
- rainwater
- sea, river, lake or dam water.

Bore water (treated and untreated)

The water supply, regardless of the source, must be capable of supplying the fire hydrant system with the minimum flow rate and minimum pressure for a minimum of four hours.

A backflow prevention device must be fitted to any fire hydrant service connected to a town water supply to protect the community's drinking water (see Figure 7.27). Some water authorities will require the installation of a double-check detector assembly to monitor low water usages; for example, a fire hose reel being used to wash plant or equipment.

FIGURE 7.27 Double-check detector assembly

GREEN TIP

Plumbers have an important responsibility in protecting our community's drinking water by installing backflow preventers.

Fire hydrant services connect to a water main by a tee and valve. They are a dedicated service for firefighting, and therefore not metered (see Figure 7.28).

The minimum flow rate for a feed fire hydrant is 10L per second with a residual pressure of 150kPa in NSW and 200kPa in all other states and territories, as per AS 2419.1.

FIGURE 7.28 Fire hydrant service connected to water main with a tee and valve

The minimum flow rate for an attack hydrant is 10L per second with a residual pressure of 250kPa in NSW and 350kPa in all other states and territories, as per AS 2419.1.

The minimum required pressure may need to be achieved using a pump if necessary.

Water that is stored for firefighting purposes in tanks must have a specific volume solely dedicated to firefighting. The water storage tanks must conform to AS/NZS 3500.1.

Where water is to be drawn from a source below a pump, such as a dam or a river, the maximum vertical lift shall be 3m.

The water must not be corrosive to the system. If sea water is used, then the fire service must be flushed with fresh water on completion of use.

Fire hose reel services usually connect to the metered service (see Figure 7.29). The system connects to the domestic supply and shall have isolating valves independent of the hose reel system so the supply is not interrupted.

FIGURE 7.29 Fire hose reel service connected to water main with a main tapping

The minimum pressure required for a fire hose reel is between 200kPa and 230kPa with flow rate of 0.33L per second for a 19mm hose.

Testing and commissioning

Testing and commissioning the system is the final process of the job prior to clean-up and the final handover. It is a vital process ensuring the system is sound and working to the correct pressures and flow rates.

Testing

Once installed the fire hydrant and fire hose reel service must be pressure tested. AS2419 states the fire service must be hydraulically (water) tested to a pressure of 1700kPa or 1.5 times the design pressure, whichever is the greater, for a minimum of two hours.

It is good practice to carry out a preliminary air test first to check for any major leaks such as open ends or open valves. Doing this could eliminate possible flooding. This test is not compulsory but highly recommended.

HOW TO

HOW TO CARRY OUT A HYDRAULIC TEST

1 Do a visual check to ensure the installation is complete.
2 If testing sections, ensure the section is isolated.
3 Connect the pump to the test point or use the permanently installed pump with a gauge and isolating valve.
4 Open the control valve and fill the system with water.
5 Bleed air from the system from the furthest outlet.
6 Start the pump and increase pressure at 200kPa intervals. Hold for 30 minutes at each interval for the system to adjust to the pressure and to allow screwed hemp joints to swell.
7 When pressure is at 1700kPa, or as specified, hold for two hours.
8 If there are any leaks, rectify the problem and re-test.
9 The owner or representative is to witness the start and finish time of the test and sign off on its successful completion.
10 Record test data on company records (quality assurance procedure).

HOW TO

HOW TO CARRY OUT AN AIR TEST

1 Do a visual check to ensure the installation is complete.
2 If testing sections, ensure the section is isolated.
3 Connect the air compressor to the test point (remove the hose cock on the booster assembly and connect the pressure gauge and isolating valve for the connection point).
4 Raise the air pressure to 300kPa.
5 Hold for 30 minutes. Air pressure should not drop more than 20–30kPa over this time.
6 If this test passes, proceed to the hydraulic test.
7 If the air test drops, locate and repair the leak. Then repeat the air test.
8 Record test data on company records (quality assurance procedure).

Commissioning

The fire hydrant and fire hose reel system must be complete, tested and inspected by the hydraulic designer, who will ensure the system meets all the design requirements and issue an inspection report.

A copy of this report would be kept by the client and the contractor.

When commissioning the fire hydrant service, all hydrants must be opened to release any trapped air, and flow rates and pressure must be checked at each hydrant. If pumps are installed, they need to be run to ensure they are working properly.

When commissioning the fire hose reel, check the following points:

- Be sure there is no leakage from the hose reel when charged and pressurised with the nozzle closed.
- Check that the hose unwinds easily from the reel in its intended direction.
- Check the required flow rate and pressure.
- Make sure the hose is rewound evenly onto the reel with the hose under pressure.
- Engage the nozzle to the interlock and close the isolating valve.
- Open the nozzle to depressurise the hose and then close the nozzle.

LEARNING TASK 7.3

1 What type of pipework must have thrust blocks installed?
2 What is the most common method of joining large-diameter galvanised steel pipe?
3 How is the town's water supply protected from contaminated water?

7.5 Cleaning up and packing up

An essential part of the job is to clean up and leave the site tidy. An important rule is to clean as you go! Clean-up should be carried out at the end of each day. There are several reasons for this. The main reason, as always, is for health and safety, as a clean workplace reduces the chance of accidents occurring, such as via trip hazards and falling objects. Another reason is that the build-up of waste and debris on a job site creates a problem for storage and access.

Offcuts and leftover material should be appropriately stored and used on another job. This not only prevents waste but also enables savings in cost and increases in profit.

At the end of the job there is usually a large amount of cardboard and paper packaging. This should be placed in the recycling waste so it can be reused and reduce landfill.

A thorough sweeping of the work area helps remove dirt, dust and rubbish. A clean and tidy work area helps the job run more smoothly, again increasing profits.

All tools and equipment must be packed up and stored safely every day. This will help to reduce the loss or misplacement of tools. All tools and equipment must be well maintained by regularly cleaning and lubricating them. Blades and cutting tools must be kept sharp or replaced if necessary to be effective in use. Any plant and equipment on hire needs to be either taken off hire or returned to the hire company as soon as possible to reduce company costs.

Company policies must be followed regarding any final documentation. This would include the final design approval certificate, certificate of compliance and any as-built drawings. Also, any written instructions for the operation and maintenance of equipment and fixtures installed must be handed over to the client, along with any warranties.

LEARNING TASK 7.4

1 Why is it important to clean as you go?
2 What documentation should be issued on job completion?

 COMPLETE WORKSHEET 3

SUMMARY

- Effective planning and preparing for fire hydrant and fire hose reel installations is important for a cost- and time-effective job.
- There are several types of fire hydrant and fire hose reel systems to be aware of.
- Several types of piping material are allowed to be used, depending whether the fire service is below or above ground and what is stated on the written specifications.
- Be sure to place accurate orders for material to avoid inconvenience and time wasting, and check the order is correct upon delivery.
- When setting out the job it is important to check plans for dimensions and levels. For underground pipework, check excavation depth and plan trench support if required.
- Install pipework in the most direct route possible with minimal bends.
- There is a range of methods used to join pipes, fittings and valves. This depends on the type of piping material used, the pipe diameter and the location.
- Be sure to use adequate fixings and brackets to support pipework. Wooden or plastic inserts are not allowed. Nails and explosive fixings cannot be used.
- There are several water sources available for firefighting.
- The minimum testing requirements must be adhered to as per AS2419.
- Clean up regularly and try to reduce waste.

GET IT RIGHT

1 Which photo shows the correct installation?

2 Which valve is missing?

3 What is the purpose of this valve?

WORKSHEET 1

To be completed by teachers

Satisfactory ☐

Not satisfactory ☐

Student name: _____

Enrolment year: _____

Class code: _____

Competency name/Number: _____

Task: Review section on '7.2 Planning and preparing work activities' and answer the following questions.

1 What are two important considerations for the efficient operation of a fire hydrant and fire hose reel service?

_____

_____

2 What are the three main points in a risk assessment?

_____

_____

_____

3 Name four appropriate PPE items for installing fire hydrant and fire hose reel systems.

_____

_____

_____

_____

4 Name four requirements included in a company quality assurance policy.

_____

_____

_____

_____

5 How often do electrical tools and leads need to be tested and tagged on construction sites?

_____

WORKSHEET 2

Student name: _____

Enrolment year: _____

Class code: _____

Competency name/Number: _____

To be completed by teachers

Satisfactory ☐

Not satisfactory ☐

Task: Review section '7.3 Identifying installation requirements' and answer the following questions.

1   What does a 'wet' system mean?

_____

2   Describe a feed hydrant.

_____

3   What is an attack hydrant?

_____

_____

4   Describe a booster inlet.

_____

_____

5   Describe a booster assembly.

_____

_____

6   What is a hardstand area?

_____

_____

7   Why is the isolating valve locked in the open position?

_____

8   Name four items of information that can be found on a block plan.

_____

_____

_____

_____

9   What is the minimum size pipe supplying a single hydrant?

_____

10 What is the minimum required flow rate and pressure for a fire hydrant?

_____

11 At what height should a fire hydrant be installed?

_____

12 At what angle is an attack hydrant outlet installed?

_____

13 Why is the hydrant angle important?

_____

14 What is the minimum size pipe supplying a fire hose reel?

_____

15 What is the minimum required flow rate and pressure for a fire hose reel?

_____

16 At what height should a fire hose reel be installed?

_____

17 At what height should a fire hose reel isolation valve be installed?

_____

18 Name three types of piping material used on fire hydrant services.

_____

_____

_____

WORKSHEET 3

Student name: _____

Enrolment year: _____

Class code: _____

Competency name/Number: _____

To be completed by teachers

Satisfactory ☐

Not satisfactory ☐

Task: Review sections '7.4 Fabricating, installing and testing the system' and '7.5 Cleaning up and packing up' and answer the following questions.

1   Name three things to consider for pipe supports.

_____

_____

_____

2   Can plastic or wooden inserts be used for fixings for fire hydrant services?

_____

3   What is the purpose of thrust blocks?

_____

_____

_____

4   Which type of piping systems require thrust blocks?

_____

5   Name two jointing methods for copper tube.

_____

_____

6   What is the most common jointing method for polyethylene pipe?

_____

7   What is the most common jointing method for large-diameter galvanised pipe?

_____

8   Is the following statement true or false? 'A maximum of 2m of galvanised steel pipe can be below ground for a fire hydrant service.' Circle the correct answer.

T   F

9   Name three water sources for fire hydrant and fire hose reel services.

_____

_____

_____

10 Is the following statement true or false? 'The water supply, regardless of the source, must be capable of supplying the fire hydrant system with the minimum flow rate and minimum pressure for a minimum of three hours.' Circle the correct answer.

T   F

11 What is the maximum vertical lift that a pump can draw water from a water source below the pump, as per AS 2419.1?

_____

12 What type of backflow prevention device is used for containment on a fire hydrant service?

_____

13 What does AS 2419.1 state as the minimum test pressure and duration requirements for a fire hydrant service?

_____

14 Why is it good practice to carry out an air test before pressurising the system with water?

_____

_____

15 Why is It important to open all fire hydrants when commissioning the system?

_____

_____

16 Name four procedures that are part of the commissioning process for fire hose reels.

_____

_____

_____

_____

17 Name three documents to be handed over on completion of a job.

_____

_____

_____

18 Why is it important to reduce waste?

_____

_____

WORKSHEET RECORDING TOOL

Learner name		Phone no.	
Assessor name		Phone no.	
Assessment site			
Assessment date/s		Time/s	
Unit code & title			
Assessment type			

Outcomes

Worksheet no. to be completed by the learner	Method of assessment WQ – Written questions PW – Practical/workplace tasks TP – Third-party reports SC – Scenarios RP – Role plays CS – Case studies RW – Report writing PF – Portfolio	Satisfactory response	
		Yes ✓	No ✗
Worksheet 1	WQ	☐	☐
Worksheet 2	WQ	☐	☐
Worksheet 3	WQ	☐	☐

Assessor feedback to the learner

Feedback method (Tick one ✓):	☐ Verbal ☐ Written (if so, attach) ☐ LMS (electronic)

Indicate reasonable adjustment/assessor intervention/inclusive practice (if there is not enough space, a separate document must be attached and signed by the assessor).

Outcome (Tick one ✓):	☐ Competent (C) ☐ Not Yet Competent (NYC)

Assessor declaration: I declare that I have conducted a fair, valid, reliable and flexible assessment with this learner, and I have provided appropriate feedback.

Assessor name:	
Assessor signature:	
Date:	

Learner feedback to the assessor

Feedback method (Tick one ✓):	☐ Verbal	☐ Written (if so, attach)	☐ LMS (electronic)

Learner may choose to provide information to the RTO separately.

	Tick one:	
Learner assessment acknowledgement:	**Yes** ✓	**No** ✗
The assessment instructions were clearly explained to me.	☐	☐
The assessment process was fair and equitable.	☐	☐
The outcomes of assessment have been discussed with me.	☐	☐
The overall judgement about my competency performance was fair and accurate.	☐	☐
I was given adequate feedback about my performance after the assessment.	☐	☐

Learner declaration:
I hereby certify that this assessment is my own work, based on my personal study and/or research. I have acknowledged all material and sources used in the presentation of this assessment, whether they are books, articles, reports, internet searches or any other document or personal communication.
I also certify that the assessment has not previously been submitted for assessment in any other subject or at any other time in the same subject and that I have not copied in part or whole or otherwise plagiarised the work of other students and/or other persons.

Learner name:	
Learner signature:	
Date:	

INSTALL WATER PUMP SETS

8

Chapter overview

This chapter addresses the general and basic installation of pumps either as part of the supply of water to a property or building or for other areas that are part of a water supply system.

The information and principles mentioned in this chapter, although directly relating to water pump sets, also apply to other areas of plumbing such as waste water removal. Only a basic knowledge of some pump categories and types are provided here, and some of the common technical issues related to them.

Learning objectives

Areas addressed in this chapter include:

- preparing for work and work health and safety (WHS)
- pump requirements
- types of pumps
- information required to select a pump
- installing and testing pump sets
- cleaning up.

8.1 Background

A water pump can be simply defined as a mechanical device that is generally driven by a motor and is used for raising water from a lower to a higher level, for pressurising a pipework system or for circulating water in a pipework system. Pumps are generally referred to according to their principle of action, manner of construction or method of operation. In addition, nearly every pump type is defined by a complementary application (such as circulating or submersible pumps) or by motive type (such as centrifugal or piston pumps) that can be used in water service applications. In this chapter we deal with the two major groups of pumps: rotodynamic pumps and positive displacement pumps. Their names describe the method of moving a fluid (in our case, water).

- Rotodynamic pumps are based on bladed impellers that rotate within the fluid to accelerate the fluid and cause a consequent increase in the energy of the fluid. Figure 8.1 shows the types of rotodynamic pumps in use today.
- Positive displacement pumps cause the fluid to move by trapping a fixed volume of fluid then forcing (displacing) that trapped volume into the discharge pipe. Figure 8.2 shows the types of positive displacement pumps in use today. These are further discussed in the section '8.3 Types of pumps' later in this chapter.

A pump can be used for many reasons on a water supply system. The three main purposes of a pump are to:

- increase pressure in the pipeline that is attached to its outlet

- move water from a lower elevation to a higher elevation
- help circulate water in a continuous loop around a water supply system.

For large industrial, commercial or residential systems, the design engineer normally plans and specifies the size and type of pump system, whereas on smaller systems this may be left for the plumber to do. The plumber's task is to order and install a pump that is appropriate for the job it has to do, and to connect the pipework and ancillary equipment to and from the pump assembly.

While plumbers are generally involved in the installation of pumps, pump maintenance and repairs are normally left to those who specialise in these areas. However, as the tradesperson most concerned with water supply, plumbers should have a good basic understanding of pumps – their function, how they operate, and when and why they are used in water supply.

FROM EXPERIENCE

Having the knowledge to choose the correct pump for a particular job is an important plumbing skill.

Situations in which pumps are required

Here are some contexts where water pumps are generally required as part of the water supply.

Source: © Pump Industry Australia Incorporated, *Australian Pump Technical Handbook 5e*, http://www.pumps.asn.au.

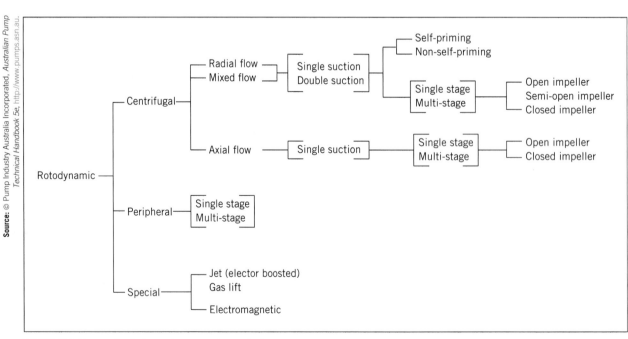

FIGURE 8.1 Types of rotodynamic pumps in use today

Source: © Pump Industry Australia Incorporated, *Australian Pump Technical Handbook 5e*, http://www.pumps.asn.au.

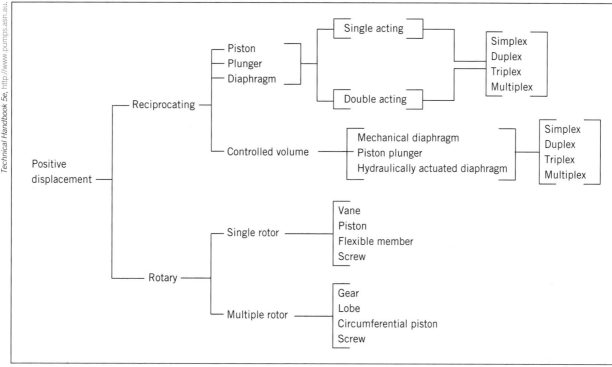

FIGURE 8.2 Types of positive displacement pumps

Cities and towns

- In industrial areas, for:
  - circulating water through machinery for cooling purposes
  - circulating hot and cold water
  - food processing
  - air-conditioning and other mechanical services
  - irrigation.
- In commercial buildings, for:
  - raising water to storage tanks
  - circulating hot and cold water
  - air-conditioning
  - other mechanical services.
- In residential buildings, for:
  - raising water to storage tanks
  - circulating hot and cold water
  - air-conditioning
  - pressurising water from rainwater tanks.
- In schools, churches and various other community establishments, for:
  - air-conditioning
  - circulating hot and cold water
  - irrigation.
- In hospitals, for:
  - raising water to storage tanks
  - circulating hot and cold water
  - air-conditioning
  - other mechanical services.
- For fire services:
  - hydrants
  - hose reels
  - sprinklers for any or all of the above building types.

- In other areas, for:
  - irrigation (both domestic and urban)
  - moving water into and out of household appliances (washing machines and dishwashers)
  - circulating water in domestic spas
  - water services – plumbers use a test pump to raise the pressure within a water service when testing the service for soundness.

Rural areas

- Pumping to dwellings and other buildings from:
  - dams
  - storage tanks/rainwater tanks
  - wells
  - streams.
- For irrigation, by pumping from:
  - rivers
  - creeks
  - dams
  - bores
  - other sources.
- For supplying water for stock:
  - by pumping from rivers, creeks, dams, bores and other sources
  - via windmills and other devices.

LEARNING TASK 8.1

1 What is an impeller?
2 Name a pump that is of the positive displacement type.

COMPLETE WORKSHEET 1

8.2 Pump requirements and basic pumping terminology

The plumber has to:

- know how to read plans and specifications in order to determine the type and location of the pump assembly
- liaise with the client to ensure that the pump installation meets their needs
- liaise with the pump manufacturer and/or the supplier to ensure that the correct pump unit is selected for the process
- liaise with the local water utility and complete the necessary documentation during the installation process
- safely install the pump assembly, associated pipework and ancillary items in accordance with the relevant codes, the manufacturer's instructions and the job specification
- liaise with other trades to ensure not only that the pump installation is completed satisfactorily, but also that the pump and ancillary equipment are maintained in the future.

It should be noted here that in most large pump installations there is normally more than one pump installed. This is to allow for such things as:

- helping to prevent an individual pump from overworking (that is, to share the load on the pump assembly)
- allowing the system to continue operating while maintenance takes place on part of the pump assembly.

In some areas there may be alternative power sources for the pump assembly. For example, fire services have both diesel and electric power, in case there is a fire and the electricity supply is interrupted or shut down. In such a case, the pump is still able to provide water to help control the fire.

In cities and towns, the local water supply utility normally has an obligation to provide water at a minimum of 150kPa (approximately 15m of head) at any part of its water supply. Obviously, the pressure within the main varies depending on different sources of water supply and different types of land features within towns and cities, and is normally much higher than the guaranteed minimum.

GREEN TIP

Water utilities normally maintain a conservative approach to their water supply. This is mainly to allow for future growth in particular areas, even if the current available water pressure might appear to be more than adequate.

For this reason, when supplying water to properties that require large volumes of water (such as fire services, some industrial areas or where there are multistorey buildings), the water utility may insist that the property owner takes responsibility by using either pumps or a combination of both pumps and a storage tank, particularly if the main supply has insufficient volume or pressure to feed all parts of the property during peak periods.

Current Australian Standards also mention that the most disadvantaged point of a drinking water supply system should have a *minimum head no less than 50kPa* (approximately 5m of head) and that fixtures connected to the supply require minimum flow rates. Therefore, if the pressure or volume of water required is insufficient, it can be seen that there is a need for pumps to be installed and it is the plumber's job to install them according to current rules and regulations.

 AS/NZS 3500.1 PLUMBING AND DRAINAGE: WATER SERVICES

Head (pressure)

The International System of Units (SI) units of measurement that are used to express head (pressure) throughout this unit include:

- bar
- kilopascals (kPa)
- millimetres (mm head)
- megapascals (MPa)
- metres (m head)
- metres head of water (H)
- pascals (Pa).

To choose the proper pump for a particular system the plumber has to know what size pump is required, or how many pumps are required. The plumber also has to be aware of how much water needs to be moved, how far and how fast it needs to be moved, and how much resistance may be encountered while it is being moved.

Pump manufacturers sum up all of these factors into two basic measurements and rate their pumps accordingly. The measurements are flow rate and delivery head.

Flow rate relates to the amount of water that can be moved in a given period of time and is generally measured in the number of litres per second (L/s) or cubic metres per hour (m³/hr).

Delivery head is the vertical height in metres that the pump has to deliver water to from the pump outlet, which would be the highest outlet from the pump. It can be further classified depending on whether the resistance is encountered on the suction side of a pump (*suction head*) or the delivery or discharge side (*discharge head*); whether it is caused by the standing (*or stationary*) weight of the water (*static head*) or by the movement of water through the system (*dynamic head*); or whether the resistance is brought about by friction as the water passes through the pipes, fittings and valves installed on the system (*friction head*).

The plumber needs to be aware of the discharge head against which the pump has to operate and also whether there is a positive suction head or a negative suction head available, as this is part of the information that has to be provided to the pump supplier when selecting the appropriate pump for a specific application. A positive suction head is where the available water source for the pump is above the pump inlet; see **Figure 8.3** (a). A negative suction head (or suction lift) is where the pump has to draw the water from a source lower than the pump inlet; see **Figure 8.3** (b). Note that the pump does not 'suck' but creates a partial vacuum, allowing atmospheric pressure to push water up into the inlet pipe.

When discussing pumps and pumping, the term 'head' is used for the pressure generated by the pump or the pressure the pump needs to overcome. The way we calculate pressure, therefore, is to measure the vertical height (head) of a column of water and for each metre multiply by 9.81 (a 1m vertical column of water exerts a pressure of 9.81kPa at its base).

The common formula that is used to calculate the pressure in kPa is:

$$P = H \times 9.81$$

where

P = pressure (kPa)
H = head (height) in metres
9.81 = conversion factor

One simple way to do this is by approximation (in other words, saying that 9.81 is approximately equal to 10). For example, if a building is 30m high, then at the bottom of the building we can assume that the pressure is approximately 300kPa; for 40m, approximately 400kPa; for 50m, approximately 500kPa and so on.

EXAMPLE 8.1

CALCULATE THE PRESSURE FOR A 30M-HIGH BUILDING

Formula: $P = H \times 9.81$

where: P = pressure (kPa) (to be determined)
H = head (height) in metres (30)
9.81 = conversion factor (9.81)
Workings: $P = 30 \times 9.81$
Answer: P = 294.3kPa (which is quite close to 300kPa, as approximated above).

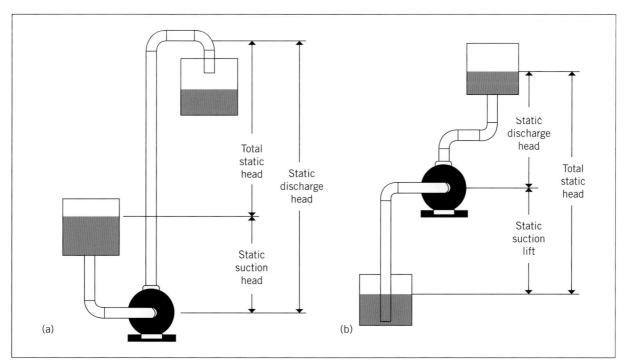

FIGURE 8.3 (a) Positive suction head; (b) Negative suction head/lift

Examples of technical data related to pump installations can be seen in **Figure 8.4**.

The plumber also has to be aware of suction lift limits when preparing for and installing a pump. In theory, atmospheric pressure (approximately 100kPa or 1 bar) can support a water column of some 10m. The suction lift of a pump is limited by the height to which the atmosphere will force water against the partial vacuum formed by the action of the pump.

A diagrammatic sketch indicating various terms related to suction lift and suction head can be seen in **Figure 8.5**.

For most pumps it is preferable to have a positive suction head rather than a negative suction head or suction lift. Simply stated, it is better, where possible, to position the pump below the water source. However, in a situation where there is a negative suction head, it is advisable to ensure that the water supply level on the suction side of the pump is kept to a minimum

Source: All rights reserved. Grundfos Pumps Pty Ltd.

Temperature conversions

Degrees Centigrade to degrees Fahrenheit $= \dfrac{°C \times 9}{5} + 32$

Degrees Fahrenheit to degrees Centigrade $= (°F - 32) \times \dfrac{5}{9}$

Kilowatts to horsepower
kW to HP = kW $\times$ 1.341
HP to kW = HP $\times$ 0.746

Circle formulas
Area = $\pi\, r^2$
Circumference = $2\,\pi r$
Volume of cylinder = $\pi\, D^2 H \times .25$

Volume of sphere = $0.166 \times \pi\, D^3$ where:
$\pi = 3.14$
r = radius
D = diameter
H = height

Velocity of flow past
SP motor

$V = \dfrac{Q \times 353}{Db^2 - Dm^2}$

Where:
V = velocity metres/sec   Db = borehole diameter (mm)
Q = flow m³/hr   Dm = motor diameter (mm)

Torque
Ft lbs to Nm = Ft lbs $\times$ 1.355
Nm to Ft lbs = Nm $\times$ 0.73757

Weight
Pounds to kilograms = lbs $\div$ 2.2
Kilograms to pounds = kg $\times$ 2.2
Gallons to pounds = gallons $\times$ 10
Litres to kilograms = litres $\times$ 1

Affinity laws
(how changes in pump speed affect pump flow, head and input power)
$Q = Q_1 \times N/N_1$
$H = H_1 \times (N/N_1)2$
$P = P_1 \times (N/N_1)3$

where:
N = new speed
N_1 = old speed
Q = flow @ N
Q_1 = flow @ N_1

H = head @ N
H_1 = head @ N_1
P = input power @ N
P_1 = input power @ N_1

NOZZLE FLOW RATES (litres/sec)

Pressure		Diameter of nozzle (inches*) mm								
		5/64"	5/32"	1/4"	5/16"	25/64"	15/32"	5/8"	25/32"	1"
kPa	psi	2	4	6	8	10	12	16	20	25
200	29	0.06	0.25	0.56	1.01	1.57	2.26	4.02	6.29	9.82
300	43	0.08	0.31	0.69	1.23	1.93	2.77	4.93	7.70	12.03
400	58	0.09	0.36	0.80	1.42	2.22	3.20	5.69	8.89	13.89
500	73	0.10	0.40	0.89	1.59	2.48	3.58	6.36	9.94	15.53
600	87	0.11	0.43	0.98	1.74	2.72	3.92	6.97	10.89	17.02
700	102	0.12	0.47	1.06	1.88	2.94	4.23	7.53	11.76	18.38
800	116	0.12	0.50	1.13	2.01	3.14	4.53	8.05	12.58	19.65
900	131	0.13	0.53	1.20	2.13	3.33	4.80	8.54	13.34	20.84
1000	145	0.14	0.56	1.26	2.25	3.52	5.06	9.00	14.06	21.97

*Approximate imperial equivalent

UNITS OF FLOW RATE

Convert to / Convert from	Imperial gallons per min	US gallons per min	Cubic metres per hour	Litres per second	Litres per minute
	Multiply				
Imperial gallons per min	1	1.2	.273	.076	4.546
US gallons per min	.833	1	.227	.063	3.787
Cubic metres per hour	3.666	4.4	1	.278	16.66
Litres per second	13.19	15.85	3.6	1	60
Litres per minute	.219	.264	.06	.016	1

UNITS OF VOLUME

Convert to / Convert from		Litre	Kilolitre	Cubic metres	Imperial gallon	US gallon
		Multiply by				
Litre	lt	1	0.001	0.001	.220	.264
Kilolitre	klt	1000	1	1	220	264
Cubic metre	m³	1000	1	1	220	264
Imperial gallon	gal	4.546	0.00454	0.00454	1	1.201
US gallon (US)	gal	3.785	0.0038	0.0038	0.833	1

UNITS OF PRESSURE

Convert to / Convert from		Kilopascal	Metrehead	Bar	lbs per sq. inch	Feet of water	Atmosphere
		Multiply by					
Kilopascal	kPa	1	0.102	.01	0.145	0.335	–
Metrehead	m	9.804	1	0.098	1.42	3.28	0.098
Bar	Bar	100	10.20	1	14.5	33.45	1
lbs/sq. inch	psi	6.895	0.704	0.069	1	2.307	0.069
Feet of water	ft	2.98	0.3048	0.03	0.4335	1	0.03
Atmosphere	At	100	10.33	1.0	14.7	33.9	1

UNITS OF LENGTH

Convert to / Convert from		Millimetre	Centimetre	Metre	Kilometre	Inch	Foot	Mile
		Multiply by						
Millimetre	mm	1	0.1	0.001	–	0.0394	0.0033	–
Centimetre	cm	10	1	0.01	–	0.394	0.0328	–
Metre	m	1000	100	1	0.001	39.37	3.281	.000621
Kilometre	km	–	–	1000	1	–	3281	0.621
Inch	in	25.4	2.54	0.0254	–	1	0.083	–
Feet	ft	304.8	30.48	0.305	–	12	1	–
Mile	mile	–	–	1610	1.61	–	5280	1

Note: The information provided on these pages is for guidance only. Grundfos Pumps Pty Ltd accepts no responsibility for the misuse or misapplication of this information.

FIGURE 8.4 Typical sheet showing relevant engineering data relating to pumps

Source: © Pump Industry Australia Incorporated, *Australian Pump Technical Handbook 5e*, http://www.pumps.asn.au.

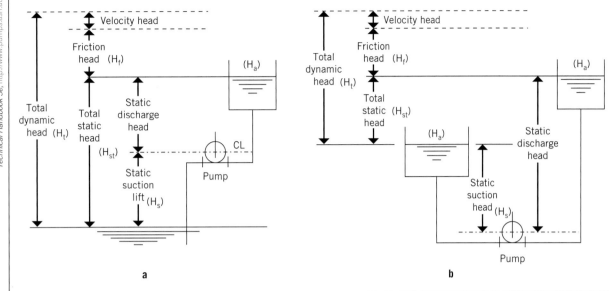

FIGURE 8.5 (a) Suction lift; (b) Suction head

(normally no more than 6m below the centre line of the pump, and in extreme cases no more than about 8m). This is because of friction losses caused by the pipe, fittings, valves and so on, and can vary depending on site conditions and the manufacturer's recommendations. (In all cases it is recommended to refer to the pump manufacturer's installation instructions before connecting the pump set so as not to void any warranty. If there is any doubt at all, contact the pump manufacturer.)

LEARNING TASK 8.2

1 What are the two main factors required when choosing a pump?
2 What force pushes water into the pump inlet?
3 How is flow rate measured?

COMPLETE WORKSHEET 2

8.3 Types of pumps

Figure 8.6 shows examples of domestic pumps. The plumber's role is to install not only the pump but also the associated pipework, ancillary valves and equipment required for the pump assembly.

It is not essential for the average plumber to be able to identify all of the different types of pumps available. This is the role of the pump manufacturer or supplier, who can provide advice as long as the plumber can provide the relevant information relating to the pump installation. This chapter therefore addresses the basic principles related to the more common types of water pumps and does not discuss the specifics about all of the pumps. If you wish to find out more about pumps, the internet is a good place to start. See the references at the end of the chapter.

Rotodynamic pumps

A rotodynamic pump is a pump that uses the rotation of an impeller or propeller to create a pressure in the liquid it is moving. Pumps that use rotation to move a liquid are commonly referred to as centrifugal pumps.

Centrifugal pumps

Owing to its flexibility, the centrifugal pump is the most common and widely used pump in water supply systems, particularly where automatic control is necessary or where hot water is to be pumped (see Figure 8.7).

Construction of a centrifugal pump

Larger pumps generally have a cast iron or cast metal casing. Domestic pumps are generally plastic. Centrifugal pumps have a volute, or spiral fin, that guides the water to the outlet of the casing. Inside the volute is an impeller with vanes or blades. The impeller, coupled with the pump driving mechanism, creates the energy to pump the water.

Vanes have a backward curve to the direction of rotation to minimise frictional resistance offered to the water as it passes from the centre (eye) of the impeller along the blade to its tip by centrifugal force.

Source: All rights reserved. Grundfos Pumps Pty Ltd.

Self-priming
Built for long trouble-free life, the self-priming pumps with excellent suction capacity are suitable for a wide variety of water supply and transfer duties in home, garden and hobby applications as well as in agriculture and horticulture—wherever a reliable household water supply is needed.

Pressure systems
This range of Grundfos pumps are designed for any pumping application involving clean and non-aggressive water in household applications and small scale irrigation as well as in booster applications.

Submersible
Grundfos submersible pumps are easy to handle and suitable for operation for a variety of applications such as domestic water supply, irrigation and pressure boosting.

Wastewater and drainage
Grundfos wastewater and drainage pumps are designed to make pumping from household applications as simple and efficient as possible.

Hot water and heating
The Grundfos range of circulators for heating and hot water re-circulation are available in a variety of materials and finishes to cover even the most exceptional tasks.

FIGURE 8.6 Examples of domestic pumps and their applications

How does a centrifugal pump operate?

This type of pump uses the principle of centrifugal force for its operation. The pump normally has a motor, attached to the pump, that drives the impeller to revolve at high speed. Water flows from the inlet pipe into the eye or centre of the impeller either by suction or by gravity. Water then passes through the eye of the impeller and is flung through to the tip of the vanes (blades), and thus into the volute casing, by centrifugal force (this can be likened to a bucket of water being swung around 360°; no water is spilt because centrifugal force pushes the water to the bottom of the bucket). When it leaves the impeller blades, the water collects in the casing and is forced through the outlet, towards the point of discharge, at a constant pressure.

Pump connections

Centrifugal pump sets are normally driven by a motor that is either close coupled or long coupled. As can be seen in **Figures 8.8** and **8.9**, a close coupled connection is where the pump is directly connected or bolted to the motor, and a long coupled connection is where the pump and motor are a distance apart, each having its own shaft that is connected by a coupling. The advantage of a long coupled motor is that it is easier to service and maintain and therefore more common on larger installations, whereas the close coupled motor is much more compact and is most commonly used in domestic applications. In some cases centrifugal pump sets also can be belt-driven.

Source: *Basic Training Manual 11–3 Water supply 3*, Australian Government Publishing Service, from July 1970 to 1997.

OUT

Impeller

Housing

Drive shaft

IN

Cross-section through pump and drive

Impeller rotation

Water acceleration on vane blade surface

Vane blade

Cross-section through impeller

FIGURE 8.7 Cross-section of a centrifugal pump set

Source: © Pump Industry Australia Incorporated, *Australian Pump Technical Handbook 5e*, http://www.pumps.asn.au.

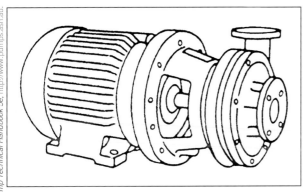

FIGURE 8.8 Close coupled motor connection of a centrifugal pump set

Source: © Pump Industry Australia Incorporated, *Australian Pump Technical Handbook 5e*, http://www.pumps.asn.au.

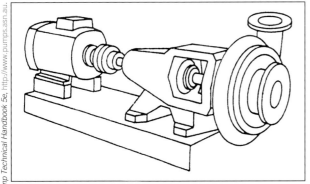

FIGURE 8.9 Long coupled motor connection of a centrifugal pump set

Pump efficiency

Because of its design and principle of operation, the centrifugal pump gives its best performance when used at the rate of discharge and head for which it is designed. The term used for this is 'best efficiency point' (BEP). The efficiency of the pump falls off when these conditions are varied.

For any given pump, the delivery, head and power input will increase as the speed increases. Thus, to make a pump deliver more water, or pump against a greater head, the speed must be increased. However, it is undesirable to drive simple pumps of this type at very high speeds; therefore, in such cases, a larger-capacity pump, or a number of pumps, should be installed.

Types of motors

Centrifugal and multi-stage centrifugal pump sets may be driven by any form of motor or engine. However, the electric motor is the most common and generally the most convenient method of driving a pump.

In Australia, electricity is supplied in two voltages: 240V and 415V. Both have a frequency of 50Hz. The electricity supply is often referred to as alternating current (AC) or direct current (DC). If the motor is wired up incorrectly, which sometimes happens, it can force the pump in the wrong direction. This can be dangerous and can damage the pump unit. If wiring has been done incorrectly, it must be corrected immediately to avoid serious damage.

 It is important to have a licensed electrician connect any hardwire electric motor for a pump and be certain the pump is operating in the right direction.

Controlling the pump

The power control of the pump unit can involve simply plugging it into a 240V power point (such as a pump unit connected to a domestic water storage tank) or hardwiring it through a control panel using either a 240V or a 415V power source (such as when pumping water from a low-level source to an elevated water storage tank).

The flow control of the water can also vary. The pump could simply operate each time a tap or valve is opened, but this can cause unnecessary wear and tear on the pump as well as inconvenience to the user because of noise transmission, hydraulic shock and water hammer. The more common type of control is therefore by automatic operation (for example, by float switches or level controls) in water storage tanks that are connected back to the pump control panel, or by flow switches or pressure switches located in a pipeline, which are again linked back to a pump control panel.

Float or level controls are generally used when filling storage tanks by means of a pump. The devices themselves can range from simple plastic floats, which incorporate a ball that acts as a switch, to more complex rods that incorporate sensors that control the flow of water.

In order to meet the varying demands that are required in certain situations, multi-stage pumps can be used. These pumps gradually increase in power and capacity as demand increases, thus building up in stages. This system helps to prevent undue stress on the pump system and ancillary pipework and equipment.

A typical example of pump information can be seen in Figure 8.10.

Source: All rights reserved. Grundfos Pumps Pty Ltd.

Applications

The MQ booster pump is a self-priming compact unit designed for the distribution and boosting of clean domestic water. Its integral tank avoids the need for a cumbersome additional water tank and enables quick installation in confined spaces.

Features

- Quiet operation
- Protection against dry running
- Suction lift: up to 6.5 metres (see operating instructions for maximum suction pipe length)
- Built-in non-return valve
- Supplied with an electric cable (2 m) and plug
- Motor with thermal protection
- Discharge outlet adjustable by +/–5° to facilitate hose attachment

Construction

- Stainless steel pump housing
- Integrated 0.4 litres diaphragm tank
- IP54 motor
- Insulation Class B
- Voltage: 1 x 240 V – 50 Hz

Performance

Pumps	Discharge pressure (kPa/psi)					
	150	200	250	300	350	400
	22	29	36	44	51	58
	Output (L/min)					
MQ3-35	61	50	39	24	-	-
MQ3-45	66	58	49	39	27	13

As a guide 1 tap = 10 litres per minute

1 sprinkler = 15 litres per minute

Technical features

Pumps	P2 (W)	In (A)	Water temp	Connections		Weight (kg)
				Inlet	Outlet	
MQ3-35	550	4.0	0/35 °C	1"M	1"M	13
MQ3-45	670	4.5	0/35 °C	1"M	1"M	13

Dimensions / installation

Dimensions in mm

The MQ booster pumps integrated control unit automatically protects against:
- dry running
- motor overheating
- motor overload

In the event of a fault, the MQ stops automatically, then tries to restart after half an hour.

FIGURE 8.10 Typical example of pump information

Positive displacement pumps

Positive displacement pumps have two main types: reciprocating and rotary. They have been used for a very long time and in fact were used extensively for water supply prior to centrifugal pumps becoming more popular. They were used to pressurise domestic services, domestic feeds to storage tanks, fire services and so on, as well as to supply water from rivers and dams, or from underground, to a storage tank in rural areas. The design of this type of pump allows it to obtain extremely high pressures and it should therefore be fitted with a safety valve on the outlet side to prevent over-pressurisation of the water service system.

Reciprocating pump

A reciprocating pump can be defined as a type of pump that operates with a constant motion, employing a straight-line path with a to-and-fro motion, moving backwards and forwards or up and down (as opposed to a circular motion). It may be of the piston, plunger or bucket type, unlike a centrifugal pump. Reciprocating pumps are generally made up of three moving elements for their operation and consist of:

- an inlet valve
- a piston or plunger
- an outlet valve.

One of the more common types of positive displacement pump can be seen on a windmill. The windmill makes effective use of the suction and lift pump, as can be seen in Figure 8.11. Multi-blade windmills traditionally pump water by directly operating a pump cylinder with a drive rod. The pump cylinder is submerged in the well attached to the end of the delivery pipe. It is a very simple pump, similar to a hand-operated bicycle pump. The drive rod is operated directly by the windmill rotor through the drive gearing,

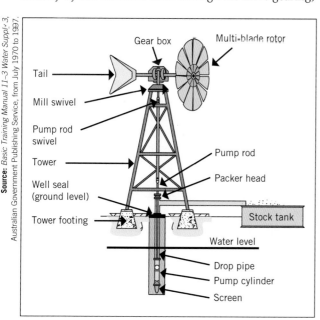

FIGURE 8.11 Windmill driving a positive displacement pump

which translates the rotating motion to the up-and-down reciprocating motion. Two one-way valves in the pump direct water through the pump, as illustrated in Figure 8.12.

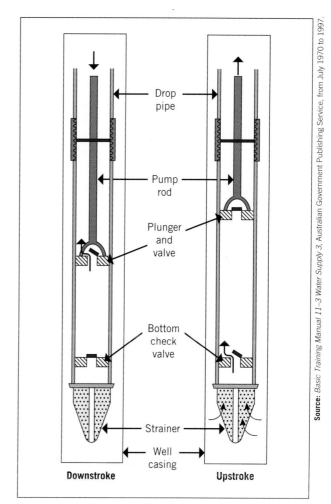

FIGURE 8.12 Operation of a typical windmill pump cylinder

The bottom check valve is fixed in position and opens to allow water to be drawn into the bottom chamber during the upstroke of the plunger and valve, which is attached to the pump rod. As the pump rod moves down again, during the downstroke, the water in the lower chamber flows through the upper check valve into the upper chamber. This water is then lifted during the upstroke towards the point of discharge and so on (hence the term 'reciprocating', because it moves backwards and forwards).

The hydrostatic test bucket, which is used by plumbers when testing water services, operates on a similar principle to that mentioned above (see Figure 8.13).

Rotary pump

The rotary or geared pump relies on cogs (gears) turning together and displacing fluid. This type of pump is used to move highly viscous (thick and gooey) fluids such as oil, paint, ink and chocolate.

Source: *Basic Training Manual 11–3 Water Supply 3, Australian Government Publishing Service, from July 1970 to 1997.*

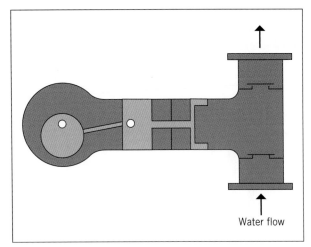

FIGURE 8.13 Positive displacement (reciprocating) pump

As the gears turn, the cogs come apart on the inlet side creating a vacuum that is then filled by the fluid and forced to the pump outlet by the rotating gears. At this point the fluid is displaced at a higher pressure (see Figure 8.14). The robust design with tight tolerances of the gears allows the ability to pump highly viscous fluids at very high pressures.

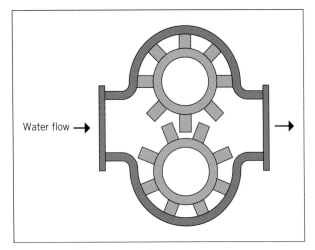

FIGURE 8.14 Rotary or geared pump

LEARNING TASK 8.3

1 Why is the centrifugal pump the most commonly used?
2 What is the name of the casing in which the impellor rotates?
3 What does the acronym 'BEP' stand for?
4 Name two devices that control the pump operation.

COMPLETE WORKSHEET 3

8.4 Information required to select an appropriate pump

The plumber should be aware of what information is required by the pump manufacturer or supplier so that the right pump is chosen to suit the task. This may include the:

■ source of water supply (such as a water main, storage tank or dam)
■ temperature of the water being pumped
■ purpose for which the pump assembly is required
■ highest outlet above the pump
■ pipework material and the type of valves being used
■ material from which the pump is manufactured
■ flow rate required at the point of use (amount of outlets)
■ pressure required at the point of use
■ pressure losses to be overcome through the piping system
■ type of power source required
■ specific location of the pumps
■ number of pumps required.

Typical examples of a manufacturer's pump selection sheets can be seen in Figures 8.15 and 8.16.

Quite often this information has already been catered for during the design process, particularly on larger installations, and is normally covered in the documentation. The plumber's role is to check the details on-site to ensure that all information is correct.

If there are any major variations it is always advisable to consult the client and obtain the necessary approvals before making any changes. Variations in the site conditions can also cause a variation to the installation of the pump assembly, so check all details relating to the pump before actually ordering and installing the unit/s.

Selecting the right pump

Once the pump supplier has been provided with all the necessary relevant information, they will select a pump using what is called a pump curve. A pump curve, also known as a performance curve, is a graph that pump manufacturers use to describe the performance of their products. This graph basically compares flow rate produced by the pump to the total head developed by the pump. From this graph the supplier will provide options based on the best efficiency point (BEP) for specific pump types. See Figure 8.17 for examples of pump curves.

Pump suppliers will readily provide pump curves to their clients to reinforce the decisions made when selecting the appropriate pump for the task at hand.

Source: All rights reserved. Grundfos Pumps Pty Ltd.

What pump do I need?			1
Application		Household water supply	
		Drainage	
		Irrigation	
		Water transfer	
		Other	
Water source		Above ground tank	
		Underground tank	
		River	
		Dam	
		Other	
Power supply		240V single phase	
		415V three phase	
Water requirement	Household		
		House only	
		House and garden	
		Showers (number)	
		Sprinklers (number)	
		Sprinklers (type)	
		Evaporative airconditioner connected	
	Irrigation		
		Sprinklers (number)	
		Sprinklers (type)	
		Automatic operation	
		Manual operation	
	Drainage and water transfer		
		Lift from pump (A) to point of discharge (B)	
Details of existing pipeline		Size (mm)	
		Type (polyethylene/PVC/copper/steel)	
		Length (m)	

How much flow (Q)?		2
Water pressure systems	✓	
Weekend cottage		10 to 20 L/min
Small home		20 to 30 L/min
Average home		30 to 50 L/min
Large home		50 to 90 L/min
Average water consumption	No.	
Standard shower head		15 L/min
Water saving shower head		6–7 L/min
Household standard tap		10–15 L/min
Tap with an aerator of flow restrictor		4–6 L/min
Lawn sprinkler		10–15 L/min

Calculate the flow rate	3

$$Q = (\quad) \text{ L/min}$$

FIGURE 8.15 Selection sheet to help calculate flow rate

FROM EXPERIENCE

The experienced plumber will be familiar with pump curves and should be able to extract the relevant information from them to choose the best pump for the job.

Selecting materials for the pump components

When selecting a suitable material for a pump component, several factors have to be considered by the pump manufacturer. The choice for each component will often depend on cost, availability and performance. Other factors affecting the materials selection include:

- the type of pump
- delivery (discharge) head in metres (how high the water needs to be pumped)
- flow rate in litres per minute (the amount of water required for all fixtures)
- liquid being pumped (consider issues such as temperature, corrosive qualities, abrasive qualities and the possibility of erosion)
- manufacturing method (casting or machining properties)
- initial cost
- economic life
- specific material properties such as strength, durability and flexibility.

General-purpose water pumps for low- and medium-pressure applications usually have a cast iron casing, high-tensile steel shafts, cast iron or bronze impellers and cast iron or bronze wearing parts. In some smaller pump units certain plastic components may be used. For higher pressures the casing is usually made from cast, forged or welded steel with the shaft, impellers and diffusers in stainless steel.

Source: All rights reserved. Grundfos Pumps Pty Ltd.

How much pressure (P)?

P = pump head

Hd = height difference between the pump and the highest point of use

Hs = pressure already available at the pump level (tank example with positive suction head). If you pump water under the level of the pump (well, river, underground tank), contact your dealer in order to calculate the suction lift and to select the right pump.

Hf = friction loss or pipe resistance to water flow (see chart at right for poly pipe friction loss)

Pr = residual pressure, i.e. the required pressure at the tap, shower or sprinkler. As a guide, shower head, standard ½" tap or sprinkler requires approx. 150 kPa (15 m or 21 psi)

Flow Rate			Friction loss - PN12.5 high density polyethylene pipe (m/100 metres of pipe)					
L/min	m³/hr	L/sec	20 mm	25 mm	32 mm	40 mm	50 mm	63 mm
12	0.7	0.2	10.9	3.7	1.1	0.4	0.1	-
24	1.4	0.4		13.4	3.9	1.3	0.4	0.1
36	2.2	0.6			8.3	2.8	0.9	0.3
48	2.9	0.8			14.2	4.8	1.6	0.5
60	3.6	1				7.2	2.4	0.8
72	4.3	1.2				10.1	3.3	1.1
84	5.0	1.4				13.5	4.4	1.5
96	5.8	1.6					5.7	1.9
108	6.5	1.8					7.1	2.3
120	7.2	2					8.6	2.8

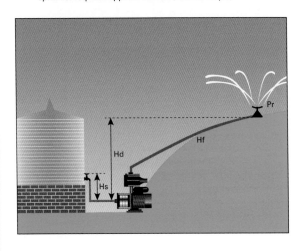

Calculate the pressure

5

$$P = (\quad) Hd - (\quad) Hs + (\quad) Hf + (\quad) Pr$$

Example

Q (Flow rate) = **60 L/min** = **4** (sprinklers) × **15 L/min**

Hs = 2 m

Hd = 15 m

Hf = 3.6 m (50 m of 40 mm poly pipe — see friction loss chart above)

Pr = 15 m (150 kPa)

P = 15 − 2 + 3.6 + 15 = 31.6 m = 310 kPa = 44 psi

Useful conversion

From \ To		Flow conversion						
		litres per minute		litres per second		cubic metres per hour		gallons per min
1 L/min	=	1	=	0.017	=	0.06	=	0.22
1 m³/hr	=	16.7	=	0.28	=	1	=	3.7
1 L/sec	=	60	=	1	=	3.6	=	13.2
1 gpm	=	4.5	=	0.076	=	0.27	=	1
Example								
60 L/min	=	60 L/min	=	1 L/sec	=	3.6 m³/h	=	13.2 gpm

From \ To		Pressure conversion						
		Metre of head		Kilopascal		Pounds/inch²		Bar
1 m	=	1	=	9.8	=	1.4	=	0.1
1 kPa	=	0.1	=	1	=	0.14	=	0.01
1 psi	=	0.7	=	6.9	=	1	=	0.07
1 bar	=	10.2	=	100	=	14.5	=	1
Example								
200 kpa	=	20 m	=	200 kPa	=	28 psi	=	2 bar

FIGURE 8.16 Selection sheet to help calculate pressure

The plumber should be aware of the pump components and select pipe materials and ancillary equipment for connection to the pump set that are compatible with the pump components and complement the specific application. When installing the pipework and ancillary equipment, care should be taken to ensure that there is no strain on the pump set that could, in turn, cause the pump unit to malfunction.

Source: © Pump Industry Australia Incorporated, *Australian Pump Technical Handbook 5e*, http://www.pumps.asn.au.

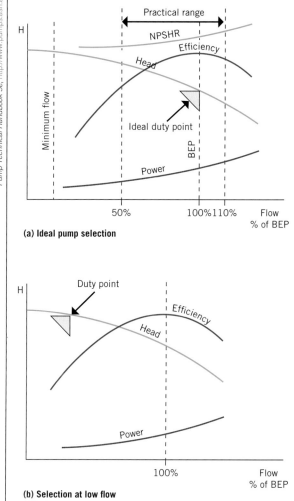

FIGURE 8.17 Examples of pump curves

Connection to the water utility's main system (towns and cities)

The designer and/or plumber must comply with the relevant water utility requirements when connecting a pump to the water supply system. This is important because connecting a pump could have an adverse effect on the rest of the water supply system. If the water supply system cannot provide sufficient water for the pump to operate, the pump assembly will also be affected.

The minimum guaranteed water pressure in a water utility's main is 150kPa (15m head). It should be remembered that there will be a pressure loss brought about by friction losses through pipes, valves, fittings and other ancillary equipment – all of which have to be factored in when considering the pump installation.

The information required by the plumber would be similar to that needed by the supplier when you are ordering the pump (see above), but would normally include extra information such as:

■ a written document from the utility stating the available water pressure in the mains located near the building/property
■ the site details including nearest cross streets (for confirmation of the location).

Connection in other situations

Rural plumbers are often involved in installing water storage tanks and associated pump systems. With the move towards rainwater harvesting and water reuse in towns and cities (due to requirements for more efficient water usage), urban plumbers also require knowledge of such installations.

> **GREEN TIP**
>
> It is generally a standard requirement for new homes to store a minimum of 2000L of rainwater for flushing toilets, irrigating the garden and washing the car.

Whenever purchasing water tanks, a variety of suitable pumps are normally suggested to accompany them. The plumber should complete all necessary research before attempting to install the pump assembly. Consult HB 230 – *Rainwater Tank Design and Installation Handbook.*

An important issue to be aware of, particularly in urban situations, is that of cross-connections. These may be direct (for example, a tank supply and a water main supply linked together by pipework so that the client can switch between the two supplies) or indirect (for example, where a client uses a hose to top up a rainwater tank from the main supply). In these situations there is a potential for the tank water to enter the water main supply. This may be a problem due to the possibility that the tank may contain what is considered to be non-drinking water, so a suitable approved backflow prevention device may need to be installed. Before installing or connecting in these situations, consult your local water utility for installation requirements. (Backflow prevention and tank water usage are covered in more detail in Chapters 3, 4 and 9.)

 The plumber is directly responsible for identifying and preventing cross-connection to avoid contamination of the community's drinking supply.

> **LEARNING TASK 8.4**
>
> 1 Name the two main requirements necessary to choose the correct pump.
> 2 How may a potential cross-connection occur with a pump installation?

8.5 Installation requirements

It is important for the plumber to have the knowledge and skills to determine the most suitable location for a pump and to install the correct valves, fittings and pipework required for efficient pump operation. This next section explains the location and installation requirements for pumps.

Location of pumps

Before installing a pump assembly, the plumber should check that the installation will meet the current standards and regulations. The pump assembly is normally located as close as possible to the source of water supply in order to minimise potential problems with the suction side of the pump.

In the majority of situations plumbers have the advantage of being able to locate a pump according to a plan. It is therefore vital that they can interpret the information provided on a plan and, normally, on the accompanying specification. Occasionally, however, plumbers may have to use their own discretion to provide and install a pump (for example, for water supply from a rainwater tank), so they must be aware of any specific installation requirements and details when locating the pump.

Installing the pump assembly

Pump assemblies, when located within a building, are usually located in a plant room or service duct. This room or duct must provide good accessibility and noise-proofing. Noise and vibration can be transferred from the pump assembly through the connecting pipework and/or the support structure into the rest of the building. This may not be too much of a problem in an industrial situation, but in residential or commercial buildings the noise can become quite disturbing. Continual vibration caused by the pump assembly not only may be a disturbance, but also may have a harmful effect on the connecting valves, pipework and surrounding equipment over a period of time.

The pump installer should be aware of noise generation and take the necessary steps to prevent noise transfer and vibration. This can be achieved by installing anti-vibration flexible connections. Installed next to the pump unit on both the suction and the discharge sides of the pipework, flexible connections will prevent noise transmission and protect the pipework from being damaged when the pump motor starts and stops. If the pump is mounted on the floor area, it may also require rubber mountings or anti-vibration springs on the pump base.

On larger pump sets (for example, long coupled centrifugal pumps) the motor and pump set are assembled in the factory and delivered to the site. The pump itself is connected to the motor shaft, sometimes with a combination of flexible couplings and spacer pieces. Because of its size, this pump set must be mounted on either a frame or a specially designed pump base (sometimes called an inertia block). These pump bases have to be carefully levelled, fixed and grouted into the building foundation if installed inside the building.

The foundation for the pump set should preferably be a rectangular concrete block with a mass approximately equivalent to five times the weight of the complete pump unit. This foundation needs to support the pump and motor assembly and, if necessary, prevent vibration and noise transmission through to the building structure.

The final positioning of the complete pump unit is essential so that correct alignment of moving parts is maintained and the pump warranty is not voided. Therefore, the plumber should pay particular attention to following the accepted trade practice when grouting in and/or fixing the pump unit. This information can be obtained from the pump supplier or manufacturer if the plumber is unsure of what the acceptable method is.

When the pipework is installed for the pump assembly, consideration must be given to both the isolation and the disconnection of the pump so that it can easily be maintained or its parts replaced. This is usually achieved by installing an isolating valve on both the inlet and the outlet, and also by using flanges or unions for connection purposes. The type of isolation valve used depends on the type and size of pump assembly and is normally a full way type of valve (for example, a gate or ball valve). A non-return (check) valve must be installed on the outlet upstream of the isolation valve so head pressure exerted on the pump outlet is avoided. The use of flanges or unions (for removing or maintaining parts of the pump assembly) normally depends on the size and type of pipe being used.

Assembly

A centrifugal pump is used in a reticulated system such as a water heating or water cooling system (see Figure 8.18). The assembly consists of:

- the suction side:
  - suction full way (gate) valve
  - suction strainer
  - foot valve (check valve and strainer)
  - flexible coupling
  - suction gauge
  - unions/flanges
- the discharge side:
  - discharge gauge
  - flexible coupling
  - non-return (check) valve
  - full way (gate) valve
  - unions/flanges.

FIGURE 8.18 Chilled water circulating pump. Note the protective guard over the drive shaft, the flanged flexible connections on the pipework and the pump mounted on springs (vibration pads)

AS/NZS 3500.1 PLUMBING AND DRAINAGE: WATER SERVICES

Pipe systems for pumps

Many of the problems experienced with pump installations are brought about by poor pipeline design or the improper application of the pump itself. If care is taken when designing the layout of the system and if good pumping practice is used, there is normally a marked improvement in the pump's overall performance.

Piping systems for pumps, irrespective of the type of pump used, can be considered as two separate sections: the suction pipework and the discharge pipework.

Suction pipework

The suction side is the side between the source of the liquid to be pumped and the pump itself. As mentioned earlier, when the pump is below the liquid source it is said to have a positive suction head (see **Figure 8.19**), and when it is above the liquid source it is said to have suction lift (negative suction head; see **Figure 8.20**).

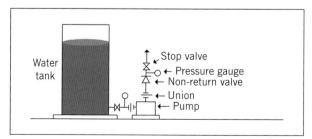

FIGURE 8.19 Positive suction head installation and valves

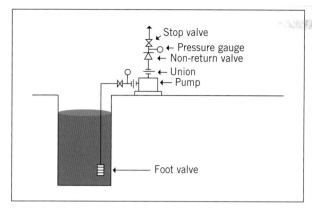

FIGURE 8.20 Negative suction head/lift installation and valves

Issues to consider with suction pipework include the following:

- Keep the suction pipe as short as possible.
- Where possible, try to provide positive suction head to the pump. (Install pump below water source.)
- Ensure that air pockets are eliminated by allowing for a continual rise to the pump (if there is a potential for air pockets to form, consider redesigning the pipe layout or allow for an air release valve to be installed).
- When selecting the ancillary equipment, such as foot valves and strainers, select those that will provide minimal friction loss and resistance to flow.
- If a reducer needs to be installed in the pipework adjacent to the pump, use an eccentric (tapered) reducer rather than a concentric (bush) reducer to avoid the possibility of forming an air pocket and/or causing cavitation.
- The suction pipe cannot exceed 6m in length if the water source is below the pump.

Discharge pipework

Careful consideration should be given to the types of pipe, fittings and ancillary equipment used when installing the discharge pipework, so that friction losses are kept to a minimum.

It is important to install a check valve (non-return valve) on the discharge side of the pump so that no undue backpressure is placed on the pump assembly, which could cause the pump to malfunction. It is good practice to install the check valve no closer than 15 times the pipe diameter to the pump outlet in order to prevent turbulence acting on the check valve.

When using reducers on the discharge pipework it is preferable to create a gradual taper using a tapered reducer piece, rather than using a reducer with a stepped shape, taking the larger diameter to a smaller diameter. This helps to prevent air being trapped and the possibility of the pump cavitating.

Ancillary equipment

Examples of ancillary equipment used with pumps include the following:

- Check valve – a non-return valve installed in the pipework. It is always installed on the discharge side and sometimes on the suction side of the pump. Avoid using a swing-type check valve as these have the potential to cause water hammer.
- Foot valve – a type of check valve with a built-in strainer. Used at the point of the liquid intake (suction side) to retain liquid in the system, it prevents the loss of prime when the liquid source is lower than the pump (generally when a suction lift is required).
- Strainer – a device installed in line on the suction side of the pump. The strainer is designed to prevent foreign matter from getting into the main body of the pump and possibly causing damage to the pump.
- Reducer – pipe reducers are used for changing pipe sizes. It is important to use an eccentric reducer on the suction side of the pump so that air does not become trapped in the pipework and possibly contribute to cavitation.
- Pipe supports – these are a very important part of the pump assembly. Some pipe supports are designed only to take the load off the pipe assembly. Others are designed to support both the suction and the discharge line and may include noise suppressors as part of the support to help prevent pump noise being transmitted into the building.
- Drains – some larger pumps have seals that are designed to leak normally and may include a packing gland. For this type of pump, a drain has to be installed to drain water away from the pump assembly to an approved point of discharge.
- Pressure gauges – these should be installed and can be used on both the suction and the discharge sides of the pump pipework assembly. They provide a ready check to see if the system is operating satisfactorily.
- Air cells – hydro-pneumatic pump assemblies include an air (pressure) cell (see Figure 8.21) on the discharge side. This cell has an internal bladder that is pressurised with air. When the pump cuts in, it initially fills the pressure cell with water, as well as pumping water through the system as it is required. The cell is designed to store water for intermittent (irregular) use so that the pump does not continually shunt. Figure 8.22 shows a small hydro-pneumatic (air/pressure cell) installation.
- Vibration eliminators – installed in line on both the suction and the discharge sides of the pump and as close as possible to the pump itself. These are designed to prevent damage to both the pump and the pipework, particularly when the pump starts and stops.

FIGURE 8.21 Pump with pressure cell designed to boost water pressure

LEARNING TASK 8.5

1 Why is the pump located as close as possible to the water source?
2 What valves and fittings are connected to the pump inlet (suction) side?
3 What valves and fittings are connected to the pump outlet (discharge) side?
4 Why is it important that no air enters the inlet (suction) pipe?

8.6 Testing the pump system

When the pump assembly has been completed, it is important to check that the system complies with the drawings and specifications, that it is free from leaks and defects, and that the flow through the system is correct. To check that the system is free from leaks, the plumber will need to hydrostatically pressure test the piping system in accordance with the job specifications or current relevant standards. Water services must be tested to the requirements of AS/NZS 3500.1.

The pump system should now be ready to start so that the plumber can check whether it is operating correctly. Before starting the pump, however, there are several points that must be checked (failure to do this could result in damage to the pump):

1 Electric wiring – this must be checked by a qualified electrician and must comply with all current and relevant codes.
2 Direction of rotation – depending on the type of pump this can be checked by removing the coupling between the motor and the pump and briefly starting the motor. Confirm that the direction of rotation is the same as that indicated on the pump casing. If it is correct, refit the coupling and check for soundness.

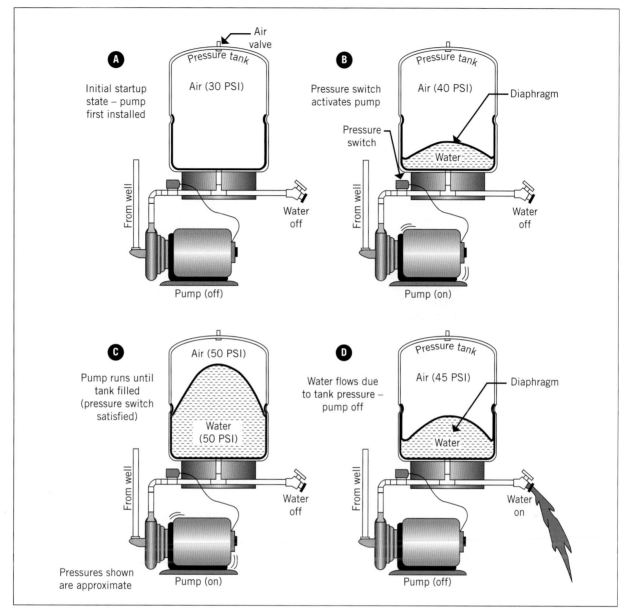

FIGURE 8.22 Diagram of a pump and bladder (air cell) assembly

3 *Pump gland* – there are two types of shaft seals used on pumps: mechanical seals and packed glands. Mechanical seals do not drip, whereas packed glands (stuffing boxes) require a drip rate of approximately 20 drips per minute. If it does not drip, the gland instantly overheats and destroys itself and the pump shaft.

4 *Priming the pump* – do not operate a pump dry. Before operating the pump, the system must be flushed to remove any debris and then it must be primed with clean water. The pump and the suction line must be completely filled with water before the pump is switched on.

5 *Other considerations* – carefully read the manufacturer's installation and starting instructions before switching on the pump.

The plumber will now need to test the pump set and record the test data. This test also needs to be done in accordance with the specification and relevant codes.

(Testing the pump flow requires specialised knowledge and is not covered in this unit. However, if the plumber has all the information listed below, they will be able to contact the pump manufacturer and then use the information to check the flow rate of the pump.)

To test the performance of the pump for correct flow and operation, the plumber will need to record the:
- operating pressures at both the suction and the discharge sides of the pump
- flow rate required and its application
- pump data from the name plate.

All the relevant information should then be recorded by the plumber for their records. Depending on the nature of the installation, this same information is usually given to the facility's manager (or equivalent) for maintenance purposes. It is also usual for this information to be permanently affixed next to the pump assembly or the pump control panel.

8.7 Work health and safety (WHS)

Be sure to identify all the hazards of the work area and the tasks involved before starting the job. Then write a job safety analysis (JSA) stating the control measures that will reduce the risks.

 Before commencing work on the pump assembly and associated pipework, the overall task should be risk assessed in accordance with current WorkCover requirements.

Some risks/hazards associated with a pump installation are now discussed.

Manual handling/lifting

Pump units are normally quite heavy and difficult to manoeuvre. Care should be taken when locating the pump in the correct position. If necessary, use lifting equipment to locate the pump, associated valves and equipment.

Working at heights

Although pumps are not usually located in high positions, the associated pipework may be. The plumber must therefore ensure that suitable safe equipment is used while working at heights. If items such as mobile platforms and scissor lifts are used, ensure that the person operating this equipment has the appropriate current qualification and has received adequate training in its use. Care should be taken to protect the safety of fellow workers and the public by the use of barricades and so on.

Fumes

If there is a potential for fumes to be present during the installation process, ensure that suitable respiratory equipment is worn and that adequate ventilation is provided and used. Fumes could be generated during welding processes or by other equipment operating in the surrounding work area. Attention should also be paid to correct ventilation of the area if the pump is operated on diesel fuel.

Moving parts

Pumps have moving parts and may therefore need to have suitable guards installed. Most times these protective guards come as part of the pump assembly; however, they may have to be made to suit particular site conditions. Pump guards should be designed and installed to prevent injury to all persons, including those who are working near the pump. It is also a good idea to install some warning signs near the pump assembly.

Electrical work

All electrical work should be carried out by a licensed electrician. When working on the pipework on the pump assembly, the plumber should test that pipework with an electrical test lamp or probe to check whether it is live as a result of stray electricity. It is also advisable to bridge any portion of the pipework (or ancillary equipment) that is to be removed with an approved 'bridging conductor' before starting any maintenance work. If you have any doubts at all, have the system checked by a licensed electrician.

Tools and equipment

Some of the tools and equipment that a plumber may require when working on a pump installation can be found in Chapter 7 of *Basic Plumbing Skills*.

LEARNING TASK 8.6

1 A packing gland on a shaft seal is meant to leak. What is the recommended drip rate?
2 Explain why pumps need to be primed.
3 Why is it important to check the rotation direction?
4 Name two hazards that could be present when installing pumps.

Troubleshooting

In most cases troubleshooting should be left to those who are well experienced in pump installations. Different pump installations may have different problems. Before making any alterations to the pumps or the pump installation, it is advisable to consult with the pump manufacturer, or their agent, so that any warranty is not voided.

There are, however, some simple checks that can take place when faced with a problem on-site, as shown in Table 8.1.

Troubleshooting is, of course, not limited to the items listed in Table 8.1. Where there is a problem, it is advisable to contact the manufacturer of the pump. They, in turn, may:
■ refer you to the literature provided with the pump
■ arrange for their representative to meet you on-site
■ refer you to their website to access the necessary information to assist you.

TABLE 8.1 Troubleshooting for pumps

Problem	Possible causes and/or remedies
No water is being pumped or the pump is not delivering enough water.	The suction line and pump casing are not properly primed (air in the system).
	Ensure that the system is completely flushed prior to commissioning the installation.
	The discharge head is too high (check the friction head).
	The mechanical seal may be worn and leaking air or water. Check the weep hole.
	Suction lift is too great. If pumping from a well or pit the maximum practical distance between the water level and the pump inlet is about 5m (see the discussion on suction lift earlier in the chapter).
	The impeller, suction pipe or foot valve is clogged (also check the strainer in the suction line if fitted).
	There is an air leak in the suction line.
	The suction and/or discharge piping is too small in diameter.
The pump operates for a while then stops pumping.	The pump is not properly primed.
	There is an air pocket in the suction line.
	There is an air leak in the suction line.
	The suction strainer is blocked.
The pump is noisy when operating.	The pump and motor are out of alignment. Check the pump alignment.
	Foreign matter is rotating with the impeller. Visually inspect the pump suction inlet port.
	The pump is cavitating (for example, the pump suction lift is too great or the pipework is incorrectly installed).
The pump vibrates during operation.	The pump and motor are out of alignment.
	The foundation is insufficiently rigid.
	The suction or discharge piping may be unsupported.
	The stuffing box becomes hot.
	The packing is too tight or not dripping enough water (approximately 20 drips per minute are required on packed gland seals).
The pump does not prime properly.	Make sure the pump casing is filled with water.
	Look in the suction line or fittings. Check to see that all fittings are tight in the suction line and make sure there is no leak in the pipe itself.
	The mechanical seal may be worn and leaking air.
	The inlet valve rubber may be frozen to the seat.
	The pump may be running too slowly.
	Suction lift may be too high. Keep the pump as close to the water source as is safely possible.
	The suction line or suction strainer may be clogged.

Good planning and design of a water pump installation minimises the long-term costs of the overall installation by:

- ensuring that the pump operates efficiently
- ensuring that the right type of pump is installed to achieve the required flow rates and pressures
- ensuring that the life of the pump is maximised
- minimising hydraulic (friction) losses
- helping minimise energy costs
- ensuring easy pump operation
- ensuring that foundation and structural requirements are met.

Figure 8.23 shows a typical pump assembly with dual pumps, a pressure cell and pump control panel.

FIGURE 8.23 Completed pump assembly with control panel

8.8 Cleaning up

Having a general tidy-up in the work area at the end of the day is a good practice. The work area is more productive when clean and tidy. It is easier and safer to move about and the thought process is more direct without the distraction of clutter.

Any leftover material, pipe and fittings should be stored and reused on another job. All recyclable rubbish should be disposed of in the proper bins.

All tools and equipment must be cleaned and stored so they are ready for use next time.

All company quality assurance procedures must be followed. These include the clean-up process, the pack-up and storage process, and issuing the required final documentation.

LEARNING TASK 8.7

1 What could cause the pump to not deliver water or to stop pumping?
2 What could prevent the pump from priming properly?
3 What should happen to leftover material?

 COMPLETE WORKSHEET 4

SUMMARY

- Water pumps have three purposes: to increase water pressure, to move water from a lower place to a higher place, and to circulate water in a continuous loop.
- The two major groups of pumps are rotodynamic and positive displacement.
- Centrifugal pumps are the most commonly used and come under the rotodynamic group.
- There are several different mechanisms that control the pump operation, including float or level controls and pressure controllers.
- It is preferable that the water source is above the pump (positive suction head) because the pump will self-prime and it doesn't have to draw (suck).

- A pump cannot draw water that is more than approximately 6m below it, regardless of its size.
- It is preferable to install the pump as close to the water source as possible.
- Always install a full-way isolation valve, line strainer and union on the pump inlet.
- A foot valve must be installed on the suction pipe when the water source is below the pump (negative suction).
- Always install a union, non-return valve and full-way isolation valve on the pump outlet.
- Installing an air cell gives the pump a longer life because it reduces the pump's operating time and saves energy.

REFERENCES

Davey: **http://www.davey.com.au**

Grundfos: **http://www.grundfos.com**

GET IT RIGHT

1 Which pump set is correctly installed?

2 Why is it correct?

WORKSHEET 1

To be completed by teachers

Satisfactory ☐

Not satisfactory ☐

Student name: _____

Enrolment year: _____

Class code: _____

Competency name/Number: _____

Task: Review the sections '8.1 Background', and 'Situations in which pumps are required' and answer the following questions.

1 Name the three main functions of a pump.

_____

_____

_____

2 Describe at least four situations where pumps may be used within a water supply system.

_____

_____

_____

_____

3 What is the definition of a water pump?

_____

_____

4 State the two basic categories that pumps fall into.

_____

_____

5 List two examples of each of the pump types from the two categories above (see Figures 8.1 and 8.2).

Type 1   Type 2

_____   _____

_____   _____

6 Which type of water pump is the most commonly used?

_____

_____

WORKSHEET 2

To be completed by teachers

Satisfactory ☐

Not satisfactory ☐

Student name: _____

Enrolment year: _____

Class code: _____

Competency name/Number: _____

Task: Review the section '8.2 Pump requirements and basic pumping terminology' and answer the following questions.

1   When selecting and installing pump sets for particular applications, the plumber has a number of responsibilities. List four of these responsibilities.

_____

_____

_____

_____

2   On large pump installations it is normal for more than one pump to be installed. Give two reasons why this is done.

_____

_____

3   Explain the term 'delivery head'.

_____

_____

4   Explain the term 'friction head' and state examples of situations that cause friction head.

_____

_____

_____

5   Explain the formula $P = H \times 9.81$

where P = _____

H = _____

9.81 = _____

6   Using the above formula, determine the pressure supplied to outlets on the 6th, 14th and 29th floors
    of a high-rise building that has a water tank located on the roof of the 31st floor that is 2m in height.
    (Note that each floor is 3m high.) What delivery pressure would be required of a pump at ground level
    to fill the water tank?

    6th floor:

    _____

    14th floor:

    _____

    29th floor:

    _____

    Pump delivery:

    _____

7   Explain the term 'flow rate'.

    _____

    _____

    a   In relation to the term 'suction head', what is the theoretical maximum height that a pump can be
        above the water source?

        _____

    b   However, in reality what is the maximum height a pump should be positioned above the water
        source?

        _____

    c   Give a reason why the maximum theoretical height cannot be achieved.

        _____

8   Define the terms 'positive suction head' and 'negative suction head'.

    _____

    _____

    _____

    _____

9   Why is it preferable to have positive suction head?

    _____

    _____

WORKSHEET 3

Student name: _____

Enrolment year: _____

Class code: _____

Competency name/Number: _____

To be completed by teachers

Satisfactory ☐

Not satisfactory ☐

Task: Review the section '8.3 Types of pumps' and answer the following questions.

1   On what basic principle does a centrifugal pump operate?

_____

_____

_____

2   Name two major parts of a centrifugal pump.

_____

_____

3   Name two fuel types available to run pumps.

_____

_____

4   What is the basic principle of operation for a multi-stage pump?

_____

_____

5   When testing water services for soundness, plumbers use a test bucket. Name the type of pump that is used on the test bucket.

_____

6   Explain the differences between a close coupled pump and a long coupled pump.

_____

_____

_____

_____

7 Identify the various parts of a windmill pump cylinder on the diagram below.

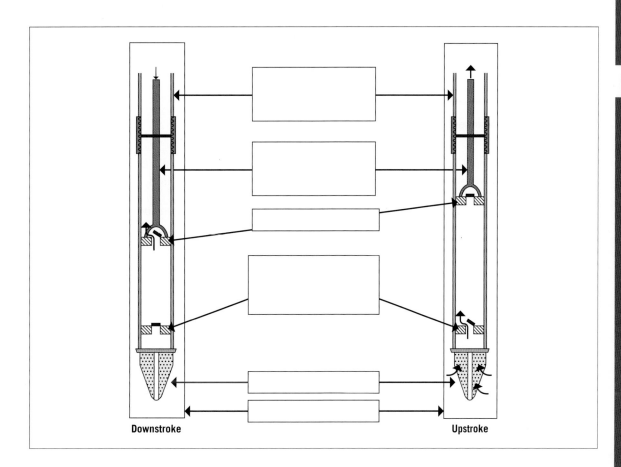

Downstroke   Upstroke

WORKSHEET 4

Student name: _____

Enrolment year: _____

Class code: _____

Competency name/Number: _____

To be completed by teachers

Satisfactory ☐

Not satisfactory ☐

Task: Review the sections '8.4 Information required to select an appropriate pump', '8.5 Installation requirements' and '8.6 Testing the pump system' and answer the following questions.

1 Before contacting the pump manufacturer or supplier, the plumber must obtain certain facts about the pump installation so that the right pump is chosen. Name five pieces of information that may be required in selecting the right pump.

_____

_____

_____

_____

_____

2 What is the function of the pump (performance) curve?

_____

_____

_____

3 Why is it necessary to use flexible connections and vibration eliminators when installing pump sets?

_____

_____

_____

4 Why should a non-return valve be installed on the discharge side of a pump?

_____

_____

_____

5 Explain how the hydro-pneumatic pumping system works.

_____

_____

_____

_____

6 Why should a pump never be run dry?

_____

_____

7 What does the term 'cavitation' mean?

_____

_____

8 Name two risks/hazards associated with installing pumps.

_____

_____

📋 WORKSHEET RECORDING TOOL

Learner name		Phone no.	
Assessor name		Phone no.	
Assessment site			
Assessment date/s		Time/s	
Unit code & title			
Assessment type			

Outcomes

Worksheet no. to be completed by the learner	Method of assessment WQ – Written questions PW – Practical/workplace tasks TP – Third-party reports SC – Scenarios RP – Role plays CS – Case studies RW – Report writing PF – Portfolio	Satisfactory response	
		Yes ✓	No ✗
Worksheet 1	WQ	☐	☐
Worksheet 2	WQ	☐	☐
Worksheet 3	WQ	☐	☐
Worksheet 4	WQ	☐	☐

Assessor feedback to the learner

Feedback method (Tick one ✓):	☐ Verbal ☐ Written (if so, attach) ☐ LMS (electronic)
Indicate reasonable adjustment/assessor intervention/inclusive practice (if there is not enough space, a separate document must be attached and signed by the assessor).	
Outcome (Tick one ✓):	☐ Competent (C) ☐ Not Yet Competent (NYC)
Assessor declaration: I declare that I have conducted a fair, valid, reliable and flexible assessment with this learner, and I have provided appropriate feedback.	
Assessor name:	
Assessor signature:	
Date:	

Learner feedback to the assessor

Feedback method (Tick one ✓):	☐ Verbal	☐ Written (if so, attach)	☐ LMS (electronic)

Learner may choose to provide information to the RTO separately.

Learner assessment acknowledgement:	Tick one:	
	Yes ✓	No ✗
The assessment instructions were clearly explained to me.	☐	☐
The assessment process was fair and equitable.	☐	☐
The outcomes of assessment have been discussed with me.	☐	☐
The overall judgement about my competency performance was fair and accurate.	☐	☐
I was given adequate feedback about my performance after the assessment.	☐	☐

Learner declaration:
I hereby certify that this assessment is my own work, based on my personal study and/or research. I have acknowledged all material and sources used in the presentation of this assessment, whether they are books, articles, reports, internet searches or any other document or personal communication.
I also certify that the assessment has not previously been submitted for assessment in any other subject or at any other time in the same subject and that I have not copied in part or whole or otherwise plagiarised the work of other students and/or other persons.

Learner name:	
Learner signature:	
Date:	

9

CONNECT AND INSTALL STORAGE TANKS TO A DOMESTIC WATER SUPPLY

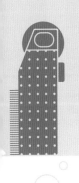

Chapter overview

This chapter looks at the connection and installation of storage tanks to a reticulated water supply pipe system, which may be a new work site or an existing structure being renovated, extended, restored or maintained.

Learning objectives

Areas addressed in this chapter include:
* basic design requirements for the location of storage tanks
* basic water storage tank installation requirements
* connecting mains water supply to a storage tank
* cleaning up.

9.1 Background

The content of this chapter relates to the basic principles of the connection and installation of above-ground water storage tanks that are 'open' to the atmosphere. The water supply is connected to the tank via a physical air gap and is therefore subjected to atmospheric pressure. This is a gravity feed supply system, as compared with a 'closed' or 'pressurised' system, such as a mains pressure supply to a water heating storage tank that is directly connected to the drinking water supply, as discussed in Chapter 5.

All connections made to fixtures, tanks or service pipes must have the relevant authority's permission. The supply of water to fixtures must be carried out according to strict regulations. These regulations apply whether the fixtures are connected directly to the drinking water supply or indirectly through storage tanks or flushing devices. Water from any fixture, storage tank or flushing device cannot flow back into the supply pipes. Cross-connection must always be prevented, and is to be achieved by fitting a backflow prevention device or by providing a minimum air gap of at least 20mm between the lowest level of the water inlet and the highest part of the overflow pipe or rim of the fixture.

Planning and preparation

Before commencing any work in relation to connecting or installing storage tanks, it is important to comply with the most current Australian Standard/s, any relevant overseas standards (if they apply), the relevant authority's requirements and the manufacturer's instructions. So that proper planning for the task can take place, the following is also important:

- Plans and specifications relating to the task must be obtained and carefully reviewed (when provided) to ensure that the system complies with current rules and regulations.
- All tasks must be planned and sequenced in conjunction with others involved in or affected by the work.
- Sustainability principles and concepts must be applied to work preparation and application.
- Safety and workplace environmental requirements associated with connecting static storage tanks to a water supply system must be followed.

The information in this chapter will show the vast range of storage tank installations. It will explain some of the complexities related to the installation of storage tanks so there is a better understanding of what is seen on various work sites. Refer to the following chapters for more detailed support information and awareness relating to this topic: Introduction and Chapters 4, 5, 8 and 10, along with the online appendix. It is also recommended that you visit various websites relating to tanks and tank manufacturing to supplement the contents of this chapter.

Storage tanks

A storage tank is defined as a container or vessel open to the atmosphere at the top, connected to a drinking (potable) water supply and filled with water to be used for various purposes. It is hard to find a clear definition of what constitutes small- and large-capacity storage tanks, but in this chapter small-capacity storage tanks are deemed to be those up to 500L and large-capacity those in excess of 500L.

> ✔✔ **AS/NZS 3500.0 PLUMBING AND DRAINAGE: WATER SERVICES**

The following illustrate the range of storage tanks available:

- A simple cistern for a water closet (see **Figure 9.1**), where all the plumber has to do is ensure that the cistern is installed properly, connect it to the water supply and check that it operates correctly. With these installations the cistern is manufactured with an approved air gap and the water level is controlled by a float valve and an overflow, which is integral to the unit.

FIGURE 9.1 Water supply to a cistern

- A header tank or make-up tank (see **Figure 9.2**) or similar, where a considerable amount of planning is required to manufacture the tank, locate it on-site and carry out the installation. The tank is sized according to the specific site requirements. It is manufactured with the necessary connection points and placed in its nominated position, and then all connections are made to and from the tank.
- A large-capacity water storage system, which consists of an extremely large vessel that may contain several thousand litres of water, required for a variety of reasons, on a building site(see **Figure 9.3**). The tank is sized according to the specific site requirements. Obviously, more detailed planning and approval must be carried out before installation can take place, as it requires input from other trades. This type of system may use tanks that are either prefabricated or built in situ. *Note:* In some installations where it is important to maintain a dedicated supply of water at all times, such as a fire

FIGURE 9.2 Header tank or make-up tank, showing the outlet connection from the tank, the sludge valve, safe tray and safe waste, and tank support (note the bracket holding the tank and support material)

Source: Shaun Kristen.

FIGURE 9.3 Large-capacity storage tank used for a fire service

supply service, it is advisable to have more than one tank available for supply (see **Figure 9.28** later in this chapter). This helps avoid major disruptions to supply or possibly unsafe situations due to breakdown or maintenance of the system.

Storage tanks may be used to serve the following purposes:

■ *To provide a supply of water that is physically disconnected from the mains supply* (for example, see **Figures 9.1, 9.2** and **9.3**). This includes all items that have the potential to cause a cross-connection and where a registered air gap (RAG) or registered break tank is required for containment, zone or individual protection purposes. Other examples of services that may be coupled with these include, but are not limited to, sanitary flushing (flusherette supply), air-conditioning services, 'make-up' water, contingency reserve, ablutions and combined systems (for example, rainwater harvesting, where rainwater storage is backed up by a mains pressure supply).

■ *To store and distribute water in buildings at a level higher than that which the supply authority can provide through its water mains.* This is generally the case for multistorey buildings where pump sets would be required to supply water to a storage tank located at or near the top of the building (see Chapter 8).

■ *To provide a large reserve of water (such as for fire services) where the potential demand exceeds the available mains supply.* Storage tanks are also required where required flow rates are insufficient or there is a lack of suitable water pressure available. Pumps are used in conjunction with these tanks.

GREEN TIP

It is very common today and compulsory for new houses (depending on local council regulations) to install rainwater tanks with mains water back-up to supply toilets, washing machines and garden taps.

COMPLETE WORKSHEET 1

Licensed plumbers have to consider many factors when selecting and installing a storage tank for specific purposes. In this chapter, we look at two main considerations:

1 Basic design requirements for locating a storage tank, including:
   ■ tank materials
   ■ the shape, construction method and location of the tank.
2 Basic tank installation requirements, including:
   ■ water supply piping both to and from the tank
   ■ parts of the storage tank
   ■ support for the storage tank.

Other factors may also have to be considered, but these are not discussed here as they relate more to design than installation. An example would be whether wind and earthquake design considerations are required to enable water tanks to survive seismic and high-wind activity.

9.2 Basic design requirements for locating a storage tank

The location of a storage tank influences the choice of material that is most appropriate for the design and construction of the tank. Factors affecting this may include atmospheric conditions, surrounding equipment or storage, and pedestrian or vehicular traffic. The tank shape is an important consideration, as is the access available to get the tank into position.

Tank materials

The tank material chosen should comply with plans, specifications and current Standards, such as AS 3855 Suitability of plumbing and water distribution systems products for contact with potable water. Materials may include, but are not limited to:

- cast iron (sectional)
- concrete (cast in situ or precast)
- copper (both in sheet form and corrugated)
- fibreglass
- plastic
- stainless steel
- steel (galvanised, Colorbond® or sectional).

Shape and construction method

Storage tanks come in a variety of shapes and sizes, and this chapter deals with the main shapes used today: tanks rounded in shape – circular tanks; and tanks

FROM EXPERIENCE

Plumbers use their knowledge and experience to choose the correct tank material for the most efficient application.

square or rectangular in shape – rectangular tanks. The shape chosen generally complements the tank's location with the available space during construction and after completion, as space is a valuable resource on any premises.

Tanks must be fitted with a close-fitting cover. Figure 9.4 indicates the various parts of a storage tank installation, including the cover.

Circular tanks

Circular tanks are available in a range of approved materials. The size of the tank may influence the material chosen and the choice of construction method. The advantage of round tanks is that there are fewer corners and crevices for mould and mildew to form in.

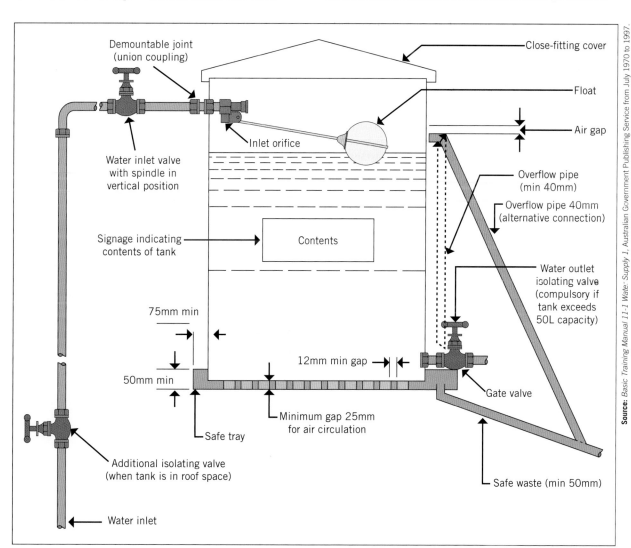

FIGURE 9.4 Parts of a storage tank installation

Source: Basic Training Manual 11-1 Water Supply 1, Australian Government Publishing Service from July 1970 to 1997.

Also they are easier to transport. The main disadvantage is that round tanks use more space. See **Figure 9.5**, which indicates a tank with a close-fitting (removable) cover. Other points to note in this photograph are:

- the water supply pipe to the tank
- the supply pipe from the tank
- that each pipe is connected with a demountable joint (threaded joint and union) to enable easy removal if required
- the timber support for the tank
- the safe tray.

FIGURE 9.5 A corrugated copper tank with a close-fitting cover

Smaller tanks are generally made of lighter materials; for example, thin sheet metals (such as copper and Colorbond®) fibreglass and UV-resistant plastics.

In larger applications there is greater demand for site space, and along with this increase in size come heavier construction materials. Larger tanks quite often incorporate a different range of materials from those used for smaller tanks. They can be manufactured off-site using materials as mentioned in the tank materials list above, and then delivered to the site. They can also be assembled on-site in sectional panels or cast in situ using reinforced concrete for the floor and walls and sheet metal for the roof.

Rectangular tanks

Rectangular tanks are available in similar materials to those mentioned above, such as concrete, sectional steel or cast iron, and have similar influences on material type.

Figure 9.6 shows a bolted together, sectional tank construction. Such a design allows for variations in shape and size if site conditions alter, although standard four-sided shapes are the preferred economical choice. Sectional tanks are obviously made up in sections – the more common sizes generally being 600mm × 600mm square and 1000mm × 1000mm square – which are bolted together on-site. One of the main advantages of

Source: Shaun Kristen.

FIGURE 9.6 Rectangular (sectional construction) storage tank; note the bollards located near the corners to help provide protection

sectional tanks is that they can be put together within a confined area; for example, a tank built in situ in a plant room or car park area. They can be of the internally or externally bolted type of construction, depending on specific needs.

Tank location

Current regulations state that storage tanks must be constructed and installed entirely above ground level, *unless* permitted otherwise by the relevant authority.

The location of any tank will be dictated by the site conditions and the required usage. As stated above, the manufacture and location of the tank depend upon the tank's environment – that is, indoors, outdoors, above ground or below ground (considering the temperature of the area where water will be stored with regard to freezing). Suitable access to the tank and its controls must also be considered (see **Figure 9.7**). If a tank is

FIGURE 9.7 A concrete storage tank located in a plant room at a hospital; note where the water supply enters (top left-hand side), the water level indicator (clear plastic tubing on left), the panel to gain access to the inside of the tank (with a danger sign indicating a confined space) and the warning sign indicating the contents of the tank and that no chemicals are to be added

located internally, sufficient headroom above the tank and space around the tank should be provided to allow access for maintenance. Another point to consider is whether a pressurised (pumped) supply is required to deliver water to the tank and therefore what type of controls are necessary for the installation.

Tanks located externally may have some restrictions placed upon them. Some of these restrictions include, but are not limited to:

■ a minimum distance between the tank and surrounding buildings or structures; for example, fire brigade requirements

■ minimum clearances from heavy vehicular traffic to avoid possible damage to the tank.

■ tanks that store drinking water cannot be located directly beneath any sanitary plumbing or pipes carrying non-drinking water.

Note: These restrictions are normally mentioned in the specifications for the project.

Storage tanks can be placed in a variety of places around a building or property. In a domestic residence they may be placed in a ceiling space, on a roof, on the outside ground or concrete, or on the floor within a building (see Chapter 5).

On larger building sites, tanks may be placed in plant rooms, on rooftops, in car park areas or in other approved locations.

Remember that working inside a tank is classed as working in a confined space. Be sure to follow the correct safety procedures.

Plant rooms

The location of a plant room will govern the source of supply and connection to the tank/s. The plant room may be located at a level where the tank/s can be directly fed from the authority's water main and reticulated supply and so may not require pumps for the inlet supply, or it may be located at a high level with a pumped water supply as part of the reticulation system and therefore use large-capacity storage tank/s to provide water for the rest of the building.

A plant room may contain small-capacity and/or large-capacity storage tanks.

■ Small-capacity storage tanks (see **Figures 9.5, 9.8** and **9.9**) may serve a variety of purposes, including heated water, sanitary flushing, air-conditioning services, ablutions and make-up water.

– If the water supply to the plant room is fed directly from a water main and reticulated supply, then each individual tank may be connected directly to that system in accordance with the relevant authority's requirements.

– If the water supply is from a larger storage tank, then a separate pump system, located within the plant room, is normally installed to provide water

FIGURE 9.8 A small-capacity tank with signage indicating that the tank is used for heating water make-up supply

FIGURE 9.9 Storage tanks and a hydro-pneumatic system located in a plant room

to these points. The pump system used for this purpose is normally a hydro-pneumatic system (see **Figure 9.9** and Chapter 8).

■ Large-capacity storage tanks located in a plant room within a high-rise building often will have a greater need for a dual water supply. These tanks generally consist of separate units, but may be a single tank partitioned to contain both the domestic and fire supply (hydrants and hose reels) in the one tank. *Note:* Tanks supplying fire sprinkler systems generally require a separate tank.

- Tanks located in these areas are generally square or rectangular in shape due to the limited space available. Materials used may vary (such as concrete construction; see Figure 9.7), but they are normally made from either cast iron or steel and are generally of a sectional construction.
- These larger tanks are normally built on engineered elevated platforms, helping to provide adequate clearance and strength in supporting their weight (see the section 'Support for storage tanks' later in the chapter).

Rooftops and car park areas

Tanks located on rooftops or in car park areas are designed similarly to those located in plant rooms and are manufactured from the wide range of approved materials. Both small- and large-capacity storage tanks can be placed in these locations. Tanks placed within car park areas may require additional protection, such as bollards placed at the corners to prevent vehicular damage (see Figure 9.6); consideration should also be given to eliminating vandalism and mechanical damage (see Figure 9.10).

FIGURE 9.10 This tank installed within a car park is fenced in within a recessed area, providing protection from vehicular and pedestrian traffic and vandalism, and displaying efficient use of a limited space

Suitable access must be provided to any tank in any location, and some may require a ladder, trafficable walkways or provisions to eliminate fall hazards (see Figures 9.27 and 9.28 and 9.30 later in the chapter).

LEARNING TASK 9.1

1 What is a bollard?
2 Explain the purpose of a sectional tank.
3 What is the purpose of a tight-fitting cover on a water tank?
4 Name four materials that tanks are made from.
5 List two hazards involved in working inside a tank.

 COMPLETE WORKSHEET 2

9.3 Basic tank installation requirements

It is important that plumbers have the skills and knowledge required to install and connect piping to a storage tank. Selecting the correct valves and fittings of the correct material ensures a professional job. Having a water tank installation fail inside a building could cause serious flooding and costly damage.

Water supply to and from the storage tank

Connections to the inlet and outlet of a storage tank are much the same regardless of the tank's size. However, the type and size of piping materials, tank construction materials and installation procedures for the tank may vary considerably depending on the requirements of the client, various authorities, manufacturers and suppliers.

The materials used for water supply systems to static storage tanks throughout Australia will vary depending on the area. The plans and specifications generally identify the materials chosen to supply water to the storage tank and state any specific requirements for the installation. Although the materials are selected from those listed in the Standards, they must also comply with the local authority's codes and regulations. These details would be dealt with at the development application stage. It is essential that an approved air gap is provided on the inlet supply. This air gap may have to be carefully calculated, recorded and checked on a regular basis once the tank has been installed (for example, RAG or registered break tank). AS/NZS 3500.1 explains how the air gap is calculated.

The connections to and from the tank will use either union couplings or flanges adjacent to the tank itself to allow for easy disconnection

All piping materials must be adequately supported according to current regulations.

 **AS/NZS 3500.1 PLUMBING AND DRAINAGE: WATER SERVICES**

Water supply piping materials

The most common piping materials used for water supply to and from storage tanks are plastics and copper. Other materials – such as ductile iron, cast iron–cement lined or galvanised mild steel – may be used depending on the purpose of the tank and its specific water supply source.

Plastics must be installed and joined according to the manufacturer's recommendations, and may include:
- acrylonitrile-butadiene-styrene (ABS)
- cross-linked polyethylene (PE-X)

- polybutylene (PB)
- polyethylene (PE)
- polypropylene (PP-R)
- unplasticised polyvinyl chloride (PVC-U).

Copper must also be installed in approved locations and according to the manufacturer's recommendations. Various methods may be used for jointing copper pipe, including:

- silver solder
- push-fit
- press-fit
- flanged
- roll-grooved.

Refer also to Chapter 1.

AS/NZS 3500.1 PLUMBING AND DRAINAGE: WATER SERVICES

FROM EXPERIENCE

It is important for the plumber to have the knowledge and skill to choose the correct material for a specific application.

Valves and ancillary items

There is a variety of valves and ancillary items that may be required for tank installation, depending on the size, location and purpose of the tank itself. Different valves may be used for the inlet supply, outlet supply and other parts of the tank (for example, sludge valves). Refer also to Chapter 4.

Inlet supply to a water storage tank

The more common valves for the inlet supply are approved stop taps/valves; for example, loose valves (jumper valves; see **Figure 9.11**), control valves, gate valves (see **Figures 9.12** and **9.13**) and float valves (see **Figure 9.14**). *Note:* It is recommended that the control valves be installed with the valve stem in a vertical position.

- *Loose valves (jumper valves), control valves*. These are designed with a loose valve (jumper valve) inside and fitted loosely to the spindle, thus making them a one-way flow valve. (Check the direction of flow carefully when installing this kind of valve, as no flow will occur if these are installed back to front.)
- *Gate valves*. Designed as a full flow valve, they open and close by having a sliding gate that operates laterally inside the valve.

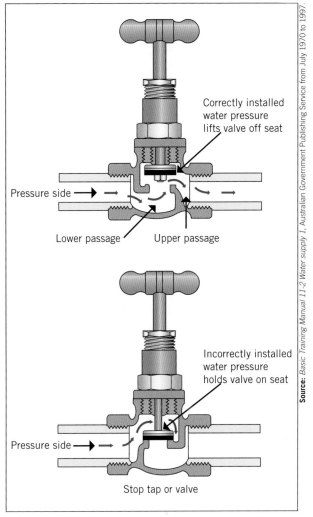

FIGURE 9.11 Stop tap with a loose valve (jumper valve)

- *Float valves*. The level of water within the tank is normally controlled by an approved float valve or other flow control device. These devices are sized to provide a calculated amount of replenishment water to the Standards and system design requirements. The size of the inlet float valve affects the size of the overflow outlet. For some installations (such as a pumped supply) the flow control may be regulated by a float control device, or devices, coupled with a control panel and pumps (see Chapter 8).

Outlet supply from a water storage tank

According to current regulations, any storage tank with a capacity in excess of 50L requires an isolating valve on the outlet, and the outlet connection is to be situated at least one pipe diameter (that is, of the outlet pipe itself) from the bottom of the tank.

Source: *Basic Training Manual 11-2 Water supply 1*, Australian Government Publishing Service from July 1970 to 1997.

Source: Zetco Valves Pty Ltd.

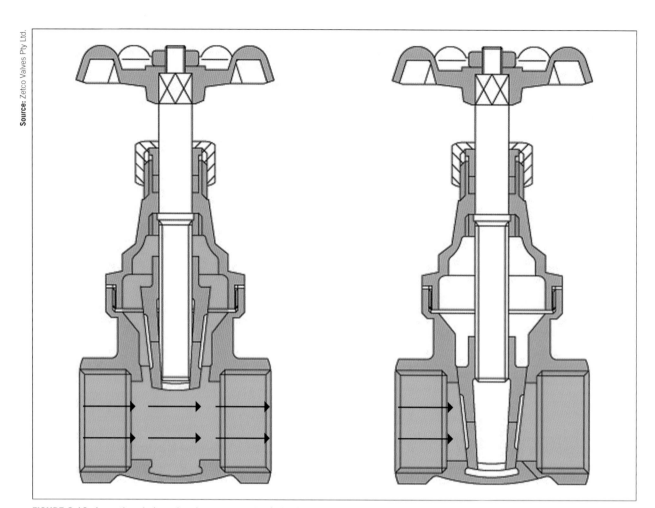

FIGURE 9.12 A sectional view showing a gate valve in both the open and closed position (refer to Figures 9.13, 9.16 and 9.18, which show a gate valve installed)

Source: Shaun Kristen.

FIGURE 9.13 A gate valve used on a fire service

Source: Terry Bradshaw.

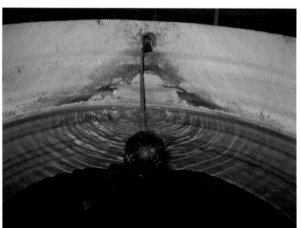

FIGURE 9.14 A float valve inside a tank

The actual connections to the tank outlet are normally full-way flow valves such as gate valves or ball valves (see Figures 9.12 and 9.15). These are used so as not to restrict the flow of water from the tank. These valves may be used as sludge valves where the tank capacity is in excess of 500L, and in these cases it is advisable to plug or cap the open end of the valve to help prevent any wastage of water.

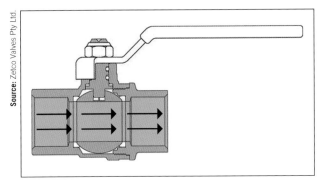

Source: Zetco Valves Pty Ltd.

FIGURE 9.15 A sectional view showing a typical ball valve in the open position

In cases such as fire service supply, the valve may have to be locked in the open position to ensure that supply is maintained at all times. **Figure 9.16** shows a fire service with an outlet using a 150mm gate valve strapped in the open position, with an anti-tamper device fitted to the isolating valve.

Source: Shaun Kristen.

FIGURE 9.16 Outlet from a fire service tank

The tank shown in **Figure 9.17** appears to have a capacity greater than 50L. The photo shows that it does not comply with current regulations because it does not have a valve on the outlet. It would therefore have to be completely drained to allow any work to take place on its outlet side.

FIGURE 9.17 The outlet point from a storage tank without an outlet valve

Ancillary items

A number of ancillary items are used on the actual connection of the water service to the tank. They may include, but are not limited to, check valves (non-return valves), union couplings and flange connections.

In some situations, an additional valve such as a check valve (non-return valve) may be required downstream of the outlet valve (see **Figures 9.18** and **9.19**). These are designed to prevent the reversal of flow from the downstream section of pipe. Depending on the design of the non-return valve, the position of its installation may affect its performance. Always check the manufacturer's recommendations for installation requirements.

Source: Terry Bradshaw.

FIGURE 9.18 Threaded union coupling connection, gate valve and check valve on the outlet of a tank

Regulations state that the inlet supply and outlet connection to a water storage tank must have an allowance for disconnection. This can be in the form of a union coupling or a flange, and is dependent on the size of the service connecting to the tank.

- *Union couplings* are defined as a demountable connection between two pipes, or fittings, that can be separated without damaging the pipework or components for reassembly. Union couplings are normally used on pipe sizes up to and including 50mm. Unions also have the advantage of allowing some minor deflection between the hard piping and the tank connection, which can assist during installation.
- *Flanges* are also a demountable joint, but have much less flexibility than a union coupling. Flanges are normally used on pipe sizes in excess of 50mm. The type of flange will vary, but must always match the valve/tank connection to which it will be joined. Appropriate gaskets are also used to ensure that the joint is watertight. It is extremely important when installing flange connections to ensure that the alignment of the faces of the flanges and their bolt holes are accurate, as there is no margin for error.

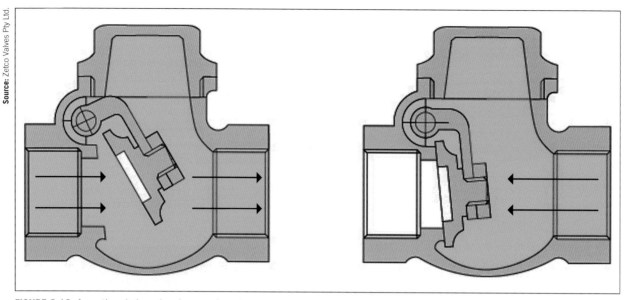

Source: Zetco Valves Pty Ltd.

FIGURE 9.19 A sectional view showing a swing check valve in both the open and closed position – these must be installed in a horizontal and upright position (refer to **Figure 9.18**)

Other examples of ancillary items that may be used (but are not discussed here) include:

- alarm valves
- butterfly valves
- flow switches
- pressure-limiting valves
- pressure switches
- strainers.

Parts of the storage tank

Plumbers need to understand several key parts of the storage tank.

The overflow pipe

The overflow pipe should be designed according to current regulations such as AS/NZS 3500.1 and the specification (if provided). The minimum size for any storage tank overflow is DN 40. Sizing the overflow outlet is normally based on the current relevant Australian Standards.

AS/NZS 3500.1 PLUMBING AND DRAINAGE: WATER SERVICES

Materials used for the overflow must match the specification requirements and be of a type that conforms to current regulations. The overflow pipe will normally discharge into the safe tray, but on occasions may discharge onto the floor or other approved area.

Figures 9.20 and **9.21** show an overflow pipe on a storage tank located externally. This can drain to the storm water drainage system (if approved).

FIGURE 9.20 Typical overflow connection to a storage tank

Safe tray and safe waste

The safe tray and safe waste must be designed according to current regulations such as AS/NZS 3500.1 and the specification (if provided).

The safe tray is a watertight tray that is fitted under a storage tank, where required; it drains through an outlet to a safe waste (see **Figures 9.2** and **9.4**). The safe tray should be constructed from approved materials,

FIGURE 9.21 Dual storage tank overflow connection

generally sheet metal, with specific measurements that must comply with current regulations. The minimum height of the sides of the tray is 50mm, and the minimum clearance between the side of the storage tank and the safe tray is 75mm. Timber supports should be placed both beneath and inside the safe tray; the storage tank is placed on these supports. The main purpose of these supports is to help to evenly distribute the weight of the tank and its contents and to allow air movement beneath the base of the tank.

The safe waste must be of an internal diameter greater than that of the tank overflow but not less than 50mm. It must be constructed from approved materials in accordance with current regulations. The safe waste must discharge to a point where it is noticeable but does not cause any inconvenience.

AS/NZS 3500.1 PLUMBING AND DRAINAGE: WATER SERVICES

Sludge valve

The sludge valve is primarily designed to allow the tank to be cleared of silt/sediment build-up that has settled out of the water and deposited on the bottom of the tank. Sludge valves are located at the bottom of the tank with a minimum size of half the outlet diameter or DN 40 minimum. They must be fitted to any tank that exceeds 500L in capacity. As stated above, it is advisable to plug or cap off the open end of the valve to help prevent any wastage of water. An example of a sludge valve is shown in Figure 9.22.

Approved cover

The cover should enable easy access to the various controls for the tank (float valves, level controls etc.)

FIGURE 9.22 Sludge valve and drain

and be either removable (see Figure 9.5) or of an approved design and fixed in position (such as approved roofing material or a specially constructed cover that is compatible with the tank). In the case of larger-capacity tanks, approved access must be allowed both onto and into the tank itself for maintenance purposes. The removable access cover must be a minimum of 0.5m^2 in size as per AS/NZS 3500.1. Refer to Figure 9.23 and the section 'Access to all parts of the tank' later in the chapter.

Support for storage tanks

The structural supports for water storage tanks must be designed according to current regulations such as AS/NZ 3500.1 and the specification (if provided).

When support for a tank is required, the material used is generally determined by the location, size and volume of the tank. As you can imagine, a smaller tank for residential and domestic supply located externally to the building may require only a concrete pad to support the base of the tank. Larger tanks may require additional support, such as the installation of a concrete or timber plinth to support the tank (see Figures 9.24 and 9.25). These plinths may have a dual application as they also allow ventilation under the tank, thus reducing the likelihood of corrosion and thereby increasing the tank's longevity.

Tanks placed in roof spaces or in elevated locations may require the design input of an engineer to enable additional support structures to be put in place. The main thing to consider, regardless of where the tank is located (internally, externally, on the ground or elevated), is the mass of the tank, its contents and any material used to support the tank.

A structural engineer can calculate and make provision for the structural requirements for large-capacity tanks. The engineer must take into consideration the mass of the tank material itself plus the support material – for example, a cast iron tank with reinforced concrete supports – and then design the structure to support the entire installation.

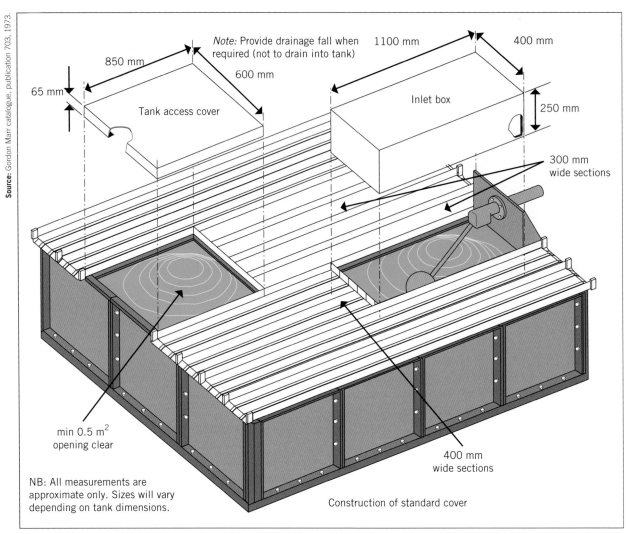

Source: Gordon Marr catalogue, publication 703, 1973.

Note: Provide drainage fall when required (not to drain into tank)

1100 mm

400 mm

850 mm

600 mm

65 mm

Tank access cover

Inlet box

250 mm

300 mm wide sections

min 0.5 m² opening clear

400 mm wide sections

NB: All measurements are approximate only. Sizes will vary depending on tank dimensions.

Construction of standard cover

FIGURE 9.23 Allowance for access to a large-capacity storage tank

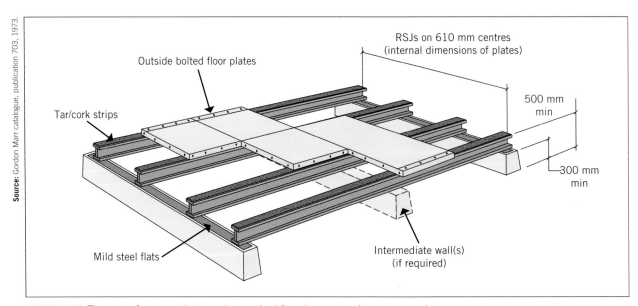

Source: Gordon Marr catalogue, publication 703, 1973.

RSJs on 610 mm centres (internal dimensions of plates)

Outside bolted floor plates

500 mm min

Tar/cork strips

300 mm min

Mild steel flats

Intermediate wall(s) (if required)

FIGURE 9.24 The type of support that may be required for a large-capacity storage tank

Source: Tank Industries, A Division of Hunt Engineering & Staff Pty Ltd.

FIGURE 9.25 This tank has been placed on a roof with additional lateral support being provided to the timber slats, giving the system the ventilation and corrosion control it needs

EXAMPLE 9.1

HOW TO CALCULATE TANK CAPACITY AND MASS

A water storage tank is 10m long, 6m wide and 4m high. Determine the volume of the tank in cubic metres (m^3) and litres (L), and the mass in tonnes (T).

Note: 1L of water has a mass of 1kg and $1m^3$ contains 1000L, so 1000L has a mass of 1000kg or 1T.

Volume in cubic metres

We can calculate the volume in cubic metres by using the formula:

$$V = L \times W \times H$$

where:
V = volume in cubic metres
L – length in metres (10)
W = width in metres (6)
H = height in metres (4)
$$V = 10 \times 6 \times 4$$
$$= 240m^3$$

Volume in litres

We know that there are 1000L in a cubic metre and that this tank contains $240m^3$ of water.

Thus, multiplying 240 (m^3) by 1000 will give the capacity in litres (and also in kilograms).

$$\text{Volume in litres} = 240 \ (m^3) \times 1000$$
$$= 240 \ 000L$$

Mass in tonnes

Mass is 240 000kg; that is, 240T.

Understanding the above method to determine a tank's water capacity and total mass is important to plumbers in everyday work activities related to the installation of tanks. For example, if a tank is to be situated in a ceiling space in a domestic residence, careful consideration should be given to the location of that tank. It is essential to locate the tank over corners of the wall below, and preferably spanning two or three walls, such as near a bathroom or hallway area.

> Severe damage and possible injury could occur if a tank was installed in the ceiling space in the middle of a room without adequate support.

Other points to be considered regarding tank installation

There are a few other points to be considered regarding tank installation, including:
- access to all parts of the tanks
- internal corrosion protection
- use of an approved membrane
- sustainable identification markings.

Access to all parts of the tank

As mentioned above, it is important to have suitable, approved access to the tank for maintenance purposes. An access ladder is required if the internal tank depth is 2m or more. Any access ladder should be designed to specifications (as per AS 1657), incorporating safe work practices at heights and allowing for access to both the tank cover and access opening for periodical maintenance and cleaning purposes. All access openings must have signage adjacent to the opening stating 'Warning – confined space'.

Figures 9.26 and 9.27 illustrate various forms of access.

Source: Shaun Kristen.

FIGURE 9.26 Trafficable access provided to a tank's maintenance cover

FIGURE 9.27 Storage tanks with approved ladder access

FIGURE 9.28 Multiple rainwater storage tanks, a service from the drinking (potable) supply – complete with a backflow prevention device and its drain – the pumping unit and the supply from the tanks to the points of usage

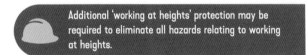
Additional 'working at heights' protection may be required to eliminate all hazards relating to working at heights.

Internal corrosion protection

Some storage tanks are required to have internal corrosion protection. One of the reasons for accessing some larger tanks is to check and possibly resurface the inside of the tank to help prevent corrosion. It is therefore important to have suitable access to the tank and for the personnel working on this process to have current confined-space qualifications.

Use of an approved membrane

In some situations, regulations and/or specifications may state that a membrane of non-corrosive insulating material may be required for all metallic tanks or other tanks as directed. It is advisable to check all documentation to determine what is required.

Suitable identification marking

In some situations, regulations and/or specifications may state that piping materials, valves, specific tanks or parts of the tank installation may require identification. It is advisable to check all documentation to determine what is required.

An example of a water storage tank installation

Currently it is very common for the installation of a storage tank system for rainwater harvesting where the stored roof water is used for activities such as:
- flushing toilets
- topping up a swimming pool
- washing clothes
- watering the garden
- washing cars.

When this type of system is used, the water storage can be supplemented by topping up the tank water from the reticulated system, and therefore an approved backflow prevention device must be installed on that inlet supply to prevent cross-connection. The water from the tank system is distributed by means of a pump unit to the various points of usage. Refer to Figures 9.28 and 9.29.

FIGURE 9.29 The roof water drainage pipe to the tanks, the drinking (potable) water supply to the tank system and the outlet supply to the points of usage for the system shown in Figure 9.29; note the approved identification stickers indicating flow

COMPLETE WORKSHEET 3

LEARNING TASK 9.2

1   What is the minimum size of a removable access cover?
2   Name three uses of storage tank water.
3   What does 1L of water weigh?
4   When must a sludge valve be fitted?
5   What type of isolation valve must be installed on the tank outlet?

9.4 Installing a water storage tank and its components

Having covered the basic background information relating to connecting and installing storage tanks to a reticulated water supply, we turn to what is required to complete the task. Much of the necessary information relating to the various elements, such as piping materials and valves, has already been covered in other chapters, and even though it seems repetitive, it is necessary to revisit some of the important parts here.

Prepare for the work

Regardless of the task, it is important to plan and prepare in readiness for that task. The following need to be carried out:

- Plans and specifications should be obtained. If these are not available, check the work site at least to ensure that all contingencies are catered for.
- Safety and workplace environmental requirements associated with connecting static storage tanks to a water supply system must be followed. If in doubt, ask your employer or instructor. It is important to be aware of any specific manual handling or mechanical lifting requirements and to make the necessary provisions for these. As with every other practical task, it is essential that a risk assessment is carried out and a safe work method statement (SWMS) is completed during a toolbox talk.
- Quality assurance requirements should be identified and followed in accordance with workplace requirements. Again, if in doubt, ask your employer or instructor.
- Tasks should be planned and sequenced in conjunction with others involved in or affected by the work. This is extremely important whether the site is a domestic residence, a larger building site or a simulated workplace such as a TAFE college workshop. On a larger building site, for example, it is necessary to liaise with the site manager and representatives from other trades that may be affected by the task, so that the installation takes place with minimum disruption to other parties.

- Tools and equipment, including personal protective equipment, should be selected and checked for serviceability. If everything is in good working order, there is less chance of an accident and/or possible injury.
- The work area should be prepared to allow for an accessible connection and installation. Again, this is extremely important whether the site is a domestic residence or a larger building site.

Identify installation requirements

If installation requirements are not checked, it is possible that the task may not be completed according to your plans, which may cost you time and money. Points to be considered include the following:

- Required materials should comply with relevant Australian Standards, and job specifications should be determined from plans and specifications. It is also important to check the availability of the materials and suitable possible alternatives should these materials be unavailable.
- Sustainability principles and concepts should be applied to work preparation and application. In today's community and climate, this is an important factor in all applications.
- Quantities of required materials should be calculated from plans and specifications.
- Materials and equipment should be ordered and collected according to workplace procedures. The magnitude of the task will determine when the items are ordered and collected and whether there is a suitable secure location to store them.
- Tank manufacturers can provide tanks in all shapes and sizes and generally have a range of tanks from which to select to suit your specific installation. In most cases tanks can be ordered directly from the manufacturer, but in some cases they need to be ordered through the local plumbing supply company. Tanks can usually be delivered directly to the site as and when required. When placing a special order for a tank (that is, one designed to suit a particular installation), it is essential to provide specific details relating to the tank's connection points. Most times, a sketch indicating any specific requirements will be helpful for the supplier.
- Materials and equipment should be checked for compliance with relevant Australian Standards, docket and order form, and for acceptable condition. If any items do not comply with the above, they should be noted and returned to the supplier for replacement.

Install and test the storage tank

As stated previously, it is important to be aware of any specific requirements related to issues such as working at heights, working in confined spaces, manual handling and mechanical lifting at this point and to make the necessary provisions as required. This should have been considered in the planning process and as part of the toolbox talk, risk assessment and subsequent SWMS. Do not place yourself or others in a dangerous position and ensure that all the necessary safety precautions are taken.

The following points should be carried out in this part of the process:

- The storage tank and associated pipework should be set out in accordance with the plans, specifications and job instructions.
- Pipe supports and fixings, compliant with relevant Australian Standards, should be installed as per the plans and the manufacturer's specifications.
- The tank, piping and materials should be installed in accordance with the plans, specifications and relevant Australian Standards.
- Jointing systems should be confirmed as compliant with relevant Australian Standards.
- The installed system should be pressure tested and commissioned in accordance with the relevant Australian Standards and job specifications. All tanks should be tested for soundness prior to, and after, installation so as not to cause any inconvenience to the user once installed.
- Test data should be recorded in the format required by the job specifications and quality assurance procedures.

HOW TO

HOW TO CLEAN AND MAINTAIN THE STORAGE TANK

Storage water tanks must be cleaned and disinfected before initial use and when the tank is being repaired or serviced. This involves draining the tank so all the sludge can be removed and all the internal surfaces can be cleaned with a high-pressure water jet, scrubbed and swept.

After the cleaning is complete, the tank is disinfected with a chlorine solution. There are two methods that can be used and they have retention times that must be adhered to. Refer to AS/NZS 3500.1 for this information.

AS/NZS 3500.1 PLUMBING AND DRAINAGE: WATER SERVICES

9.5 Cleaning up

Upon completion, the work area must be left clean and tidy. Any material left over should be stored and reused on another job or recycled, which will reduce waste.

Tools and equipment need to be cleaned before storing and kept in good working order. Any tools or equipment on hire should be returned.

All relevant documentation must be submitted to the regulatory authorities, principal contractor (builder) and the client (owner).

Refer to the corresponding section in *Basic Plumbing Skills* and to **Figure 9.30**, which shows a tank installation site with the clean-up completed.

Source: Picture courtesy of Altanks Pty Ltd.

FIGURE 9.30 A tank installation site with the clean-up completed

LEARNING TASK 9.3

1 What needs checking when the tank is under test?
2 Give an example of a quality assurance requirement.
3 What are some main considerations when installing a water tank?

 COMPLETE WORKSHEET 4

- Water storage tanks have three main purposes:
  1. to provide a water supply that is physically disconnected from the mains supply to prevent cross-connection
  2. to store and distribute water in buildings at a level higher than that which the supply authority can provide through its water mains (high-rise buildings)
  3. to provide a large reserve of water (such as for fire services) where the potential demand exceeds the available mains supply.
- The advantage of round tanks is that there are fewer corners and crevices for mould and mildew to form in, but they use more space.
- The advantage of rectangular tanks is that they can save space and can be built in sections, but they have more corners and crevices for mould and mildew to form in.
- It is important to have isolation valves on the inlet and outlet connections with unions or flanges for disconnection.
- Adequate tank support is critical to support large tanks.
- Sufficient access must be allowed in order to maintain tanks.

REFERENCES

AS 3855 Suitability of plumbing and water distribution systems products for contact with potable water
AS/NZS 3500.0 National Plumbing and Drainage Code – Glossary of terms
AS/NZS 3500.1 Plumbing and drainage, Part 1: Water services

GET IT RIGHT

1 Which photo has a correct connection?

2 Why is it correct?

WORKSHEET 1

Student name: _____

To be completed by teachers

Satisfactory ☐

Not satisfactory ☐

Enrolment year: _____

Class code: _____

Competency name/Number: _____

Task: Review the section '9.1 Background' and answer the following questions.

1 What is meant by the term 'open type' in relation to water storage tanks?

_____

_____

_____

2 Provide a definition of a water storage tank.

_____

_____

_____

_____

3 Describe the difference between a small-capacity storage tank and a large-capacity storage tank.

_____

_____

4 State the three main purposes of a water storage tank.

_____

_____

_____

5 Why is it important to have more than one large-capacity storage tank available for a water supply in some instances?

_____

_____

_____

6 Give three examples of what a storage tank may be used for.

_____

_____

_____

WORKSHEET 2

Student name: _____

Enrolment year: _____

Class code: _____

Competency name/Number: _____

Task: Review the sections '9.1 Background' and '9.2 Basic design requirements for locating a storage tank' and answer the following questions.

1   List two basic design requirements for locating a storage tank.

   _____

   _____

2   List six different materials that may be used to construct a storage tank.

   _____

   _____

   _____

   _____

   _____

   _____

3 Identify the various items shown in the diagram of a storage tank below.

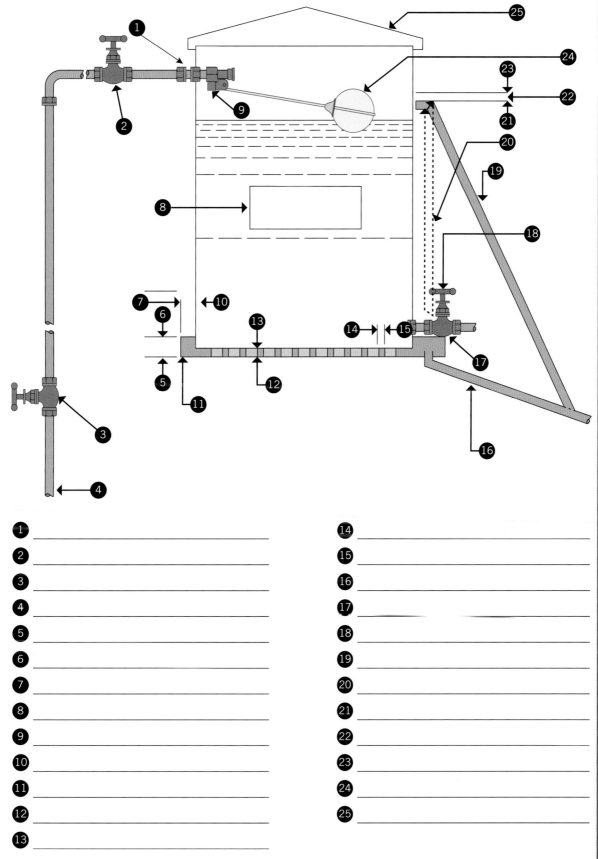

1	_____	14	_____
2	_____	15	_____
3	_____	16	_____
4	_____	17	_____
5	_____	18	_____
6	_____	19	_____
7	_____	20	_____
8	_____	21	_____
9	_____	22	_____
10	_____	23	_____
11	_____	24	_____
12	_____	25	_____
13	_____		

Source: *Basic Training Manual 11-1 Water Supply 1*, Australian Government Publishing Service from July 1970 to 1997.

4   List the two main shapes of storage tanks in use today.

_____

_____

5   Describe the main advantages of a circular tank.

_____

_____

_____

_____

6   Describe an advantage of a rectangular tank.

_____

_____

_____

_____

7   List three possible locations in which storage tanks can be positioned on large building sites.

_____

_____

_____

8   Describe how a storage tank located in a public car park area could be protected from vandalism and mechanical damage.

_____

_____

WORKSHEET 3

To be completed by teachers

Satisfactory ☐

Not satisfactory ☐

Student name: _____

Enrolment year: _____

Class code: _____

Competency name/Number: _____

Task: Review the section '9.3 Basic tank installation requirements' and answer the following questions.

1   Name the two most common piping materials used for water supply to and from a storage tank.

_____

_____

2   Describe how a float valve works.

_____

_____

_____

3   State two methods that may be used for jointing copper pipe.

_____

_____

4   Where would a loose jumper isolation valve be installed on a water storage tank?

_____

5   Complete the following statement: 'According to current regulations, any storage tank with a capacity in excess of _____ requires an isolating valve on the outlet'.

6   Describe in your own words why the actual connections to the tank outlet are normally gate valves or ball valves.

_____

_____

7   Describe the method used to allow for disconnection of a water service from a storage tank.

_____

_____

8   When would flanges be used instead of unions?

_____

9   State the minimum size for any storage tank overflow.

_____

10   State the minimum size of a safe waste from a safe tray.

_____

11   Describe the purpose of a sludge valve on a storage tank.

_____

_____

12   What is the required distance that an outlet is located from the bottom of the tank?

_____

13   State four uses for water contained in a tank used for rainwater harvesting.

_____

_____

_____

_____

14   Complete the following calculation and fill in the missing parts of the formula and other statements
     with the correct response.

     A water storage tank is 8m long, 4m wide and 3m high. Determine the volume of the tank in cubic
     metres (m^3) and litres (L), and the mass in tonnes (T).

     Volume in cubic metres

     We can calculate the volume in cubic metres by using the formula:

     $V = L \times W \times H$

     where:
     V = _____
     L = _____ (__)
     W = _____ (__)
     H = _____ (__)
     V = __ × __ × __
       = ___ m^3

Volume in litres

We know that there are 1000L in a cubic metre and that this tank contains _____ m^3 of water.

Thus, multiplying _____ (m^3) by 1000 will give the capacity in litres (and also kilograms).

Volume in litres = _____ (m^3) × 1000

 = _____ L

Mass in tonnes

Mass is _____ kg; that is, _____ tonnes.

WORKSHEET 4

Student name: _____

Enrolment year: _____

Class code: _____

Competency name/Number: _____

To be completed by teachers

Satisfactory ☐

Not satisfactory ☐

Task: Review the section '9.4 Installing a water storage tank and its components' and answer the following questions.

1 State the points to be considered when planning and preparing for a work activity related to installing storage tanks.

_____

_____

_____

_____

_____

_____

2 State the points to be considered in identifying installation requirements for a work activity related to installing storage tanks.

_____

_____

_____

_____

_____

_____

3 What is it recommended that you do when placing an order for a storage tank directly with a supplier?

_____

_____

_____

_____

_____

_____

4   State the main workplace safety points to be considered when carrying out work related to installing and testing a storage tank for a work activity.

_____

_____

_____

_____

_____

_____

5   How would a tank installation be tested?

_____

_____

_____

_____

6   Why is it important to reuse and recycle leftover material?

_____

_____

_____

_____

_____

_____

Group activities

7   Discuss with your colleagues any sustainability practices that might be considered in relation to the task of installing and testing a storage tank system. Record specific elements of the discussion below.

_____

_____

_____

_____

_____

_____

_____

8 Discuss with your colleagues each of the installations shown in the two photographs below and note any items that do not comply with AS/NZ 3500.1 Plumbing and drainage, Part 1: Water services. (For example, look closely at the jointing method used on the cold water supply inlet to the tank in photograph (a).) Note the specific clause number that relates to the non-compliant item and write it down in the space below.

(a)

(b)

_____

_____

_____

_____

_____

_____

9 Effective capacity relates to the actual amount of liquid that a tank can discharge, not the amount that it can theoretically hold. This has to be calculated carefully in relation to the design and location of the tank. It is the licensee's responsibility (not the apprentice's) to check the details prior to installation. However, to assist in the communication process when discussing storage tanks, it is helpful for you to understand some of the more specific items relating to the task.

a The diagram below shows a large-capacity storage tank and indicates some commonly applied terms in relation to determining the effective capacity of a storage tank. With the assistance of your instructor and your colleagues, research and note down the meaning of these terms and then discuss them as a group.

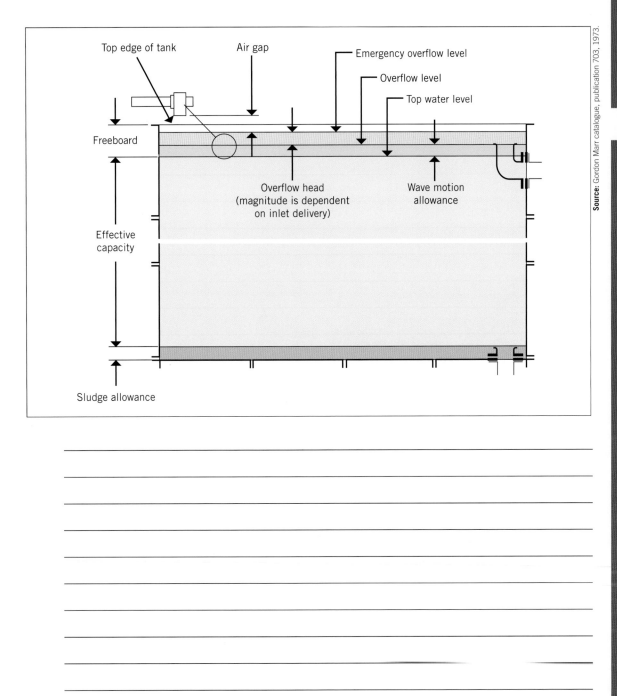

Source: Gordon Marr catalogue, publication 703, 1973.

b Referring to Chapter 6 in conjunction with the diagram in question 9(a), discuss with your colleagues the methods of controlling the water level within a tank if the tank is fed from a pumping system (for example, pump cut-in, pump cut-out, low-level and high-level alarms). Record specific elements of the discussion below.

_____

_____

_____

_____

_____

_____

_____

_____

_____

WORKSHEET RECORDING TOOL

Learner name		Phone no.	
Assessor name		Phone no.	
Assessment site			
Assessment date/s		Time/s	
Unit code & title			
Assessment type			

Outcomes

Worksheet no. to be completed by the learner	Method of assessment WQ – Written questions PW – Practical/workplace tasks TP – Third-party reports SC – Scenarios RP – Role plays CS – Case studies RW – Report writing PF – Portfolio	Satisfactory response	
		Yes ✓	No ✗
Worksheet 1	WQ	☐	☐
Worksheet 2	WQ	☐	☐
Worksheet 3	WQ	☐	☐
Worksheet 4	WQ	☐	☐

Assessor feedback to the learner

Feedback method (Tick one ✓):	☐ Verbal ☐ Written (if so, attach) ☐ LMS (electronic)

Indicate reasonable adjustment/assessor intervention/inclusive practice (if there is not enough space, a separate document must be attached and signed by the assessor).

Outcome (Tick one ✓):	☐ Competent (C) ☐ Not Yet Competent (NYC)

Assessor declaration: I declare that I have conducted a fair, valid, reliable and flexible assessment with this learner, and I have provided appropriate feedback.

Assessor name:	
Assessor signature:	
Date:	

Learner feedback to the assessor

Feedback method (Tick one ✓):	☐ Verbal	☐ Written (if so, attach)	☐ LMS (electronic)

Learner may choose to provide information to the RTO separately.

Learner assessment acknowledgement:	Tick one: Yes ✓	No ✗
The assessment instructions were clearly explained to me.	☐	☐
The assessment process was fair and equitable.	☐	☐
The outcomes of assessment have been discussed with me.	☐	☐
The overall judgement about my competency performance was fair and accurate.	☐	☐
I was given adequate feedback about my performance after the assessment.	☐	☐

Learner declaration:
I hereby certify that this assessment is my own work, based on my personal study and/or research. I have acknowledged all material and sources used in the presentation of this assessment, whether they are books, articles, reports, internet searches or any other document or personal communication.
I also certify that the assessment has not previously been submitted for assessment in any other subject or at any other time in the same subject and that I have not copied in part or whole or otherwise plagiarised the work of other students and/or other persons.

Learner name:	
Learner signature:	
Date:	

BACKFLOW PREVENTION

10

Chapter overview

This chapter provides further knowledge of the type of work that a plumber is involved in when protecting the potable water supply and installing irrigation systems.

Learning objectives

Areas addressed in this chapter include:

- cross-connection
- backflow prevention devices
- various types of irrigation systems
- installation procedures for connection of irrigation systems to a drinking supply
- planning work activities
- testing and commissioning procedures
- cleaning up.

10.1 Background

The drinking (potable) water supply for a community should never be taken for granted and should always be protected from contamination. If the water supply becomes contaminated in any way at all, the health and hygiene of the community is potentially at risk.

Procedures involved in determining the size of the connection to the water service – and, more specifically, what precautions must be taken to protect the drinking water supply from contamination that could occur due to a cross-connection between the drinking water supply and any alternative water supply, service, system or fixture – must be adhered to as per AS/NZS 3500.

The avoidance of cross-connection is called backflow prevention. This is an area of utmost importance to the water utilities, as they make every effort to provide good-quality drinking water to the community. It is for this reason that the water utilities insist that once the water leaves their supply main, through a control valve and then onto a property, it must not re-enter that water main.

 The role of the plumber in backow prevention is important, and that is why it is necessary to undergo extensive training. It is the plumber's responsibility to help to protect the community's water supply.

FROM EXPERIENCE

Plumbers should always be aware of all current rules and regulations, and safe work practices, and be aware of any new technology related to their work.

10.2 Cross-connections

A cross-connection is any connection, physical or otherwise, between any drinkable (potable) water supply system either directly or indirectly connected to a water main and any fixture, storage tank, receptacle, equipment or device through which it may be possible for any contaminated water or substance to enter a water system. Backflow can be caused by either backpressure or back siphonage. Cross-connection may take many forms, including:

- hoses left in pools
- hoses with chemical injectors attached
- outlets lower than the flood level of the fixture (submerged inlet)
- irrigation systems
- cooling towers
- bidets
- dental cuspidors
- photographic laboratories and other industrial and commercial processes.

Backpressure is the difference between the pressure of the water supply and a higher pressure within any vessel or pipework to which it is connected. If the water outlet on a property is much higher than a low-pressure water main, it is possible for backpressure to occur.

Back siphonage occurs when the water supply pressure falls below atmospheric pressure – in other words, a vacuum or partial vacuum is created. Reductions in drinkable (potable) water supply pressure occur whenever the amount of water being used exceeds the amount of water being supplied – such as during water mains flushing, firefighting or breaks in water mains – which creates back siphonage. In a water-utility potable water system, the effect of back siphonage is similar to drinking water through a straw.

Backflow is managed by the installation of a backflow prevention device. This can be a mechanical device, a registered air gap or a break tank, which is designed to automatically ensure the disconnection or separation of the drinkable (potable) water supply from a potential source of contamination or backflow.

Protection required

The type of backflow protection required within properties is defined into three different areas, as per AS/NZS3500.1.

Containment protection

Containment protection is installed immediately downstream of the meter assembly for a property that may have an area of risk. Generally, this means all commercial and industrial properties. It should be noted that the highest identified hazard on-site will also determine the hazard rating applied to the containment device.

Zone protection

Zone protection is when hazards with the same rating are identified and a device is installed on the supply line to those areas, such as a particular building in a complex, so as to protect the rest of the property from backflow or contamination of the drinking supply.

Individual protection

Individual protection is when a single hazard is identified and a backflow prevention device is installed for protection of individual fixtures, appliances or apparatus.

Selecting backflow prevention devices

Which backflow prevention device is needed for the installation is determined by the hazard rating (see Table 10.1) as per AS/NZ 3500.1. To select the appropriate backflow prevention device for the hazard rating, you need to:

- refer to the current relevant standard (AS/NZS 3500.1)
- identify the hazard rating from the protection tables
- select the valve most appropriate for the installation.

TABLE 10.1 Hazard ratings for water supply systems

Hazard rating	Definition
High hazard	Any condition, device or practice within a water supply system and its operation that could potentially cause death.
Medium hazard	Any condition, device or practice within a water supply system and its operation that could potentially endanger health.
Low hazard	Any condition, device or practice within a water supply system and its operation that could potentially constitute a nuisance but not endanger health.

 AS/NZS 3500.1 PLUMBING AND DRAINAGE, PART 1: WATER SERVICES

Backflow prevention devices

Backflow prevention devices must be manufactured in accordance with AS/NZS 2845.1 Water supply – Backflow prevention devices, Part 1: Materials, design and performance requirements. Air gaps and break tanks used as backflow prevention devices must be installed in accordance with AS 2845.2 Water supply – Backflow prevention devices, Part 2: Registered air gaps and break tanks and AS/NZ 3500.1 Plumbing and drainage, Part 1: Water services.

When the plumber tests any backflow prevention device, whether it is after the initial installation or as part of general ongoing maintenance, the testing must be done in accordance with AS/NZS 2845.3 Water supply – Backflow prevention devices, Part 3: Field testing and maintenance.

FROM EXPERIENCE

Only licensed plumbers with backflow accreditation may test and commission backflow prevention devices.

Installing backflow prevention devices

AS/NZS 3500.1 makes specific reference to the installation of backflow prevention devices. The rules are based on current 'best practice' principles and must be observed when installing the devices.

HOW TO

HOW TO INSTALL A BACKFLOW PREVENTION DEVICE

1 Do not apply heat to any backflow prevention device during installation. This is to prevent distortion or weakening of the valve body and internal workings through the application of heat; for example, from an oxy/acetylene flame.

2 Before installing the device, fit it with an approved line strainer. This is to help prevent any debris from fouling the internal workings of the actual backflow prevention device.

3 Place an approved isolating valve before the strainer and the device as well as downstream of the device. These isolation valves are required so that the device can be isolated for maintenance and repair, and to prevent draining the piping system of water during any maintenance and repair.

4 Flush all piping before connecting the device (and strainer). This is to help prevent any debris from getting into the valves and damaging areas such as the valve seats.

5 Do not install unprotected bypasses around the device. This is to protect the supply system from cross-connection. If a bypass is installed and there is no backflow prevention device on the bypass, then there is a potential for the contaminated water to flow through the bypass back into the drinking water supply.

6 Install the device according to the current Standard and following the manufacturer's written instructions. This is to protect both the consumer and the plumber and is normal best/safe practice. If unsure about the correct way to install the device, or the connecting pipework, it is advisable to contact the manufacturer so that any warranty is not voided. The local water authority should also be consulted to determine whether there are any rules or regulations specific to the local area.

7 Protect the device from damage such as freezing. Again this is a best-practice approach. Regulations state the actual installation requirements for the device, but if it is installed in a location where it can be damaged, or tampered with, necessary steps need to be taken to protect the valves. This should be considered when planning for the project and priced into the tender submission price.

8 If continuous water supply is essential, install another valve set-up 'in parallel' to the first, so that the latter can be shut down. This would be dependent on the purpose of the system, and if continuous supply is essential, then it must be catered for. This allows a valve or fitting on the line to be serviced or replaced while supply of water is maintained for the system. What it means is that if a particular backflow prevention valve is sized for a project, then another complete valve set-up of the

>>

same size should be installed next to – that is, 'in parallel' with – the other valve set-up. The valve configuration should be exactly the same. This can be achieved by rising up out of the ground with a supply pipe, fitting a 'tee' branch, completing the duplicate valve arrangements – complete with isolating valves, strainers, unions or flanges and the backflow prevention device – then connecting to another 'tee' branch and onto the pipework system.

9   Ensure that in-line devices can be removed or replaced. This can be interpreted to mean that unions or flanges are required on certain backflow prevention device installations. This is to allow for the removal or replacement of a device or other valve or fitting without damaging the rest of the installation.

AS/NZS 3500.1 PLUMBING AND DRAINAGE: WATER SERVICES

AS/NZS 3500 stipulates other regulations related to the installation of the backflow prevention device, as do the manufacturer's instructions. It is important that you familiarise yourself with both sources of instruction before attempting to install the device and any associated valves and pipework.

There are several different types of backflow prevention devices that can be installed in certain locations. As mentioned, the plumber should be aware of the current rules and regulations relating to the location and installation of the backflow prevention device. The plumber should also be aware of the current categories and terminology related to each of the categories and devices. Some of the terminology used in relation to the devices can be found in AS/NZS 3500.0 National Plumbing and Drainage Code – Glossary of terms and in the glossary of this book.

AS/NZS 3500.0 NATIONAL PLUMBING AND DRAINAGE CODE – GLOSSARY OF TERMS

FROM EXPERIENCE

It is important for the plumber to be aware of the pressure loss across the backflow prevention valve, especially in low-pressure areas.

Types of backflow prevention devices

There are quite a few different types backflow prevention devices designed for different hazard ratings. The low-hazard devices are not testable, but the medium- and high-hazard devices must be tested annually by an accredited licensed plumber.

Atmospheric vacuum breakers

An anti-siphon or atmospheric vacuum breaker (AVB) is designed to protect against back siphonage of contaminated water into the drinkable supply by allowing air into the pipe upstream of the device, so preventing a vacuum from forming. These vacuum breakers are for low-hazard situations such as Type B

irrigation installations, which are not subject to continuous pressure. They cannot be used in continuous pressure situations.

Dual check valves

A dual check valve is a low-hazard device and is installed on the outlet side of the water meter where dual water services are installed in a domestic residence to prevent the backflow of polluted water into the water utility's service. The dual check valve assembly consists of two in-line spring-loaded poppet check cartridges that require a 7kPa pressure difference to open. If a backflow situation occurs, the two valves will close, preventing reverse flow, which protects against backpressure and back siphoning.

Dual check valves with atmospheric port

A dual check valve with atmospheric port is a low-hazard device made for smaller supply lines. It is ideally suited for laboratory equipment, processing tanks, sterilisers, dairy equipment, dental consoles and similar applications with a low hazard rating where the device is under continuous pressure. It stops back siphonage or backpressure occurring and allows air into the pipe via the atmospheric port, preventing a vacuum from forming and discharging any water backflow to the atmosphere.

Hose connection vacuum breakers

Hose connection vacuum breakers (see Figure 10.1) are used on hose taps and prevent back siphonage in applications where hoses are connected for low hazards. The hose connection vacuum breaker is a low-

FIGURE 10.1 Hose connection vacuum breaker to a polyethylene drip system

hazard backflow device. It has a single check valve, which is spring-loaded to hold it in a closed position with an atmospheric vacuum breaker vent. These devices are generally attached to a hose tap and can also be used in other applications, including laboratory sinks, service sinks, wall hydrants, and commercial dishwashers and soap dispensers. They must be installed on all drinking water hose taps in dual water areas.

Double check valves

The double check valve provides protection to drinkable (potable) water supply from contamination in medium-hazard applications, such as Type C irrigation systems, where the device is under continuous pressure and backpressure. It is a testable device that must be tested annually. Double check valves consist of two spring-loaded check valves that require a 7kPa pressure difference to open. In a non-flow condition the check valves hold 7kPa differential minimum in the direction of flow. In a flow condition the check valves are open, proportional to the flow demand. In a backflow condition both checks will close, preventing reverse flow until the resumption of normal flow, when they will reopen. They are designed to protect against backpressure and back siphoning.

Pressure vacuum breakers

Pressure vacuum breakers (PVBs) are used on medium-hazard installations, such as Type C irrigation systems, where the device is under continuous pressure. PVBs stop back siphonage occurring and allow air into the pipe via the atmospheric port. This prevents a vacuum from forming and discharging any water backflow to the atmosphere. To allow them to operate and break siphon conditions, they must be installed at least 300mm above the hazard they protect and must be tested annually.

Reduced pressure zone devices

Reduced pressure zone (RPZ) devices are designed to protect against backpressure, back siphoning or a combination of both. They consist of two independently operating, spring-loaded, Y-pattern check valves and one hydraulically dependent differential relief valve. These devices provide protection to drinkable (potable) water supply from contamination in high-hazard applications and are testable devices. Under flow conditions the check valves are open with the pressure between the checks, called the zone, being maintained at least 35kPa lower than the inlet pressure and the relief valve being maintained closed at 14kPa. Should abnormal conditions arise (no flow or reversal of flow), the relief valve will open and discharge its contents through the relief port to maintain the zone at least 15kPa lower than the supply. When normal flow resumes, the zone's differential pressure will resume

and the relief valve will close. A significant pressure loss of up to 100kPa can occur through a RPZ device, so this needs to be factored into the installation to achieve the required pressure at the most disadvantaged outlet. An RPZ device is required to be tested annually by a registered backflow accredited plumber.

Registered break tanks

A registered break tank (RBT) is a storage tank that incorporates an air gap that is specifically designed to take into account the orifice size of the inlet and the size of the overflow, and using the criteria set out in AS/NZS 3500.1. It is a high-hazard testable device that is registered by, or on behalf of, a regulatory authority for the purposes of inspection and maintenance to ensure its functional requirements are maintained.

Registered air gap

A registered air gap (RAG) is an air gap in a storage tank or the air gap over a fixture, vat tanker filling point or drum of non-drinkable (non-potable) liquid (see Figure 10.2). It is a high-hazard testable device that is registered by, or on behalf of, a regulatory authority for the purposes of inspection and maintenance to ensure its functional requirements are maintained.

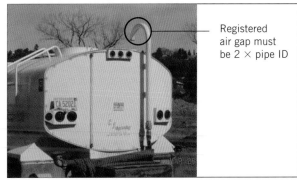

Registered air gap must be 2 × pipe ID

Source: Backflow Awareness Course @ http://www.tlch2o.com, 1 January 2004. All backflow materials are used by permission from CMB Industries, Inc.

FIGURE 10.2 Registered air gap

Most valves have limitations regarding their temperature and pressure capabilities. These factors must be taken into account when selecting backflow prevention devices. The appropriate valve for the application should be ordered.

When installing backflow prevention devices it should be remembered that no heat should be applied to them. Depending on their type, and whether they are testable or not, they should be provided with a line strainer, except when used on a fire service. Testable devices must have a resilient-seated isolating valve installed immediately upstream of the line strainer or immediately upstream of the device in cases where no integral line strainer is fitted. A resilient seated valve must also be installed on the outlet of the device. It is essential to flush the pipework before connecting the device.

LEARNING TASK `10.1`

1   What is the difference between backpressure and back siphonage?
2   Where would a pressure vacuum breaker (PVB) valve be installed?
3   Why is it important to flush the pipework before connecting a backflow valve and line strainer?
4   Where is the backflow valve installed when containment protection is required?

  COMPLETE WORKSHEET 1

10.3 Researching the irrigation system

What type of research is required before beginning the process of connecting an irrigation system?

Generally, this question should be asked before starting any form of work. If the research is not done, and the activity is not planned, then it is possible that unforeseen problems could occur. For irrigation systems, there are a number of issues that specifically need to be considered. As a guide, consider the following (although the research required is not limited to these points alone):

- the size and material for the new connection
- the layout of the irrigation system
- the type of irrigation system being used
- water pressure and volume requirements for the irrigation system
- the type of backflow prevention device required for the irrigation system
- the effect of the irrigation system on the drinking water supply
- the type of control systems required for the irrigation system
- use of an alternative water supply; will there be an alternative water source for the irrigation system/s and, if so, what effect will this have on the overall system, including the drinking water?

Some of the information required for the above points can be gathered from the plans and specifications (if any), but it is best to visit the site itself, prior to commencing work, in order to understand the requirements.

GREEN TIP

Using an alternative water supply, such as rainwater or bore water, for irrigation saves drinking water and cost.

The size and material for the new connection

The sizing and material for the irrigation system being used can be determined from the plans and specifications, if they are available. If neither is available, the size (normally no more than DN 20 pipe) will have to be determined from the literature provided by the supplier of the irrigation components. This may result in a 'hit or miss' approach because of the number of variables involved. The type of material used for the connection will depend on the material used for the main water supply and also whether the irrigation piping material is compatible with it.

If it is the plumber's job only to provide the connection point and the backflow prevention device, as per the client's instructions and in accordance with current rules and regulations, then that is where the plumber's role will finish. An experienced plumber may choose to give the client options as to what to do from here (for example, by discussing the flow dynamic of different piping materials, different valves, different layouts or fixing methods). If, however, the plumber's role is to design and complete the whole of the irrigation installation, then it is advisable to have a conservative approach and ensure that all of the facts and figures are thoroughly researched and recorded.

Layout

The layout of the irrigation system can be determined from the plans and specifications, if they are available. If not, it may be up to the plumber to suggest a basic design. It is advisable to follow normal trade practice when considering the layout. Here are some points to consider:

- If plans and specifications are available, check each of these carefully when on-site. This ensures that the task is achievable and complies with current rules and regulations. Also check that the water service that is being connected is as specified and will suit the actual design requirements.
- If plans are not available, then consult with the client about their specific needs for the irrigation service. Also refer to the manufacturer's brochures and specifications (depending on the type of service) to ensure that a service can be provided. It can then be ascertained where to provide a connection point to the water supply.
- It is not advisable to place sprinklers next to walls, as this could lead to staining of the walls. Irrigation water can also damage walls, as it may cause wood rot and/or mould problems. It is therefore advisable to keep sprinklers *at least 450–600mm away from walls* (perhaps more if the sprinklers are placed in a windy area).
- A suitable backflow prevention device must be installed on the irrigation system unless it is a Type A system.

- Ensure that the pipework system has adequate cover (if installed below ground) and that it is effectively fixed in accordance with the current rules and regulations.

FROM EXPERIENCE

Ensure client confidence by discussing the best options for the installation with professional knowledge and by addressing any foreseeable problems.

The type of irrigation system being used

Different applications on a site may require different types of irrigation system to be employed in different areas of the site. A properly designed irrigation system will cater for these various requirements. Pop-up sprinklers may be required for lawn areas, whereas (for example) a drip system, a 'tree watering' system or a trickle system may be used in various garden areas. Each of these systems may require various pipe and connection sizes.

When planning to provide water for these different types of irrigation systems, it is advisable to keep the similar types of systems on separate supply lines – in other words, do not mix the different types, as different flow rates and control methods may be required.

Designing an irrigation system

There are two main factors to consider when designing an irrigation system:
1. a plan drawn to scale showing all garden and lawn areas, buildings, driveways, pathways, retaining walls and hose tap locations
2. finding out the existing flow rate by timing how many seconds it takes to fill a bucket (a standard bucket is 9L) with the tap opened fully.

Gardens and lawns should be supplied from different zones as they require different amounts of water at different times. So keep these areas separate.

Lawns

Lawn sprinklers should be located head to head so they spray onto each other. Figure 10.3 shows minimal dry spots thanks to the overlapping of spray, therefore reducing the operating time by using six sprinklers to cover a 10m × 5m area. Figure 10.4 shows larger dry spots, so the operating time has to be longer to irrigate these dry spots because there are only two sprinklers for the same sized area.

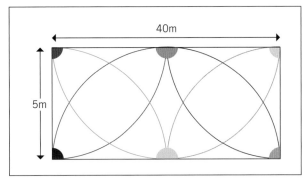

FIGURE 10.3 Efficient sprinkler spacing

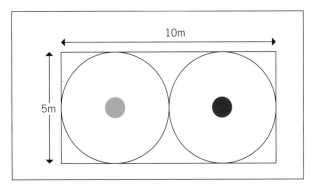

FIGURE 10.4 Inefficient sprinkler spacing

It is preferable to locate sprinklers around the edges rather than in the middle of the area (see Figures 10.3 and 10.4). If the area is too wide for the sprinkler coverage, then sprinklers would need to be placed in the middle as well.

The sprinkler type used in each zone must have a similar flow rate to ensure an even watering of that area and that an equal supply of water will occur at each outlet. For example, a fan pop-up sprinkler would use four times the amount of water than a gear-driven pop-up sprinkler, so these would not work efficiently in the same zone.

Gardens

Micro-spray sprinklers and driplines are recommended for gardens.

Micro-spray sprinklers have a radius of about 2m and a flow-rate between 50L and 70L per hour. They should be installed opposite each other like lawn sprinklers.

There are many different configurations of driplines. A common one has a flow rate of 2L per hour at 300mm spacings. These lines would be spaced at 400mm apart, therefore creating a grid pattern of driplines 300mm × 400mm. Remember it is important to calculate the flow rate required for area to be irrigated. A separate outlet (hose tap) is recommended for each zone.

Water pressure and volume requirements

As discussed earlier, the determination of water pressure and volume requirements for the irrigation system is absolutely vital in most cases. The 'full on/full off' nature of the irrigation system means that for certain systems to work properly and effectively, they have to be designed efficiently. A point to consider here is that if automatic (solenoid) valves are to be used on a system, there is potential for water hammer within the system. Provision should therefore be made to avoid this problem.

When working from plans and specifications, the pipework layout for the irrigation system is somewhat different from that for a drinking (potable) supply system. With pop-up sprinklers there is normally a trunk supply pipe feeding a number of branches or laterals to suit a particular area. As some form of zoning is normally required to ensure adequate supply to each sprinkler head, a considerable amount of piping is required to cater for each zone. It is important that provision is made for the correct flow of water through the pipes. The installation may need to be ring-fed to ensure adequate water supply.

Should pressure, volume and flow be a problem for the irrigation supply, a pumped system, a combination of a pump and tank supply system or an alternative means of water supply may be required.

Flow of water through pipes

While static pressure is a guide to the volume of water that might be expected to flow through a pipe system, other factors govern the volume that will actually flow. The most important factors are as follows:

- *The diameter of the pipe* – the larger the bore of the pipe, the more water it can deliver.
- *The length of the pipe* – the longer the pipe, the greater the friction between the water and the inside surface of the pipe, causing pressure loss.
- *Changes in direction* – sharp bends cause eddies that, in turn, cause friction and restriction of the water flow. Frictional loss varies with the angle of the bend and the radius of curvature. The sharper the angle and the shorter the radius, the greater the frictional losses will be. A radius of five times the pipe bore on the centre line gives no more friction than the same straight length of pipe.
- *The quality of jointing* – burred pipe ends, excess sealing compound and gaps between pipe ends restrict the flow of water.
- *Passage through valves.* Each time that water passes through a valve, it is restricted and usually forced to change direction once or twice, depending on the design of the valve. Allowance for pressure loss through backflow prevention valves must be considered. For isolation, full-way valves (such as ball valves) are preferred.
- *Enlargement or contraction in size.* To gain maximum flow, enlargements and contractions of pipe diameter should be made as gradually as possible.

De-burring the pipes

When the pipes have been cut, the burrs must be removed from both edges of the cut. A burr left on the outside of any pipe may foul the mating socket of the next piece of pipe or fitting. A burr left on the inside causes friction when water passes through the pipe. Outside burrs should be removed with a file. Inside burrs should be removed with a reamer, a rat-tail file or a rotary burr remover.

The type of backflow prevention device required

It is extremely important to note that there is a considerable pressure loss through most types of backflow prevention devices. These pressure losses, and the losses through other valves and fittings, must be factored into the design and installation phases of the project.

AS/NZS 3500.1 categorises irrigation systems according to the level of backflow prevention that is required. These categories consist of Type A, B, C and D systems, each with different conditions and limitations.

AS/NZS 3500.1 PLUMBING AND DRAINAGE: WATER SERVICES

Type A irrigation systems

A Type A irrigation system is any system of irrigation that has all outlets permanently open, has piping more than 150mm above ground, is not subject to cross-connection through ponding or backpressure and has no form of injection system. This type does not require any form of backflow prevention (see **Figure 10.5**).

Type B systems

A Type B irrigation system (see **Figure 10.6**) is any system of irrigation located within a domestic or a residential property that has outlets installed less than 150mm above ground level and does not have any form of injector system connected to it. These systems require a non-testable device such as a hose-type pressure vacuum breaker, atmospheric vacuum breaker or dual check valve.

Source for Figures 10.5 and 10.6: ANTA Learner Guide, BCPWT3007A Connect Irrigation Systems. This file is licensed under the Creative Commons Attribution-Share Alike CC BY 3.0 Unported licence.

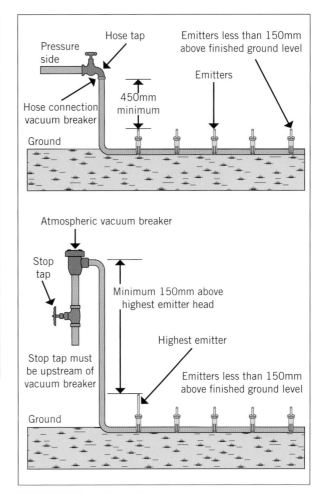

FIGURE 10.5 Type A irrigation systems

FIGURE 10.6 Type B irrigation systems

Type C irrigation systems

A Type C irrigation system (see **Figure 10.7**) is any system of irrigation in a location other than a domestic or residential property that has outlets installed less than 150mm above ground level and does not have any form of injector system connected. These systems require a testable device such as a double check valve or pressure-type vacuum breaker, depending on whether the protection required is for backpressure, back siphonage or both.

Type D irrigation systems

A Type D irrigation system (see **Figure 10.8**) is any irrigation system with a means of either injecting or siphoning chemicals, such as fertilisers, herbicides,

Source: ANTA Learner Guide, BCPWT3007A Connect Irrigation Systems. This file is licensed under the Creative Commons Attribution-Share Alike CC BY 3.0 Unported licence.

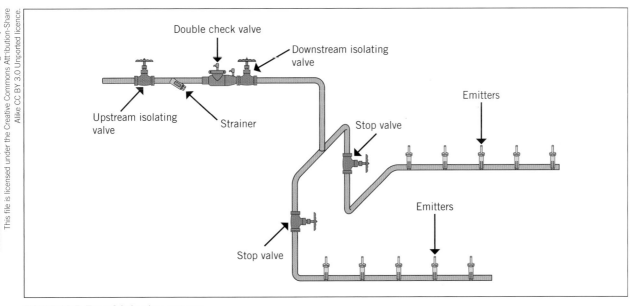

FIGURE 10.7 Type C irrigation systems

Source: ANTA Learner Guide, BCPWT3007A Connect Irrigation Systems. This file is licensed under the Creative Commons Attribution-Share Alike CC BY 3.0 Unported licence.

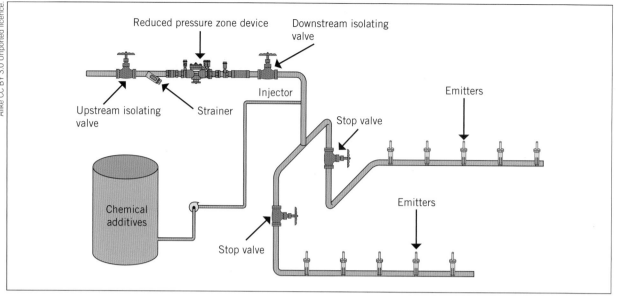

FIGURE 10.8 Type D irrigation systems

insecticides or similar, into the system. Due to the high hazard resulting from this feature of Type D systems, maximum backflow prevention must be used. The testable devices that are high-hazard rated are either a reduced pressure zone device or a registered air gap and a registered break tank.

10.4 Connecting an irrigation system to a drinking water supply

Irrigation systems, when correctly installed and maintained, are unlikely to affect the water supply system. However, if they are incorrectly installed or not properly maintained, they can have potentially dire effects on the drinking water supply system, and as a result may cause serious illness or death to the user or others in the community.

The types of irrigation system that can be used range from residential irrigation systems to urban irrigation systems. An example of a residential irrigation system is a simple 'drip' system where an above-ground hose is connected to the water supply (see **Figure 10.9** for typical connection) and simply drips onto the ground or soil. An example of an urban irrigation system is a zone irrigation system, where large areas of land or soil are divided up into zones watered by devices such as pop-up sprinklers; for example, those found in golf courses, parks and gardens.

Generally, the consumer who is using an irrigation system is trying to promote plant growth. In order to assist this plant growth, the consumer may be adding some form of fertiliser or herbicide to the irrigation water. It is extremely important to prevent such substances from getting into the drinking water supply through cross-connection. Current regulations state that

FIGURE 10.9 Typical connection for a Type A residential irrigation system

some form of backflow prevention device be installed on the water service (for example, a dual check valve at the meter) to help prevent contaminated (non-drinking) water from getting into the utility's water main or, in some cases, to other parts of the consumer's own water supply.

The plumber's role normally is to provide a connection point in the water supply for the irrigation system, but in some cases, particularly in domestic situations, the role may extend to completing the full installation for the consumer. (However, the design of urban irrigation systems is an exact science and should be left to an expert to design correctly. Specialist personnel are normally required to carry out the installation of these systems.) Should the nature of the work involve altering the water supply system itself (by physically changing the piping system), the connection to the drinking water supply must be carried out by a licensed plumber, or someone under their direct supervision, and must be done in accordance with current rules and regulations.

AS/NZS 3500.1 PLUMBING AND DRAINAGE: WATER SERVICES

Residential irrigation systems are generally Type A systems (as shown in Figure 10.5), and therefore require no form of backflow prevention. Use a hose connection vacuum breaker if connecting to a hose tap, or a dual check valve if there is a physical connection to the drinking water supply system.

Preparing for the connection

The areas of research to be completed before working on the drinking service are as follows.

The size of the available water service

The size of the available water service needs to be determined to establish whether the connection is large enough to provide water for the irrigation needs. If it is not large enough, then suitable alternatives have to be investigated before continuing the project.

The material of the available water service

The type of material used for the available water service needs to be established so that the appropriate fittings and valves can be selected. This will also help minimise the shutdown period for the connection.

The pressure and volume of water available

The pressure and volume of the available water supply is probably the most important factor when determining the type of irrigation service. The designer of the irrigation system requires this information so that they can design the irrigation system to suit their particular needs. (Remember that the plumber does not design the irrigation system and usually provides a point of connection for the supply to the system.) For a residential irrigation system, it may mean that the owner needs to rely on a smaller version of zone control for the various parts of their garden or lawn, or simply provide irrigation to different areas on different days, whereas with a larger urban irrigation system, more sophisticated methods of control and distribution may be required.

FROM EXPERIENCE

Proper planning leads to a safe and efficient installation, ensuring customer confidence and satisfaction.

The water pressure and flow rate

When discussing pressure you need to be aware of the two categories: static pressure and dynamic pressure.

- *Static pressure* is the pressure available when there is no water flowing through the pipework system. (However, this pressure alone does not indicate whether there will be enough water flow for the particular type of irrigation system.)
- *Dynamic pressure (working pressure)* is the pressure available when the water is flowing, which helps determine the sufficient flow rate required.

Determining the static pressure for a residential irrigation system may be as simple as connecting a pressure gauge to a hose tap and taking an approximate reading. Although static pressure does not indicate whether there will be enough water flow for most irrigation systems, it does help to determine whether a drip system or a simple spray system could be employed for the simple domestic system. When pressure and volume of water are essential (as they are in the more sophisticated systems), greater care and accuracy must be used to determine flow rates. Determining the dynamic pressure will indicate whether there is enough pressure and water flow combined to supply the irrigation system requirements.

When taking the static pressure reading on-site, it is important to take the reading as close as possible to the point of the connection for the irrigation system. This reading needs to be taken when there is no other flow in the system – check, for example, that a shower or washing machine is not operating, as this will give a working (dynamic) pressure reading. If these pressure readings are low, it may be necessary to take some other readings at various times throughout the day.

To design the irrigation system, the flow rate and dynamic pressure are required. Flow rate and dynamic pressure depend on the pressure available in the water main, the size of the water meter assembly and the size and length of the drinking (potable) water supply line. Elevation (height) above the water supply point also needs to be considered, as this will affect the dynamic pressure.

When pumps, and systems that require set pressures to operate, are required, a dynamic pressure/flow measuring device will have to be connected (see, for example, Figure 10.10). This will measure the dynamic pressure.

There are different methods used for calculating flow rates and pressure readings depending on the nature of the plumbing work involved. A sample of the readings taken could be recorded as shown in Table 10.2, and a worked example is shown in Table 10.3.

FIGURE 10.10 Pressure reading device

HOW TO

HOW TO ESTABLISH THE FLOW RATE

1 Connect the device to a hose tap with the hose tap adapter.
2 Ensure that the ball valve is turned off.
3 Turn the water on (slowly) at the hose tap and take the static pressure reading at the pressure gauge; take note of this pressure.
4 Carefully open the ball valve and take note of the pressure reading on the pressure gauge as the water is flowing through the device. This will give the dynamic (working) pressure.
5 While the water is flowing through the device, direct the water into a bucket and, using a stopwatch, time how long it takes to fill the bucket. This will help to determine the flow rate (generally in litres per minute or litres per second). The size of the bucket should be a minimum of 10L capacity. For a more accurate reading of the flow rate, a larger bucket should be used.

TABLE 10.2 Sample of pressure and flow information for an irrigation system

Client details	Name:
	Address:
Time of test	am/pm
Location of hose tap	
Static pressure reading	kPa
Dynamic pressure reading	kPa
Capacity of bucket	Litres
Time taken to fill bucket	Seconds
Flow rate (capacity of bucket ÷ time taken to fill it)	Litres/second
Flow rate in litres per minute (litres per second × 60)	Litres/minute

TABLE 10.3 A worked example of pressure and flow information for an irrigation system

Client details	Name: Apprentice plumber
	Address: Irrigation Road, Waterview
Time of test	10.00 am
Location of hose tap	Outside laundry
Static pressure reading	400kPa
Dynamic pressure reading	200kPa
Capacity of bucket	10L
Time taken to fill bucket	20 seconds
Flow rate (capacity of bucket ÷ time taken to fill it)	0.5L/second
Flow rate in litres per minute (litres per second × 60)	30L/minute

Remember that there will also be a pressure and flow loss through the backflow prevention device (up to 100kPa depending on the device) and any other valves and fittings that are installed as part of the connection, and this has to be factored in when considering any installation.

Once these factors have been considered, it can be decided whether the available water supply is sufficient for the type of irrigation system required, or whether a pump is required, or whether a storage tank and pump are required to provide both volume and pressure for the system. (If a pump and storage tank are required then, if set up correctly, the storage tank would fulfil any backflow prevention requirements by using a registered break tank or registered air gap.)

Locate and determine the size of the connection

It is important to determine the correct connection size to provide the required flow rate, so there is sufficient water supply to the irrigation system and fixture outlets.

Small service: residential irrigation system

When locating the connection point for the smaller residential irrigation system, the owner or plumber would normally look for the easiest point to connect (at or near an external hose tap – see **Figure 10.9**). Even so, careful consideration should be given to how to actually make the connection itself. It could be as simple as screwing a hose connection vacuum breaker onto the tap then connecting a polyethylene drip system to the breaker, as shown in **Figure 10.1**.

If, however, some variation has to be made to the piping system near the tap (such as inserting a tee and isolating valve), then more precautions and planning should be taken. This connection will involve isolating this section of pipework or shutting down the entire water supply system. Therefore, it is necessary to risk-assess this activity before commencing any work.

A residential irrigation system is likely to run at night and therefore could cause a noise problem, so it may be preferable to run a separate line to the backyard for the irrigation take-off point.

Larger service: residential or urban irrigation system

The available plans and the job specification (if any), or a site visit, will allow determination of the connection size to the water supply and the valve size. Most domestic water services are in either 20mm or 25mm copper, PVC-U or PE. The pipe material and size on-site can be verified by exposing the water service after the meter and measuring the diameter of the pipe. Again, this activity should be risk-assessed before starting any work, with careful consideration being given to other services, the client and the general public, as well as electrical safety.

 Remember when cutting into a metallic water service to check for stray current and that bonding straps should always be connected.

After determining the size of the connection, the valve size can be determined. Check the hazard rating for backflow prevention purposes, and select the appropriate backflow prevention device.

Excavation methods

The next step is to determine whether any excavation needs to be done to make the connection, what effect any excavation will have on the property and how to restore or make good the excavation area on completion. The size and location of the excavation for the connection will govern what methods will be used to make the excavation.

Any excavation should be carried out using appropriate procedures. This will involve locating and exposing any existing underground services, such as:

- power cables
- communications cables
- gas lines
- sewer lines
- stormwater pipes
- drinking or non-drinking water service.

The existing water service must be carefully exposed by using a shovel, so as not to damage it. Carefully consider those factors that contribute to the stability of the excavation. These include:

- the type of ground
- the depth and shape of the excavation
- the size and location of any spoil, or other heavy weights, such as machinery
- any buildings close to the excavation
- any vibration from traffic or other sources
- groundwater conditions
- any adjacent drains and services.

 Safe work practices must be carried out at all times while excavating to avoid damaging existing services.

The effect on the consumer of closing down the water service for the connection

During the planning phase, the plumber should be aware of, and plan for, any effect that the closing down of the water service will have on the client and the client's property. This will depend on:

- the type of establishment (residential, commercial or industrial)
- the timing of the shutdown (for example, whether it can be done in normal working hours)
- how long the system will be closed down
- what needs to be done to the rest of the water service after the connection is satisfactorily completed and the water is turned back on (for example, the effect on old joints, the hot water service and air in the line).

FROM EXPERIENCE

Effective communication with the client is important to avoid conflict when disrupting the water supply.

LEARNING TASK 10.2

1 What is the difference between water pressure and water flow?
2 Name three things to consider before excavating.
3 Where would you find an urban irrigation system?

 COMPLETE WORKSHEET 2

The effect of the irrigation system on the drinking water supply

Careful consideration should be given not only to the irrigation system but also to what effect it will have on the drinking water supply. For instance, what will happen to the domestic supply if, all of a sudden, the sprinkler system cuts in and the overall pressure drops dramatically? This type of information should be discussed with the client during the planning phase of the project.

The type of control system required

There is a range of different types of control systems used for irrigation systems. The control may consist of a simple manual control, such as a ball valve to turn the water supply on and off as required, or one that is controlled by a timer; or be more complex, such as an automated control (computer-controlled) system involving electronically controlled valves.

Figure 10.11 shows a typical control panel for an irrigation system, while Figure 10.12 shows valves on a manifold system supplying water to various parts of an irrigation system. The wiring from the control panel can be run to various parts of the property to control solenoid valves and zones located in remote situations (see Figures 10.13 and 10.14).

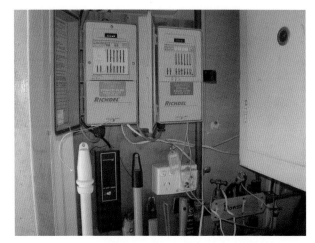

FIGURE 10.11 Typical automated control system for a residential irrigation system

FIGURE 10.12 Pipe manifold system – used to supply water to various zones

Using an alternative water supply

When using an alternative water source for an irrigation system, consideration must be given to the quality of the water and any possible cross-connection with the drinking water supply.

The supply to different areas within the irrigation system has to be designed so that each area has a water supply that is either fully on or fully off. The designer

FIGURE 10.13 Solenoid valves controlling flow to a drip system

FIGURE 10.14 Another example of remote control

has to allow for different zones within the overall system. Each zone may require different pressures and different volumes of water from that of other zones, or the system may simply be broken up, depending on the style of system the owner requires.

All of this depends, in turn, on the size of the main water supply service and the pressure and volume of water available in that main. It is for this reason, on larger projects, that a pumping system is required. This pumping system may be combined with a storage tank that feeds water to the pumps. With the current trend towards water recycling, some domestic properties rely on storage tanks to supply water for their irrigation needs (see Figure 10.15).

In some cases there may be an alternative water supply for the system, such as a dam, a creek or a river, an artesian bore or some other source. This alternative supply may either supplement the supply from the water main or be supplemented by the water main. In these cases it is very important to protect the main water supply from contamination because there is a high danger for a cross-connection to take place from

FIGURE 10.15 Storage tanks being used to supply water to a domestic irrigation system

the non-drinking water supply to the drinking water supply.

GREEN TIP

It is preferable to connect irrigation systems to an alternative water supply and conserve the drinking water supply.

 Non-drinking water cannot be used to irrigate fruit and crops that will be eaten unprocessed.

Although this chapter relates specifically to the connection of irrigation to a drinkable (potable) supply of water, this section gives a broad overview of areas where irrigation may use an alternative supply. This is becoming more common for domestic dwellings, which are increasingly using rainwater tanks for sustainability purposes (see Figure 10.15). The plumber needs to have an understanding of the protection required when dual water supplies are being used.

GREEN TIP

Plumbers are increasingly being called on to install an irrigation system connected to a non-drinking water supply to save precious drinking water.

Connecting an irrigation system to a non-drinking supply

The type of irrigation system will dictate the type of backflow prevention device to be installed at the isolation control set-up and whether a containment (on the outlet side of the water meter) backflow prevention device is to be installed (see Figures 10.16 and 10.17). This information may be found in AS/NZS 3500.1. Any testable backflow prevention device must be tested and commissioned prior to it being put into service. The local water utility may require a backflow prevention device inspection and maintenance report form to be issued by a licensed plumber with backflow accreditation.

The isolation control set-up must be installed a minimum of 300mm above the finished ground or surface level, and between 150mm and 300mm above the highest irrigation outlet, depending on the backflow prevention device used. Backflow prevention devices with an atmospheric port require a minimum height of 300mm to be maintained between the outlet port and the finished ground or surface level.

Polymer pipe may not be used to support the isolation control set-up, and all pipework must be installed in accordance with the requirements of AS/NZS 3500.1, the state or territory and any other authority having control over the installation.

When the service is connected to a non-drinking water service, the following conditions also apply:

- The pipework must be installed to the requirements of AS/NZS 3500.1, the state/territory and any other authority having control over the installation.
- All pipework, valves and outlets up to the solenoid controls must be clearly and permanently labelled with safety signs complying with the current version of AS 1319 (see Figure 10.18).

 AS/NZS 3500.1 PLUMBING AND DRAINAGE: WATER SERVICES

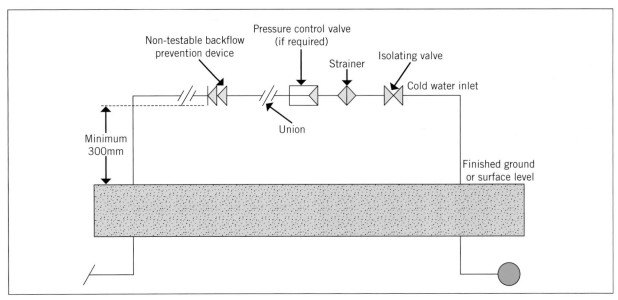

FIGURE 10.16 Standard irrigation system connected to a drinking or non-drinking water service with non-testable backflow prevention device

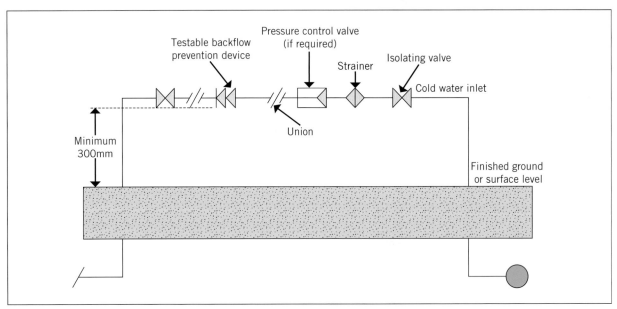

FIGURE 10.17 Standard irrigation system connected to a drinking or non-drinking water service with testable backflow prevention device

As with any water service, an irrigation service needs to be sized on the type and number of outlets. In addition:

- Application and approval is required from the local water utility prior to the installation of recycled water irrigation systems for other than domestic purposes.
- Backflow prevention devices must be installed.
- Licensed plumbers are required to contact the local water utility to arrange an inspection for all types of irrigation systems, including those for domestic use.

LEARNING TASK 10.3

1 Why can't non-drinking water be used to irrigate unprocessed fruit and vegetables?
2 What is the minimum distance that a sprinkler should be located from a wall?
3 What procedure is carried out to a testable backflow prevention device after it is installed?
4 What type of backflow prevention device is required for a Type D irrigation system?

Source: © 421 Environmental Products.

Source: (Bottom image) iStock.com/CTRPhotos.

FIGURE 10.18(a) Typical warning signs

FIGURE 10.18(b) Typical warning signs

10.5 Work preparation

The main considerations to prepare for an irrigation job have been mentioned previously in this chapter, such as plans and specifications, sequentially planning tasks, carrying out work to the statutory authorities' requirements and setting up the work area with an efficient connection to an irrigation system from a drinking water supply. Tools, equipment, quality assurance and work health and safety (WHS) requirements are as follows.

Tools and equipment

Some of the basic tools and equipment required for irrigation systems are:

- excavator for digging
- bobcat for backfilling and cleaning up
- shovels and spades for excavation
- mattock, pick and bar for breaking up ground
- tape measure, line marking paint and string lines for setting out
- hacksaw and tube cutters for cutting pipe
- press-fit tool and oxy/acetylene kit for pipe and fitting jointing
- pipe-grips (Stillsons, footprints and multigrips), shifter-spanner and spud wrench for screwed joints.

Quality assurance and WHS

Following company policy for quality assurance involves keeping all tools and equipment in good working order, carrying out the work in a professional manner, communicating effectively with the client and minimising material waste by making accurate orders.

Meeting WHS requirements, such as documenting a safe work method statement (SWMS) with an accurate risk assessment (including hazard identification, risk ranking and appropriate control measures), is part of the quality assurance policy.

10.6 Connecting and testing the system

Planning and preparation are important factors to take into consideration prior to shutting down or cutting into an existing service, as they may help avoid damage or inconvenience to the client. All services must be tested before being covered to avoid any damage.

Cutting into an existing metallic water service line

Many factors can adversely affect the standard and condition of an electrical installation. This can result in a stray electrical current passing through water pipes or other metallic components and structures. This stray electrical current can kill if the pipe is cut, breaking the electrical circuit, and the broken circuit bridges to the body. Before cutting through an existing metallic water service pipe, make sure that there is no stray electrical current by using an approved testing device and turning off the power. Bonding straps should always be used when cutting into metallic water services.

Pipes also must be buried deep enough to protect them from the weight of vehicles. AS/NZS 3500.1 specifies the depths at which pipes must be laid within footpaths, and this often also dictates the depth at which pipes must be laid on private property. Refer to the latest Standards for depth of cover.

AS/NZS 3500.1 PLUMBING AND DRAINAGE: WATER SERVICES

Laying distribution pipes

When distribution pipes are being laid, care must be taken to minimise water friction inside the pipes. This involves:
- careful reaming and de-burring of the internal surfaces of the pipe ends
- using fittings that cause the least friction (such as bends rather than elbows)
- adding as few directional changes as possible
- selecting the shortest possible route for the pipes.

Planning for future work

Stop taps should be placed in easily accessible locations so that sections of the distribution pipe installation can be isolated by shutting off the supply in order for repairs to be made or for the installation of additional piping for extensions.

Testing the installation

Small installations are usually tested as a whole, after connection to the main. Large installations are installed and tested in sections. The process for this is as follows.

HOW TO

HOW TO TEST THE INSTALLATION

1. Close off the pipe ends with plugs or blind flanges.
2. Test the section or sections that have been laid for leaks under full water pressure.
3. At the completion of the test, rectify any leaks and repeat the test until the installation is deemed satisfactory for the section under construction.
4. Turn off the water and drain it from the pipes by removing the plugs or flanges.
5. Lay the next section.
6. Repeat the testing process until the whole installation has been subjected to the necessary testing and approved by the appropriate authority.

10.7 Cleaning up

Upon completion the job must be left clean and tidy. All ground surfaces should be restored and returfed if necessary. Company quality assurance policies dictate that proper standards be maintained and that all rubbish is disposed of in the appropriate bins. Any material left over should be stored and reused on future jobs.

Tools and equipment should be packed away after they have been cleaned and maintained in proper working order. Any tools and equipment on hire must be returned as soon as possible to avoid extra cost.

All final documentation such as the certificate of compliance must be submitted to the regulatory authorities and client. Delivery dockets, invoices and order sheets should be reconciled as per company policy, and time sheets accurately completed.

LEARNING TASK 10.4

1. What tool would be used to de-burr the inside of copper tube before jointing?
2. What device is used to detect stray current?
3. Where should stop taps be located?

COMPLETE WORKSHEET 3

SUMMARY

- Plumbers have a major responsibility to protect the community's drinking water.
- Any potential cross-connection must be identified and prevented.
- Hazard ratings must be determined as per AS/NZS 3500.1 so the appropriate backflow prevention device can be identified.
- Backflow prevention devices must be located and installed as per AS/NZS 3500.1.
- Irrigation systems must be sized accordingly so the household supply is not affected.
- Available water pressure and flow rate is a major consideration for irrigation systems.
- Materials and equipment must comply with the relevant Australian Standards.
- Different irrigation systems require different backflow prevention devices.
- Alternative water supplies must be installed such that there is no possibility of cross-connection with the drinking water supply.

GET IT RIGHT

1 Which photo shows the valve correctly installed?

2 Why is it correctly installed?

3 What type of valve is it?

WORKSHEET 1

To be completed by teachers

Satisfactory ☐

Not satisfactory ☐

Student name: _____

Enrolment year: _____

Class code: _____

Competency name/Number: _____

Task: Review the sections '10.1 Background' and '10.2 Cross-connections' and answer the following questions.

1 Who is permitted to commission and test backflow prevention devices?

_____

2 What is the definition of a high hazard?

_____

_____

_____

3 What does the abbreviation 'RBT' mean?

_____

4 Break tanks used as backflow protection must comply with which Standard?

_____

5 Describe the term 'back siphonage'.

_____

_____

_____

6 Provide an example of a submerged inlet.

_____

_____

7 What is recommended when installing backflow devices if continuous water supply is required?

_____

_____

8 Describe the term 'zone protection'.

_____

_____

_____

_____

9 What does the abbreviation 'RAG' mean?

_____

10 Where should isolating valves be installed on a testable backflow prevention valve?

_____

_____

11 Provide two examples of a cross-connection.

_____

_____

12 What backflow device is used to prevent backpressure and back siphonage on a medium-rated hazard?

_____

13 When and where is a line strainer required to be installed on a backflow prevention device?

_____

_____

_____

14 What is the minimum height of the relief port on a RPZ valve from ground level?

_____

15 Name two situations where it is recommended that a dual check valve with atmospheric port be installed for protection.

_____

_____

16 List two common causes of backflow.

_____

_____

17 What pressure difference is required across the spring-loaded valves of a dual check valve to operate correctly?

_____

18   Explain why backflow protection is important to water utilities.

_____

_____

19   What is the difference between a dual check valve and a double check valve?

_____

_____

20   Registered backflow devices should be tested at intervals not longer than _____ months.

 **WORKSHEET 2**

Student name: _____

Enrolment year: _____

Class code: _____

Competency name/Number: _____

To be completed by teachers

Satisfactory ☐

Not satisfactory ☐

Task: Review the section '10.4 Connecting an irrigation system to a drinking water supply' and answer the following questions.

1   Describe the differences between a residential irrigation system and an urban irrigation system.

_____

_____

_____

2   Explain the difference between static pressure and dynamic pressure.

_____

_____

_____

3   Describe why it is important to establish both the static and the dynamic pressure when planning an irrigation system.

_____

_____

_____

4   To minimise friction loss within the system, what type of valves should be installed? Why?

_____

_____

5   Mechanical backflow prevention devices such as RPZs can dramatically reduce both flow and pressure, and this must be factored in when sizing the connection and affiliated valves and fittings for an irrigation system. Why is this important, and how would you determine the pressure and flow loss across a device?

_____

_____

_____

6   What effect could the scenario in question 5 have on the provision for backflow prevention?

_____

_____

_____

7   What is an alternative solution to provide the required pressure when the existing supply pressure is too low?

_____

_____

8   In relation to backflow prevention and connecting an irrigation system to a drinking water supply, there are three levels of hazards. Name these types of hazards and explain what they are.

_____

_____

_____

9   What backflow prevention devices can be used in a high-hazard situation?

_____

_____

_____

10   Does a Type A irrigation system require backflow prevention?

_____

WORKSHEET 3

To be completed by teachers

Satisfactory ☐

Not satisfactory ☐

Student name: _____

Enrolment year: _____

Class code: _____

Competency name/Number: _____

Task: Review the sections '10.3 Researching the irrigation system', '10.5 Work preparation' and '10.6 Connecting and testing the system' and answer the following questions.

1 Describe the safe procedure to be followed to avoid electrocution when cutting through an existing metallic water service to make a connection for the irrigation system.

_____

_____

2 What is a good practice to allow for any future work when connecting to the drinking (potable) water service?

_____

3 Describe a Type D irrigation system.

_____

_____

_____

_____

_____

4 List three alternative water supplies.

_____

_____

_____

5 Name three ways to reduce friction loss in a piping system.

_____

_____

_____

6 Which backflow prevention device is recommended for a Type D irrigation system?

_____

7   When should the installation be tested?

_____

8   Why is it recommended to use long-radius bends in irrigation services?

_____

_____

9   How does the length of an irrigation system affect the pipe size of the system?

_____

_____

10   What is the minimum distance between ground height and any atmospheric port?

_____

WORKSHEET RECORDING TOOL

Learner name		Phone no.	
Assessor name		Phone no.	
Assessment site			
Assessment date/s		Time/s	
Unit code & title			
Assessment type			

Outcomes

Worksheet no. to be completed by the learner	Method of assessment WQ – Written questions PW – Practical/workplace tasks TP – Third-party reports SC – Scenarios RP – Role plays CS – Case studies RW – Report writing PF – Portfolio	Satisfactory response	
		Yes ✓	No ✗
Worksheet 1	WQ	☐	☐
Worksheet 2	WQ	☐	☐
Worksheet 3	WQ	☐	☐

Assessor feedback to the learner

Feedback method (Tick one ✓):	☐ Verbal	☐ Written (if so, attach)	☐ LMS (electronic)

Indicate reasonable adjustment/assessor intervention/inclusive practice (if there is not enough space, a separate document must be attached and signed by the assessor).

Outcome (Tick one ✓):	☐ Competent (C)	☐ Not Yet Competent (NYC)

Assessor declaration: I declare that I have conducted a fair, valid, reliable and flexible assessment with this learner, and I have provided appropriate feedback.

Assessor name:	
Assessor signature:	
Date:	

Learner feedback to the assessor

Feedback method (Tick one ✓):	☐ Verbal	☐ Written (if so, attach)	☐ LMS (electronic)

Learner may choose to provide information to the RTO separately.

Learner assessment acknowledgement:	Tick one:	
	Yes ✓	No ✗
The assessment instructions were clearly explained to me.	☐	☐
The assessment process was fair and equitable.	☐	☐
The outcomes of assessment have been discussed with me.	☐	☐
The overall judgement about my competency performance was fair and accurate.	☐	☐
I was given adequate feedback about my performance after the assessment.	☐	☐

Learner declaration:
I hereby certify that this assessment is my own work, based on my personal study and/or research. I have acknowledged all material and sources used in the presentation of this assessment, whether they are books, articles, reports, internet searches or any other document or personal communication.
I also certify that the assessment has not previously been submitted for assessment in any other subject or at any other time in the same subject and that I have not copied in part or whole or otherwise plagiarised the work of other students and/or other persons.

Learner name:	
Learner signature:	
Date:	

GLOSSARY

A

aerator A device fitted to tapware outlets to add air to the water stream, resulting in a spray-like flow to reduce the effects of splashing and, more importantly, the flow of the water.

anneal To soften (metal).

AS/NZS 3500 A standard that provides information about product use, restrictions and specific installation requirements for plumbing and drainage in Australia and New Zealand.

atmospheric vacuum breaker (AVB) A device with a hole in the top to allow air to enter the water system if a siphon attempts to form, thus breaking the siphonage effect.

attack hydrant a hydrant that a firefighter connects a hose to extinguish the fire.

B

backpressure The difference between the pressure within any water service and a higher pressure within any vessel or pipework to which it is connected.

back siphonage The result of liquids at a lower level drawing water from a higher level.

ball valve A stop valve that is designed primarily for on/off functions.

bending spring A spiralled steel spring that is inserted into a pipe prior to bending so that pipe shape is maintained during bending.

bib tap A screw-down tap with a male threaded inlet socket and a curved spout (or bib).

block plan A permanent plan located at the booster assembly that is weather resistant and cannot fade.

blue water Discoloration of water within copper piping limited to small locations around the globe. No definitive diagnosis as to the cause is available as yet, but contributing factors include water softness and pH levels.

booster Additional heat added to maintain water temperature when the main heating source is insufficient.

booster assembly An assembly of valves on a fire hydrant or fire sprinkler service to allow the fire brigade to connect and pump water into the system at a boosted pressure.

BP/lugged Fittings that have a backplate used to screw or fix the fitting securely against a wall, post, timber or other solid object.

branch-formed joint A method of welding a branch tube to a main tube; generally called a branched tee.

breeching piece (shower) Used to connect hot and cold water from the recess bodies; allows the water to mix together prior to discharging from the shower outlet.

breeching piece (water main) Used to connect two separate water main drillings to one common line to feed water to a property; for services larger than 25mm in diameter.

Building Sustainability Index (BASIX) A sustainable planning measure introduced in NSW to ensure that homes are designed to use less drinking (potable) water and to reduce greenhouse gas emissions by setting energy and water reduction targets.

butt fusion welding A joining method in which a heated plate is placed between the squared and the cleaned/shaved ends of the pipes being joined. The plate heats the pipe ends and is quickly withdrawn. The molten pipe ends are then pushed together.

C

capillary fitting A tube fitting with a socket-like end, designed for use in a capillary joint.

cavitation A process in which cavities or bubbles form in the fluid low-pressure area of the system and collapse in a higher-pressure area, causing noise, damage and a loss of capacity.

ceramic disc tap A tap (typically single-lever or quarter-turn) that uses ceramic discs instead of conventional washers to give a quick action.

check valve A mechanical valve that permits gases or liquids to flow in only one direction, preventing flow from reversing down the line; classified as a one-way directional valve.

cistern A tank in which water is stored at atmospheric pressure.

close coupled solar system Where the solar collector panels and the storage tank of a solar water heating system are both located on the roof.

collectors Solar panels that capture energy from the sun.

combined non-return/isolating valve *See* **duo valve**.

compression joint A joint formed between two pipes using an external nut and bush with an internal compression cone (olive) to seal it.

compression sleeve A flexible sleeve placed over the pipe ends; used instead of a compression cone (olive) in a compression joint.

condensation The process in which water changes from a gas to a liquid.

conduction The transfer of thermal energy between neighbouring molecules in a substance due to a temperature gradient.

contact thermostat A thermostat attached to the outside wall of a water heater that relies on heat transfer through the wall of the tank.

continuous (instantaneous) flow water heater A water heater designed to heat water only at the time it is being used (it does not store heated water).

convection A process in which water is heated by an element installed at the bottom of the heater. The water becomes less dense and rises, leaving the colder water to fall to the bottom of the tank to be heated in turn.

copper press fit A way of joining copper tube and fittings for gas and water services without using heat.

crimp ring fitting A thin-walled fitting that involves a single O-ring. The tube section is inserted into the fitting by hand, compressing the O-ring to form the seal. The tube section meets the stop, and in this position the joint is locked by crimping the fitting sleeve so that it engages the tube section.

cross-connection Any connection (physical or otherwise) between any drinking (potable) water supply system, either directly or indirectly connected to a water main; any fixture, storage tank, receptacle, equipment or device through which it may be possible for any contaminated water or substance to enter a water system.

croxed joint Formed where the ends of the pipes are flared for a compression joint.

D

dashpot A device for cushioning or damping a movement (as of a mechanical part) to avoid shock.

deposition The process by which water changes from a gas directly to a solid.

dezincification-resistant (DR) Indicates that the fitting is made from a type of brass that resists a form of corrosion called dezincification.

Dial Before You Dig A service to help locate existing services such as water pipes and mains, communications cables and infrastructure, and electricity mains conduits (telephone 1100 or www.1100.com.au).

direct solar water heating system A system in which the water is circulated between the storage tank and the collector.

displacement principle The principle that because hot water is less dense than cold water it floats on top of the cold water, so as a hot tap is opened, cold water enters the cylinder from the bottom, forcing the hot water out. The water continues to flow into the heater until the hot tap is turned off and the cylinder is re-pressurised.

double check valve A backflow prevention device consisting of two check valves assembled in series complete with test points provided; also called a double check assembly.

dry tapping A connection to a water main that is made with the water main shut off.

dual check valve A double check valve that is non-testable.

dual water area An area that has both drinking and recycled water services.

duo valve A valve that provides for the dual functions of being an isolating valve and a non-return valve.

E

elbow bend A fitting with a sharp 90-degree change of direction.

electro fusion jointing A jointing method in which the cleaned/shaved ends of a pipe are inserted into a socket that is heated to create a bond between the socket and the pipe.

evacuated tube solar collector Where evacuated tubes are used to heat water instead of flat solar panels. They consist of two glass tubes fused at top and bottom and installed in series.

evaporation The process by which water from sources such as rivers, lakes and the ocean is heated by the sun and converted into water vapour, which then rises back into the atmosphere as a gas.

expanded joint fitting *See* **croxed joint**.

external thread male iron (MI) The externally threaded end of a fitting.

F

falling level displacement water heater A water heater that uses an off-peak electrical supply. The heated water may be used as required, but replacement water will not enter the heater until there is a power supply to operate the solenoid valve. This type of water heater may not only run out of heated water, but may also run out of water altogether, until it refills when power is available.

feed hydrant a hydrant that supplies water via a hose to the inlet of the pump on a fire truck.

flanged jointing A jointing method where each end of a pipe has been fitted with a circular flange to allow for a bolted connection.

flare-type fitting Where the end of a pipe is expanded (swaging) to allow for the end of another pipe to be inserted.

float valve A valve connected to a ball float used to control the water level within a tank or cistern.

flocculation A method of water purification in which a coagulant (such as ferrous chloride) is added to the water causing small particles to aggregate or collect as a 'floc', which then becomes heavy and settles to the bottom.

flusherette *See* **flushometer**.

flushing cistern A cistern capable of discharging a measured quantity of water automatically at intervals regulated by the rate that water is fed to the cistern, or by manual operation of the flushing mechanism.

flushometer A flushing device that uses energy from a pressurised water supply system rather than the force of gravity to discharge water.

flux A non-metallic substance used during welding, brazing or soldering to chemically clean the surface of the metal.

foot valve A type of check valve with a built-in strainer. Used at the point of the liquid intake (suction side) to retain liquid in the system, it prevents the loss of prime when the liquid source is lower than the pump (generally when a suction lift is required).

free outlet under-sink water heater *See* **push-through unit**.

frictional loss Positive head loss due to friction resistance between the pipe walls and the moving liquid.

front run The section of the water service from the outlet of the water meter to the building; it is usually buried in the ground.

full bore valve A ball valve that has an orifice with a diameter the same as that of the pipe.

G

gate valve A stop valve designed primarily to turn on or turn off water flow fully, the only application for which gate valves are recommended.

globe valve A valve designed specifically for regulating water flow in a pipeline. It consists of a movable disc-type element and a stationary ring seat, and is named after the spherical appearance of its body.

gravity unit (low pressure) *See* **low-pressure (gravity feed) hot water system**.

H

H$_2$O The chemical symbol for water: two atoms of hydrogen combined with one atom of oxygen to form one molecule of water.

hands-free infra-red tapware Taps that turn on automatically when the user's hands are within the sensor's range and turn off once they are moved out of range.

hardstand area An area for the fire brigade pumping appliance to park on and safely carry out firefighting operations.

head Another word for pressure.

heat fusion Extreme heat applied to metal surfaces to create a bond.

heat length The length of a section of tube to be bent.

heat pump air-sourced unit A form of solar hot water system. In a simple form, it is a conventional water storage tank with a heat pump attached. The heat pump converts the outside air temperature to heating energy in a similar manner to the way that an air-conditioner heats or cools a house or car.

hotplate welding *See* **butt fusion welding**.

hot water installation An installation of one or more water heaters and the required hot and cold piping system to supply hot water to a number of fixtures, appliances and outlets.

hydrant A valve that supplies water to a firefighting agency to extinguish fires.

hydraulics The branch of science that deals with fluids in motion.

hydrostatics The branch of science that deals with fluids at rest.

I

immersion thermostat A thermostat used on gas water heaters; the thermostat extends into the heater, giving a quicker response time.

indirect solar water heating system A system in which a heat exchange fluid circulates between the heat exchanger within the storage tank and the collector. Heat is transferred from the heat exchange fluid to the water via conduction in the heat exchanger.

infra-red sensing tapware Taps that use a movement sensor to detect users and provide water where required.

installation of a property service The installation of all pipework and fittings, including the copper riser pipe and the meter ball valve located at the intended meter location. A single connection to the main and single property service for each water service type (drinking and dual water systems) must be provided for each property.

internally threaded female iron (FI) The internally threaded end of a fitting.

isopropanol An alcohol-based cleaning solution.

J

joint service *See* **trunk service line**.

K

kilopascals (kPa) Metric measurement unit for pressure.

Kinco nut A nut with a hexagonal external shape with a round threaded internal shape; used for compression fittings.

L

line hammer *See* **water hammer**.

line hammer arrestor A water hammer arrestor that is installed in the line itself, as opposed to off a tee near the source of the water hammer, such as for washing machines.

line strainer A device that helps to remove solid particles from the water supply using a removable strainer.

long service A property service connected to a water main located on the other side of the street to the property.

low-pressure (gravity feed) hot water system A water heater with a cold water feed (cistern) tank fitted to the storage tank. It is designed to store water at atmospheric pressure and deliver it via gravity to the required hot water outlets.

M

mains pressure unit A water heater designed to store and deliver water at mains pressure (recommended to be above 350kPa). This provides the hot and cold water at same outlet pressure.

manifold system Where a common pipe serves multiple connection points; often referred to as a parallel system.

manipulative joint A joint formed where the shape of the tube has been altered.

mechanical compression coupling A double-ended coupling that includes one or two long-life elastomeric gaskets fitted in machined grooves; also called a double bell coupling.

melting The process by which water changes from a solid to a liquid.

multipoint unit A water heater designed with sufficient water flow capacity and thermal input to provide a consistent supply of hot water to several outlets at the same time. It may be of a storage or a continuous flow type.

N

natural water cycle The cycle created when sea water evaporates and rises to the atmosphere, where it reforms as water droplets within clouds and then returns to the ground as rain or snow.

non-ferrous materials Materials that do not contain iron.

non-manipulative joint A joint formed where the tube has not been altered.

non-return valve A valve designed to prevent reverse flow from the downstream section of a pipe to the section of pipe upstream of the valve.

O

OD Outside diameter. The term is used when describing the capillary end of a fitting.

olive A flexible cone used to seal the ends of pipes in a compression joint.

open yard A designated area greater than 500m^2 used for storing or processing combustible material.

O-ring A rubber ring fitted to the spindle of tap fittings to prevent leakage.

ovality The ability to be distorted into an oval shape.

P

performance curve *See* **pump curve**.

permanent water hardness Water hardness caused by calcium sulphates or magnesium sulphates.

pH scale A measurement of the level of acidity or alkalinity of the water. 'pH' stands for power of hydrogen.

plain fitting A fitting that has no solder ring.

Plumbing Code of Australia (PCA) Volume 3 of the National Construction Code; it contains the technical provisions for the design, construction, installation, replacement, repair, alteration and maintenance of water services. sanitary plumbing and drainage systems.

positive displacement pump A pump that operates by trapping a set amount of fluid, then forcing (displacing) it into a discharge pipe.

precipitation Rain droplets that form and fall to the ground when the condensation (water vapour) in clouds reaches dewpoint.

pressure-limiting valve A valve that controls the water pressure to a preset maximum.

pressure ratio valve A valve designed to reduce the outlet pressure of the valve by a set ratio to that of the inlet pressure.

pressure-reducing valve A valve that delivers a preset pressure, which avoids pressure fluctuations throughout the home installation.

pressure responsive control valve *See* **staging valve**.

pressure vacuum breaker (PVB) Similar to an atmospheric vacuum breaker (AVB), except that the PVB contains a spring-loaded poppet. PVBs usually have test points to which specially calibrated gauges are attached in order to ensure that they are functioning properly.

prime To have the pump suction line full of water (primed) prior to starting the pump.

property service (main to meter) The pipes and fittings used or intended for the supply of water to a property from the water main, up to and including the meter assembly, or to the stop tap if there is no meter.

pump assembly The pump in situ ready for connection of pipework.

pump curve A manufacturer's graph used to describe a pump's performance.

push lock fitting A fitting that is pushed over the tube end. The joint is made watertight by an O-ring.

push-through unit A water heater that stores a small quantity of water at atmospheric pressure. When the hot tap is opened, water is delivered at mains pressure.

push to connect fitting A connection for joining pipes such as copper, plastic, PE-X pipes and so on in any combination, without soldering, clamps, unions or glue. The pipe is inserted and the special teeth bite down and grip tight, while a specially formulated O-ring compresses to create a perfect seal.

R

radiation Any process in which energy emitted by one body travels through a medium or through space, ultimately to be absorbed by another body.

recycled water systems Non-drinking water recycled from sewage – used for flushing and irrigation.

reduced bore valve A ball valve with an orifice the diameter of which is less than that of the pipe.

reduced-pressure unit A water heater using a pressure reducing valve or an overhead storage (feed) tank connected to the cold water heater connection in order to reduce the delivery pressure to below that in the utility's water main.

reduced pressure zone (RPZ) device Two independent check valves, in series, with a pressure-monitored chamber between. The chamber is maintained at a pressure that is lower than the water supply pressure, but high enough to be useful downstream. The reduced pressure is guaranteed by a differential pressure relief valve, which automatically relieves excess pressure in the chamber by discharging to a drain.

registered air gap (RAG) An air gap in a storage tank or the air gap over a fixture, vat tanker filling point or drum of non-potable liquid.

registered break tank (RBT) A tank installed for backflow prevention that is registered by an authority for the purposes of inspection and maintenance to ensure its functional requirements.

rotodynamic pump A pump that uses bladed impellers to rotate within a fluid to increase the energy of the fluid.

rough-in The installation of any water pipework; generally required before the building has its wall sheeting fixed.

S

saddle A U-shaped piece used on the outside of a pipe so it can be secured.

safe tray A tray positioned under a storage tank or hot water heater to collect and drain away any leaks or spillage.

safe waste drain A drain pipe that runs from a safe waste, or safe tray, to a suitable discharge point.

sand bending A method of bending pipe (usually copper tube or steel pipe) that uses sand to keep the pipe round during the bending process.

short service A property service connected to a water main located on the same side of the street as the property.

silver brazing See **silver soldering**.

silver soldering A joining process whereby a filler metal or alloy is heated to melting temperature above 450°C and distributed between two or more close-fitting parts by capillary action.

single-point unit A water heater designed to supply water to one tap/outlet only. It may be of a continuous flow or a storage design.

sluice valve A valve that opens by lifting a round or rectangular gate/wedge out of the path of the fluid.

socket fusion jointing See **electro fusion jointing**.

soft soldering The process of joining metals/pipes through the application of lead-based materials with the aid of a fluxing agent and a heat source.

solar indirect system A combination hot water system where a mains pressure hot water heater passes water through an instantaneous heater mounted to the mains pressure unit to raise the water temperature if required.

solar non-return (SNR) valve A valve that prevents the backflow of heated water from a solar heater to the mains line.

solder ring fitting A fitting that has a ring of solder placed into a groove around it.

solenoid A coil of wire that creates a magnetic field when an electrical current is passed through it.

solenoid valve An electrically controlled or electromechanical valve used to control the flow of fluids (or gas) by allowing or preventing an electrical current to pass through a solenoid.

split (forced circulation) solar system A system that uses a pump to circulate the water for heating through the system.

spring-loaded tap A self-closing, spring-operated tap.

staging valve A valve designed to respond to the flow of hot water through the system by igniting each unit, depending on the flow requirements.

stand-off bracket A bracket or clip designed to fix pipework away from the surface to which it is attached.

static head The depth at any particular point when water is static.

stop tap A screw-down pattern tap with horizontal inlet and outlet connections used to control the flow of water in a pipeline. It usually incorporates a loose jumper valve, permitting flow in one direction only.

stop valve A valve used or installed to stop the water supply.

storage water heater A water heater designed to hold a useful quantity of hot water in an insulated container ready for use as required.

strainer A valve designed to prevent foreign matter from fouling the operation of devices or valves downstream.

sublimation The process in which water changes directly from a solid to a gas.

swarf Chips, turnings or filings produced during the machining of metal, wood or plastic.

T

tap A valve with an outlet used as a draw-off or delivery point.

tapping saddle A saddle through which a tapping is made into the water main and the main cock is connected.

tee A fitting that allows the connection of three pipe ends.

temperature and pressure relief valve A valve designed to relieve the pressure in the hot water heater by venting it to the atmosphere should an unsafe condition arise, such as a thermostat malfunction that may result in the temperature going beyond the specified level.

tempering valve A valve designed to mix hot water with cold water as it leaves the hot water service, to supply fixtures at a temperature of 50°C.

temporary water hardness Water hardness caused by bicarbonates of calcium or magnesium.

terminator valve A valve that mechanically shuts off the water entering an appliance when a leak is detected.

thermosiphon circulation A method of passive heat exchange based on natural convection that circulates liquid in a vertical closed-loop circuit without requiring a conventional pump.

thermostatic mixing valve (TMV) A valve that blends hot water (stored at temperatures high enough to kill bacteria) with cold water to ensure constant, safe outlet temperatures and prevent scalding. This valve is adjustable so as to provide blended water at varying temperatures.

transpiration The process in which plants take in water through their root systems and pass moisture back into the atmosphere via their leaves.

trio valve A valve with the combined functions of non-return, isolator and line strainer.

trunk service line A larger-diameter property service to accommodate local requirements.

tube expander A tool designed to expand a tube, generally made of copper, to allow another section of pipe to be inserted. The joint is then silver soldered.

tundish A funnel or other suitably shaped object that provides a connection (with an air break) for a discharge from a device or fixture to a drain. The tundish normally has a trap seal attached beneath it.

V

vacuum breaker A device designed to allow air into a water supply system to prevent a vacuum from forming.

valves Various mechanical devices that control the flow of liquids, gases or loose material through pipes or channels by blocking and uncovering openings.

W

Water Efficiency Labelling and Standards (WELS) Scheme A joint Commonwealth, state and territory regulatory scheme that requires a range of water-using products to be labelled for water efficiency. The lower the water consumption, the more stars on the label.

water hammer A noise similar to a hammer blow on the pipes that can weaken or even burst them. This occurs as a result of the pressure being rapidly increased when the liquid velocity is suddenly increased. It is usually the result of sudden starting, stopping or change in pump speed or the sudden opening or closing of a valve.

water main A pipeline that conveys drinkable (potable) water throughout the community for its use and is owned by a water supply utility.

WaterMark A labelling system that confirms that a water-using product complies with the requirements of the Plumbing Code of Australia and is fit for the purpose of installation under that code.

water pump A pump used specifically for the purpose of moving water.

water service The part of the cold water supply pipework from the water main up to and including the outlet valves at fixtures and appliances.

water storage tank A container designed to hold a specific quantity of water.

wet tapping A connection to a water main that is made with the water main under pressure.

witness mark Generally, an intentional spot, line, groove or other mark that serves as an indicator of depth. It might be found on a pipe or tube to indicate how far it should protrude into a fitting.

Z

zeolite process A method of removing permanent water hardness (usually in a domestic water softener) by exchanging magnesium or calcium salts for sodium salts that have no detrimental effects.

INDEX

TRUE COLLABORATION

LET'S JOIN FORCES

You bring the pen, we'll bring the sticky notes

Believe it or not, there are people who spend a lot of time thinking about textbooks — like instructors, and us.

At Cengage, we live and breathe textbooks.

We're obsessed with top-quality content; we're fanatical about design thinking and innovation that helps you learn.

We talk with instructors and students to find out what you like about textbooks, and what would make them better. Our UX designers crave your input!

Want to join the conversation and see how you could help create better resources for students? We'd love to hear from you. Scan the code to get started.